NEW DEVELOPMENTS IN MOLECULAR CHIRALITY

Understanding Chemical Reactivity

Volume 5

The titles published in this series are listed at the end of this volume.

New Developments in Molecular Chirality

edited by

Paul G. Mezey

Department of Chemistry and Department of Mathematics, University of Saskatchewan, Saskatoon, Canada

KLUWER ACADEMIC PUBLISHERS
DORDRECHT / BOSTON / LONDON

Library of Congress Cataloging-in-Publication Data

New developments in molecular chirality / edited by Paul G. Mezey.
p. cm. -- (Understanding chemical reactivity ; v. 5)
Includes index.
ISBN 0-7923-1021-7 (HB : acid free paper)
1. Chirality. I. Mezey, Paul G. II. Series.
QD481.N465 1990
541.2'252--dc20 90-48190

ISBN 0-7923-1021-7

Published by Kluwer Academic Publishers,
P.O. Box 17, 3300 AA Dordrecht, The Netherlands.

Kluwer Academic Publishers incorporates
the publishing programmes of
D. Reidel, Martinus Nijhoff, Dr W. Junk and MTP Press.

Sold and distributed in the U.S.A. and Canada
by Kluwer Academic Publishers,
101 Philip Drive, Norwell, MA 02061, U.S.A.

In all other countries, sold and distributed
by Kluwer Academic Publishers Group,
P.O. Box 322, 3300 AH Dordrecht, The Netherlands.

Printed on acid-free paper

Printed in the Netherlands

TABLE OF CONTENTS

PREFACE

Molecular chirality is one of the fundamental aspects of chemistry. Chirality properties of molecules have implications in a wide variety of subjects, ranging from the basic quantum mechanical properties of simple systems of a few atoms to molecular optical activity, asymmetric synthesis, and the folding pattern of proteins. Chirality, in both the geometrical and the topological sense, has also been the subject of investigations in various branches of mathematics. In particular, new developments in a branch of topology, called knot theory, as well as in various branches of discrete mathematics, have led to a novel perspective on the topological aspects of molecular chirality. Some of the mathematical advances have already found applications to the interpretation of new concepts in theoretical chemistry and mathematical chemistry, as well as to novel synthetic approaches leading to new molecules of exceptional structural properties. Some of the new developments in molecular chirality have been truly fundamental to the theoretical understanding and to the actual practice of many aspects of chemistry.

The progress in this field has been very rapid, even accelerating in recent years, and a review appears more than justified. This book offers a selection of subjects covering some of the latest developments. Our primary aim is to clarify some of the basic concepts that are the most prone to misinterpretation and to provide brief introductions to some of those subjects that are expected to have further, important contributions to our understanding of molecular properties and chemical reactivity.

I am indebted to all of our authors for their contributions. Reading their chapters has been very rewarding for me, and I hope that they, as well as all of our readers, will enjoy the book as much as I did.

Paul G. Mezey
Department of Chemistry and
Department of Mathematics
University of Saskatchewan,
Saskatoon, Canada, S7N 0W0

Fundamental Symmetry Aspects of Molecular Chirality

Laurence D. Barron
Chemistry Department, The University, Glasgow G12 8QQ, U.K.

1 Introduction

Scientists have been fascinated by molecular handedness ever since the concept first arose as a result of the discovery of natural optical activity in the early years of the last century. This concept spawned major advances in physics, chemistry and biology and continues to catalyse scientific progress even today.

In 1811 Arago observed colours in sunlight that had passed along the optic axis of a quartz crystal placed between crossed polarizers. Subsequent experiments by Biot established that the colours originated in a rotation of the plane of polarization of linearly polarized light (optical rotation) with different rotations for light of different wavelengths (optical rotatory dispersion). Biot also discovered a second form of quartz which rotated the plane of polarization in the opposite sense. Then in 1815 Biot discovered that certain organic liquids such as terpentine could also show optical rotation, which indicated that optical activity could reside in individual molecules and may be observed even when the molecules are arranged in random fashion: in contrast, the optical activity of quartz is a property of the crystal structure since molten quartz is not optically active.

Fresnel's discovery of circularly polarized light in 1824 enabled him to formulate optical rotation in terms of different refractive indices for the coherent right and left circularly polarized components of equal amplitude into which linearly polarized light can be resolved. This provided an important insight into the symmetry requirements for an optically active molecule or crystal. Quoting Fresnel:[1]

> "There are certain refracting media, such as quartz in the direction of its axis, turpentine, essence of lemon, etc., which have the property of not transmitting with the same velocity circular vibrations from right to left and those from left to right. This may result from a peculiar constitution of the refracting medium or of its molecules, which produces a difference between the directions right to left and left to right; such, for instance, would be a helicoidal arrangement of the molecules of the medium, which would present inverse properties accordingly as these helices were dextrogyrate or laevogyrate."

P. G. Mezey (ed.), New Developments in Molecular Chirality, 1–55.

The resolution by Pasteur in 1848 of a sample of the optically inactive (racemic) version of tartaric acid into equal numbers of molecules giving equal and opposite optical rotations in solution provided a dramatic confirmation of Fresnel's insight that optically active molecules must be handed. This discovery emphasised that molecules must be pictured in three dimensions, and led eventually to the concept of tetrahedral valencies for the carbon atom and to the subject of stereochemistry.

In 1846 Faraday discovered that optical activity could be induced in an otherwise inactive sample by a magnetic field. He observed optical rotation in a rod of lead borate glass placed between the poles of an electromagnet with holes bored through the pole pieces to allow a linearly polarized light beam to pass through. This effect is quite general: a Faraday rotation is found when linearly polarized light is transmitted through any crystal or fluid in the direction of a magnetic field, the sense of rotation being reversed on reversing the direction of either the light beam or the magnetic field. At the time, the main significance of this discovery was to demonstrate conclusively the intimate connection between electromagnetism and light; but it also became a source of confusion to some physicists and chemists who failed to appreciate that there is a fundamental distinction between magnetic optical rotation and the natural optical rotation associated with molecular handedness.

In 1848, Pasteur[2] introduced the word *dissymmetric* to describe objects "which differ only as an image in a mirror differs from the object which produces it." A finite cylindrical helix provides a good example since reflection reverses the screw sense and so prevents the superposition of the mirror image on the original. The two distinguishable forms of a dissymmetric system are called enantiomers (from the Greek *enantios morphe* meaning opposite shape). Dissymmetric figures are not necessarily asymmetric, meaning devoid of all symmetry elements, since they may possess one or more proper rotation axes (the finite cylindrical helix has a twofold rotation axis through the mid-point of the coil, perpendicular to the long helix axis). Specifically, dissymmetry excludes improper rotation axes; that is, a centre of inversion, reflection planes and rotation–reflection axes. Pasteur attempted to extend the concept of dissymmetry to other aspects of the physical world.[3,4] For example, he thought that a magnetic field, since it can induce optical rotation (the Faraday effect), generates the same type of dissymmetry as that possessed by an optically active molecule. As we shall see, this idea is quite wrong and has been the source of much confusion. However, Pasteur was correct in thinking that the combination of linear motion with a rotation does generate the same type of dissymmetry as an optically active molecule.

In recent years the word dissymmetry has been replaced by *chirality*, meaning handedness (from the Greek word *chir* for hand) in the literature of stereochemistry. This word was actually first coined by Sir William Thomson,

Professor of Natural Philosophy in the University of Glasgow, in his famous Baltimore Lectures delivered as long ago as 1884. These lectures were not published until 1904,[5] by which time Sir William had been elevated to the peerage, taking his title Lord Kelvin from the name of the river which runs through Glasgow University park.

To be specific, the word chiral is introduced in the Baltimore lectures to describe a geometrical figure "if its image in a plane mirror, ideally realized, cannot be brought into coincidence with itself." Although this definition is essentially the same as that used by Pasteur for dissymmetric, the two words *chiral* and *dissymmetric* are not strictly synonymous in the broader context of modern stereochemistry and theoretical physics. Dissymmetry means the absence of certain symmetry elements, these being improper rotation axes in Pasteur's usage. Chirality is a more positive concept in that it refers to the possession of the attribute of handedness which, as shown later, has a physical content: in molecular physics this is the ability to support time-even pseudoscalar observables; whereas in elementary particle physics chirality is defined as the eigenvalue of the Dirac matrix operator γ_5.[6]

This article applies some principles of modern physics, especially fundamental symmetry arguments, to develop a deeper description of chirality than that usually encountered in the literature of stereochemistry in order to facilitate a proper understanding of the structure and properties of chiral molecules, and of the factors involved in their synthesis and transformations. A central result is that, although dissymmetry is sufficient to guarantee chirality in a stationary object such as a finite helix, dissymmetric systems are not necessarily chiral when motion is involved. I have introduced the concept of 'true' and 'false' chirality to draw attention to this distinction[7-10] but do not intend this to become standard nomenclature; rather, I have suggested that the word *chiral* be reserved in future for systems that I call truly chiral. It will be appreciated from what follows that true and false chirality correspond to time-invariant and time-noninvariant enantiomorphism, respectively. We shall see that, as intimated above, the combination of linear motion with a rotation does indeed generate true chirality, but that a magnetic field does not (in fact it is not even false chirality). Examples of dissymmetric systems showing false chirality include a stationary rotating cone, and collinear electric and magnetic fields.

The recent triumph of theoretical physics in unifying the weak and electromagnetic forces into a single 'electroweak' force has provided a new perspective on chirality which has been a source of inspiration for the new synthesis reviewed in this article. Since the weak and the electromagnetic forces have turned out to be different aspects of the same, but more fundamental, unified force, the absolute parity violation associated with the weak force is now known to infiltrate to a tiny extent into all electromagnetic phenomena so that free atoms, for example, show very small optical rotations; and a tiny energy difference exists between the enantiomers of a chiral molecule. We shall see that the distinction between true and false chirality hinges on the symmetry operations that interconvert enantiomers, and that parity violation provides a cornerstone for the identification of true chirality. It is remarkable that parity violation provides a scientific basis for the general cosmic dissymmetry that Pasteur sensed over a century ago.[11,12]

2 Symmetry Principles

The symmetry arguments required for a proper understanding of molecular chirality go beyond the purely spatial aspects with which most chemists are familiar and which usually employ point group symmetry arguments to deduce qualitative information about molecules. In addition to conventional point group symmetry arguments, the fundamental symmetries of space inversion, time reversal and even charge conjugation have something to say about chirality at several levels including the experiments that show up optical activity observables, the objects generating these observables and the quantum states that these objects must be able to support. We therefore start by reviewing these fundamental symmetries.

2.1 Non-Observables and Symmetry Operations

Most symmetry arguments depend on the symmetries inherent in the laws which determine the operation of the physical world,[13] and these in turn depend on certain intrinsic uniformities that we percieve in the universe. Point group symmetry arguments, for instance, are based on the behaviour of a molecule under proper and improper rotation operations: the first of these operations relies on the fact that a physical experiment and an equivalent rotated experiment will produce equivalent results; and the second on the fact that, provided the weak force is not involved (which excludes processes such as β—decay), an experiment and an equivalent spatially-inverted experiment will produce equivalent results.

There are many important properties of molecules that depend in one way or another on motion. Manifestations of molecular magnetism, for example, depend on the orbital and spin motions of the constituent electrons and protons and on how these motions are influenced by an external magnetic field, which itself may be generated by the motion of electrons around a coil. A major feature of this article is the analysis of the motion-dependent aspects of certain enantiomorphous systems and the identification of true chirality. Arguments based on spatial symmetries alone often lead to wrong conclusions about motion-dependent aspects of molecular behaviour because they are unable to accommodate the additional symmetry implications that motion engenders.

The additional symmetry operation required to deal with properties that depend on motion is called time reversal. This name is rather unfortunate because it has mysterious connotations of travelling backwards in time, which is not the intended meaning. Sachs[14] has stressed that time reversal has nothing to do with the concept of the 'arrow of time' which, as Boltzmann pointed out, has its origins in the ability of living beings to distinguish transitions from less probable to more probable states of small regions of the universe.[15] It has to do with the universe making no distinction between equivalent motion in opposite directions in space. So it is best to think of time reversal as motion reversal, which in fact is the terminology used by Wigner[16] in his fundamental work on time reversal in quantum mechanics.

To place these considerations onto a more formal footing, it is helpful to associate symmetries in physical laws with the impossibility of observing certain basic quantities.[17] Three such non-observables appear to be particularly fundamental: absolute chirality (meaning absolute right- or left-handedness), absolute direction of motion, and absolute sign of electric charge. A non-observable implies invariance of physical laws under an associated transformation and usually generates a conservation law or selection rule.

The transformation associated with absolute chirality is space inversion, represented by the parity operator $\hat{P}$ which inverts the system through the origin of the space-fixed axes. Most physical laws (but not those describing processes which involve the weak force) are unchanged by space inversion. If replacing the space coordinates (x,y,z) by (-x,-y,-z) everywhere in equations describing physical laws leaves those equations unchanged, the corresponding physical processes are said to conserve parity.

The transformation associated with absolute direction of motion is time reversal, represented classically by the operator $\hat{T}$ which has the effect of reversing the motions of all the particles in the system. If replacing the time coordinate (t) by (-t) everywhere in equations describing physical laws leaves those equations unchanged, the corresponding physical processes are said to conserve time reversal invariance (or to have reversibility).

The transformation associated with absolute sign of electric charge is charge conjugation, represented by the operator $\hat{C}$ which interconverts particles and antiparticles. Although this exotic operation might appear to have no relevance to chemistry, we shall see that it has conceptual significance in studies of molecular chirality.

It is intriguing to note that invariance under the combined operation of $\hat{C}\hat{P}\hat{T}$ (see Section 5.3 below) is in fact the quantal and relativistic extension of $\hat{T}$ invariance,[15] which serves to emphasise the essential unity of space, time and matter and that, at least from a cosmic standpoint, the operation of time reversal is not so completely divorced from the more familiar spatial symmetry operations as might appear at first sight.

An important consequence of the existence of symmetries in the laws which determine the operation of the physical world is that, if a *complete* experiment is subjected to an associated symmetry operation such as space inversion or time reversal, the resulting experiment should, in principle, be realizable.[13,18] A detailed consideration of the natural and magnetic optical rotation experiments shows that they do indeed conserve parity and reversality;[19,20] and such arguments can be used to predict or discount possible new effects (such as an electric analogue of the Faraday effect) without recourse to mathematical theories.[19-22] It is also possible to reach the same conclusions using a distinct but complementary approach based on photon selection rules.[23,24]

2.2 Physical Quantities

It is important to classify physical quantities according to their behaviour under various symmetry operations. Physical quantities are first classified as *scalars, vectors* or *tensors* depending on their directional properties. A scalar such as temperature has magnitude but no associated direction; a vector such as velocity has magnitude and one associated direction; and a tensor such as electric polarizability has magnitudes associated with two or more directions. Scalars, vectors and tensors are then further classified according to their behaviour under $\hat{P}$ and $\hat{T}$.

Vectors such as position $\mathbf{r}$, velocity $\mathbf{v}$ and linear momentum $\mathbf{p}$ which change sign under the inversion operation $\hat{P}$ are called *polar* or true vectors. A vector such as angular momentum $\mathbf{L} = \mathbf{r} \times \mathbf{p}$ whose sign is not changed by $\hat{P}$ is called an *axial* or pseudo vector: $\mathbf{L}$ is defined relative to the sense of rotation by a right-hand rule, and $\hat{P}$ does not change the sense of rotation. A *pseudoscalar* quantity is a number with no directional properties but which changes sign under $\hat{P}$. A pseudoscalar is generated by taking the scalar product of a polar and an axial vector.

Physical quantities are classified as *time-even* or *time-odd* depending on whether they are invariant or change sign under the time reversal operation $\hat{T}$. This behaviour is usually immediately obvious from a consideration of the motions of the constituent particles. Many physical quantities do not involve motion and so of course are time-even, examples being the energy scalar W, position vector $\mathbf{r}$, electric field vector $\mathbf{E}$ and electric dipole moment vector $\boldsymbol{\mu}$. Many other quantities do involve motion and are time-odd: for example the velocity vector $\mathbf{v}$, linear momentum vector $\mathbf{p}$, angular momentum vector $\mathbf{L}$, magnetic field vector $\mathbf{B}$ and magnetic dipole moment vector $\mathbf{m}$. The elusive magnetic monopole, which has never been observed, transforms as a time-odd pseudoscalar.

Pseudoscalar quantities are of central importance in the discussion of molecular chirality because, as shown in Section 3.1 below, the natural optical activity phenomena supported by chiral molecule are characterized by time-even pseudoscalar observables such as optical rotation angle, rotational strength, Raman circular intensity difference, etc.

This is an appropriate point to introduce Neumann's principle[25,26] which states that any type of symmetry exhibited by the symmetry group of a system is possessed by every physical property of the system. In fact it is Curie's restatement of this principle, rather than the original, that is most helpful here:[27] "dissymmetry makes the phenomenon". Thus no dissymmetry can manifest itself in a physical property which does not already exist in the system. In this instance the pseudoscalar natural optical activity observables change sign under space inversion, which parallels the interconversion under space inversion of the distinguishable enantiomeric chiral molecules which make up the distinguishable enantiomeric bulk isotropic samples.

2.3 Symmetry in Quantum Mechanics

When symmetry operations are formulated in quantum mechanics, some important new features arise that are not present in the classical discussion.

Parity

The starting point for parity considerations is the invariance of the conventional (*i.e.* parity-conserving) Hamiltonian for a closed system of interacting particles to an inversion of the coordinates of all the particles. $\hat{P}$ is now interpreted as a linear unitary Hermitian operator that changes the sign of the space coordinates in the Hamiltonian and in the wavefunction.

Consider first the wavefunction:

$$\hat{P}\,\psi(\mathbf{r}) = \psi(-\mathbf{r}). \tag{1}$$

If $\psi(\mathbf{r})$ happens to be an eigenfunction of $\hat{P}$ we can write

$$\hat{P}\,\psi(\mathbf{r}) = p\,\psi(\mathbf{r}). \tag{2}$$

The eigenvalues p are found by realizing that a double application amounts to the identity so that

$$\hat{P}^2\psi(\mathbf{r}) = p^2\psi(\mathbf{r}) = \psi(\mathbf{r}), \tag{3}$$

from which we obtain

$$p^2 = 1, \quad p = \pm 1. \tag{4}$$

Thus even (+) and odd (−) parity wavefunctions are defined according to whether they are invariant or simply change sign under $\hat{P}$:

$$\hat{P}\psi(+) = \psi(+), \quad \hat{P}\psi(-) = -\psi(-). \tag{5}$$

Turning now to the Hamiltonian, its invariance under space inversion means we can write

$$\hat{P}\hat{H}\hat{P}^{-1} = \hat{H}, \quad \text{or} \quad [\hat{P},\hat{H}] \equiv \hat{P}\hat{H} - \hat{H}\hat{P} = 0. \tag{6}$$

Since $\hat{P}$ does not depend explicitly on time and commutes with the Hamiltonian, we can say that, if the state of a closed system has definite parity, that parity is conserved (the law of conservation of parity).[28]

If two eigenfunctions $\psi(+)$ and $\psi(-)$ of opposite parity have energy eigenfunctions that are degenerate, or nearly so, the system can exist in states of mixed parity with wavefunctions

$$\psi_1 = \frac{1}{\sqrt{2}} [\psi(+) + \psi(-)], \tag{7a}$$

$$\psi_2 = \frac{1}{\sqrt{2}} [\psi(+) - \psi(-)]. \tag{7b}$$

Clearly these two mixed parity states are interconverted by $\hat{P}$:

$$\hat{P}\psi_1 = \psi_2, \quad \hat{P}\psi_2 = \psi_1. \tag{8}$$

It follows from (6) that an important property of definite parity states is that they are true stationary states with constant energy W(+) or W(−), i.e.

$$\psi(\pm) = \psi^{(0)}(\pm)\, e^{-iW(\pm)t/\hbar}, \tag{9}$$

but mixed parity states are not. We shall see in Section 6 below that mixed parity states can become quasi-stationary when $W(+) \approx W(-)$, and true stationary states if $\hat{H}$ contains a parity-violating term.

All observables can be classified as having even or odd parity depending on whether they are invariant or change sign under space inversion. Even and odd parity operators $\hat{A}(+)$ and $\hat{A}(-)$ associated with these observables are therefore defined by

$$\hat{P}\hat{A}(+)\hat{P}^{-1} = \hat{A}(+), \quad \hat{P}\hat{A}(-)\hat{P}^{-1} = -\hat{A}(-). \tag{10}$$

Since integrals taken over all space are only nonzero for totally symmetric integrands, the expectation values of these operators in a mixed state such as (7a) reduce to

$$\langle \psi_1 | \hat{A}(+) | \psi_1 \rangle = \frac{1}{2}[\langle \psi(+) | \hat{A}(+) | \psi(+) \rangle + \langle \psi(-) | \hat{A}(+) | \psi(-) \rangle], \tag{11a}$$

$$\langle \psi_1 | \hat{A}(-) | \psi_1 \rangle = \frac{1}{2}[\langle \psi(+) | \hat{A}(-) | \psi(-) \rangle + \langle \psi(-) | \hat{A}(-) | \psi(+) \rangle], \tag{11b}$$

from which it follows that the expectation value of any odd-parity observable vanishes in any state of definite parity, i.e. a state for which either $\psi(+)$ or $\psi(-)$ is zero. This means that measurements on a system in a state of definite parity can reveal only observables with even parity, examples being electric charge, angular momentum, magnetic dipole moment, etc. Measurements on a system in a state of mixed parity can reveal, in addition, observables with odd parity, examples being magnetic monopole, linear momentum, electric dipole moment, etc.[20,29] The optical rotatory parameter, being a pseudoscalar, has odd parity: this leads to the important deduction that *resolved chiral molecules exist in mixed-parity quantum states.* The detailed structure of these mixed-parity states is elaborated in Section 6 below.

Time Reversal

The classical time reversal operator $\hat{T}$ introduced above does not translate into a satisfactory quantum-mechanical operator. Instead, the operator

$$\hat{\Theta} = \hat{T}\hat{K}, \tag{12}$$

where $\hat{T}$ again represents the transformation $t \rightarrow -t$ and $\hat{K}$ is the operator of complex conjugation, is taken as the time reversal operator in quantum mechanics.[20,29,30] Although it is possible to classify time-even and time-odd Hermitian operators $\hat{A}(+)$ and $\hat{A}(-)$ and their associated observables according to whether they are invariant or change sign under time reversal,[20]

$$\Theta \hat{A}(+) \Theta^{-1} = \hat{A}(+)^{+}, \qquad \Theta \hat{A}(-) \Theta^{-1} = -\hat{A}(-)^{+}, \tag{13}$$

it is not possible to classify a quantum state as being even or odd under time reversal because $\hat{\Theta}$, unlike $\hat{P}$, does not have eigenvalues. (However, the operator $\hat{\Theta}^2$ *does* have eigenvalues, these being +1 for an even-electron system and -1 for an odd-electron system).

A simple illustration is the effect of $\hat{\Theta}$ on a general atomic state $|J, M\rangle$ where both orbital and spin angular momenta can contribute to the total electronic angular momentum specified by the usual quantum numbers J and M. Using a particular phase convention, it is found that[20]

$$\hat{\Theta} |J, M\rangle = (-1)^{J-M+q} |J, -M\rangle, \tag{14}$$

where q is the sum of the individual orbital quantum numbers of all the electrons in the atom. Thus application of the time reversal operator has generated a new quantum state, orthogonal to the original, corresponding to a reversal of the sense of the total angular momentum of the atom. Since they are interconverted by time reversal, such states can be loosely regarded as having 'mixed reversality' analogous to the mixed parity states (7), even though associated states of definite reversality do not exist. One important property of mixed reversality states is that they can support time-odd observables.[20,29] Notice, however, that the states $|J,M\rangle$ do have definite parity since they are eigenstates of $\hat{P}$:

$$\hat{P}|J,M\rangle = (-1)^q|J,M\rangle. \tag{15}$$

This follows from the behaviour under space inversion of the spherical harmonics together with the standard convention that the 'intrinsic parity' of an electron spin state is +1.[31]

Charge Conjugation

A discussion of the effect of the charge conjugation operator $\hat{C}$ in quantum mechanics requires a formulation in terms of relativistic quantum field theory which is beyond the scope of this article.[17,32] All we need to appreciate here is that a charged particle is not in an eigenstate of $\hat{C}$ because charge conjugation generates a different quantum state corresponding to the associated antiparticle.

3 True Chirality

It is important to realize that optical activity is not necessarily the hallmark of chirality. In this section it is shown how a proper symmetry classification of the associated observables leads to a more precise definition of a chiral system than that usually encountered. The distinction between natural and magnetic optical activity, which is often a source of confusion in the literature of both chemistry and physics, provides a good example; and a careful analysis leads to the required new definition of chirality.

3.1 Natural and Magnetic Optical Activity

The following statement from Lord Kelvin's Baltimore Lectures[5] shows that it was recognized in the last century that the symmetry aspects of natural and magnetic optical activity are quite different:

"The magnetic rotation has neither left-handed nor right-handed quality (that is to say, no chirality). This was perfectly understood by Faraday, and made clear in his writings, yet even to the present day we frequently find the chiral rotation and the magnetic rotation of the plane of polarized light classed together in a manner against which Faraday's original description of his discovery of the magnetic polarization contains ample warning."

He probably had Pasteur in mind who, judging from his writings, was a persistent offender. Lord Kelvin's viewpoint was reinforced much later by Zocher and Torok,[33] who discussed the space-time symmetry aspects of natural and magnetic optical activity from a general classical viewpoint and recognized that quite different asymmetries are involved. A recent note of discord, however, is Stedman's classification of natural and magnetic optical activity together as 'chiral effects' in a general formalism he has developed for describing optical interaction processes:[23,34] he dismisses the present viewpoint as reflecting 'older chemical nomenclature'.[34] But far from being 'older chemical nomenclature', the sharper perception of the word chiral espoused in this article actually accords with the usage of modern elementary particle physics and provides productive analogies between the physics of chiral molecules and elementary particles. It also accords with the way in which chirality is treated in group theoretical statistical mechanics (Section 3.5). Furthermore, as shown in Section 4 below, the insistence on an absolute distinction between natural and magnetic optical activity, which must of course be reflected in the nomenclature, has resolved the confusion that has plagued considerations of absolute asymmetric synthesis for more than a century!

We obtain the required symmetry classification of the natural and magnetic optical activity observables by comparing the results of optical rotation measurements before and after subjecting the sample plus any applied field to space inversion and time reversal[20] (notice that this is a different procedure to that mentioned at the end of Section 2.1 in which the *complete* experiment, including the probe light beam, is subjected to symmetry operations in order to demonstrate conservation of parity and reversality).

Consider first the natural optical rotation experiment. Under space inversion, an isotropic collection of chiral molecules is replaced by a collection of the enantiomeric molecules, and an observer with a linearly polarized probe light beam will measure equal and opposite optical rotation angles before and after the inversion. This indicates that the observable has odd parity, and it is easy to see that it is a pseudoscalar (rather than, say, a polar vector) because it is invariant with respect to any proper rotation in space of the complete sample. Under time reversal, an isotropic collection of chiral molecules is unchanged, so the optical rotation is unchanged. Hence *the natural optical rotation observable is a time-even pseudoscalar.*

We now turn to the Faraday effect, where optical rotation is induced in an isotropic collection of achiral molecules by a static uniform magnetic field collinear with the light beam. Under space inversion, the molecules and the magnetic field direction are unchanged, so the same magnetic optical rotation will be observed. This indicates that the observable has even parity, and we can further deduce that it is an axial vector (rather than a scalar) by noticing that a proper rotation of the complete sample, including the magnetic field, through π about any axis perpendicular to the field reverses the relative directions of the magnetic field and the probe beam and so changes the sign of the observable. Under time reversal, the collection of molecules (even if they are individually paramagnetic) can be regarded as unchanged provided it is isotropic in the absence of the field, but again the relative directions of the magnetic field and the probe light beam are reversed and so the optical rotation changes sign. Hence *the magnetic optical rotation observable is a time-odd axial vector.*

The same conclusions are obtained from a more fundamental approach in which operators are defined whose expectation values generate the optical activity observables.[20,35] It is found that the natural optical rotation observable is generated by a time-even odd-parity operator, and the magnetic optical rotation observable by a time-odd even-parity operator. Another approach is to look at the associated molecular property tensors: it is found that all the contributions to natural optical rotation are generated by time-even tensors, and all the contributions to magnetic optical rotation are generated by time-odd tensors.[36,37] Last, but by no means least, these conclusions accord with those obtained from the new technique called group theoretical statistical mechanics (Section 3.5).

This analysis tells us that the nature of the quantum states of molecules that can support natural optical rotation is quite different from that of the quantum states that can support magnetic optical rotation. From the discussion in Section 2.3 above, it is clear that the former must have, among other things, mixed parity and the latter mixed reversality. The former is associated with spatial dissymmetry and corresponds to true chirality; whereas the latter originates in a different type of dissymmetry associated with lack of time reversal invariance (in some earlier articles I have called this an example of false chirality but now prefer to reserve the term for systems exhibiting time-noninvariant enantiomorphism). From Pasteur onwards these two types of optical activity have often been confused, with the magnetic field being thought of as a source of chirality. Recent examples include the assertion that an achiral molecule in a pure rotational state can be regarded as chiral,[38,39] and that there exists a chiral discrimination in the intermolecular forces between co- and counter-rotating pairs of such molecules.[40] It is certainly correct to call an achiral molecule in a pure rotational quantum state $|J,M\rangle$ (or $|J,K,M\rangle$ for a symmetric top) optically active, since it will induce optical rotation in a light beam travelling parallel to the space-fixed quantization axis. An equal and opposite optical rotation is induced by $|J,-M\rangle$. But this is equivalent to magnetic optical activity since a magnetic field, or some other time-odd influence such as a rotation of the bulk sample,[41] is required to lift the degeneracy of $|J,M\rangle$ and $|J,-M\rangle$.

3.2 A New Definition of Chirality

By now it should be clear that the hallmark of a chiral system is that it can generate time-even pseudoscalar observables. This leads to the following definition that enables chirality to be distinguished from other types of dissymmetry:[7-10,20,35]

> *True chirality is exhibited by systems that exist in two distinct enantiomeric states that are interconverted by space inversion, but not by time reversal combined with any proper spatial rotation.*

This means that the enantiomorphism shown by truly chiral systems is time-invariant. Enantiomorphism that is time-noninvariant has different characteristics that I call false chirality in order to emphasise the distinction.

It is easy to see that a stationary object such as a finite helix that is chiral according to the traditional stereochemical definition is accommodated by the first part of this definition: space inversion is a more fundamental operation than the mirror reflection traditionally invoked but provides an equivalent result; and the second part of the definition is irrelevant for a stationary object. However, the full definition is required to identify more subtle sources of chirality in which motion is an essential ingredient. The following examples will make this clear.

3.3 Translating Spinning Cones, Spheres and Elementary Particles

Consider a cone spinning about its symmetry axis. This system certainly supports enantiomorphism because the space-inverted version is not superposable on the original (Fig. 1a), so it might be thought that a spinning cone is a chiral object. However, according to the definition above, it is false chirality because time reversal followed by a rotation $\hat{R}_\pi$ through 180° about an axis perpendicular to the symmetry axis generates the same object as space inversion (Fig. 1a). But if the spinning cone is also translating along the axis of spin, time reversal followed by a 180° rotation now generates a different system to that generated by space inversion (Fig. 1b). Hence a *translating* spinning cone exhibits true chirality.

In fact the translating spinning object does not need to be a cone. A sphere translating along the axis of spin also shows true chirality. This can be appreciated by looking at just the pattern of arrows in Fig. 1b and ignoring the cone.

The molecular equivalent of a stationary spinning cone is a symmetric top in a rotational quantum state $|J, K, M\rangle$. The parity operation transforms $|J, K, M\rangle$ into $|J, -K, M\rangle$, which therefore has mixed parity,[42] so that these two states correspond to the two non-superposable cones in Fig. 1a. And just as the two cones can be interconverted by time reversal followed by a rotation through 180° about an axis perpendicular to the symmetry axis, so this sequence of operations interconverts $|J, K, M\rangle$ and $|J, -K, M\rangle$.

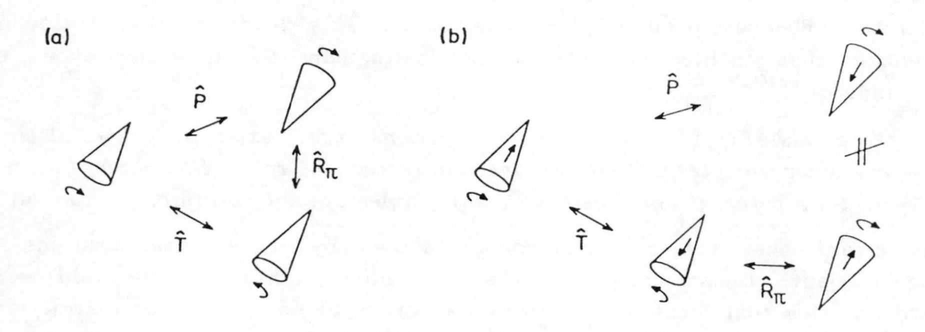

Fig. 1. The effect of $\hat{P}$, $\hat{T}$ and $\hat{R}_\pi$ on (a) a stationary spinning cone, and (b) a translating spinning cone.

This shows that mixed parity in a molecular quantum state is not necessarily sufficient to generate chirality, despite the fact that it can result in two enantiomeric objects. Although mixed parity is a necessary condition for any odd-parity observable, further characteristics are required for different types of odd-parity observable. In this instance the mixed parity characteristic of the rotational quantum state $|J, K, M\rangle$ results in a symmetric top with $K \neq 0$ showing a space-fixed electric dipole moment (an odd-parity observable transforming as a polar vector) and hence a first-order Stark effect provided the top is dipolar to start with; but in order for the top to be dipolar there is the additional requirement of mixed parity internal (vibrational-electronic) quantum states associated with a molecular framework of C_n or C_{nv} symmetry. On the other hand natural optical rotation in isotropic samples is an odd-parity observable transforming as a pseudoscalar and so requires mixed parity vibrational-electronic quantum states associated with a chiral molecular framework (symmetry C_n, D_n, O, T or I), but there is no requirement for mixed parity rotational states. The origin of these mixed parity internal states is discussed in Section 6 below.

The photons in a circularly polarized light beam propagating as a plane wave are in spin angular momentum eigenstates characterized by a spin quantum number $s = 1$ with quantum numbers $m_s = +1$ and -1 corresponding to projections of the spin angular momentum vector parallel and antiparallel, respectively, to the propagation direction. The absence of states with $m_s = 0$ is connected with the fact that photons, being massless, have no rest frame and so always move with the velocity of light (the usual $2j + 1$ projections for a general angular momentum vector are defined in the rest frame).[32] In the usual convention, the electric vector of a right-circularly polarized light beam rotates in a clockwise sense when viewed towards the source of the beam, so right- and left-circularly polarized photons have spin angular momentum projections $-\hbar$ and $+\hbar$, respectively, along the propagation direction. Considerations analogous to those above for a translating spinning sphere then show that a circularly polarized photon exhibits true chirality.

The case of a spinning electron ($s = 1/2$, $m_s = \pm 1/2$) is rather different to that of a circularly polarized photon because an electron has rest mass. From the foregoing it is clear that, whereas a stationary spinning electron is not a chiral object, an electron translating with its spin projection parallel or antiparallel to the propagation direction exhibits true chirality, with opposite spin projections corresponding to opposite handedness. Indeed, beams of spin-polarized electrons impinging on targets composed of chiral molecules are expected to show effects analogous to the polarization effects in the light beams used as probes in conventional optical activity experiments.[43–45] One such effect has recently been observed: an asymmetry in the attenuation of beams of right- and left-handed spin-polarized 5 eV electrons on passing through camphor vapour.[46] A central aspect of such experiments is that, all other things being equal, the magnitudes of the optical activity observables should increase with increasing electron velocity because electron chirality is velocity-dependent. This suggestion is reinforced by a mechanism proposed for asymmetric decomposition of enantiomeric chiral molecules by longitudinally spin-polarized electrons that is a function of v/c,[47] and by the discussion in Section 7 below of the relativistic aspects of chirality and the associated velocity-dependence of the amplitude of parity violation in the weak interaction. The fact that such a large effect was observed in the experiment just mentioned using electron beams of such low energy is probably due to resonance associated with a temporary negative ion state generated by capture of an electron into the π^* orbital of the carbonyl group.[48]

Similar experiments have been proposed for beams of spin-polarized neutrons,[49–52] which are also spin–1/2 particles with rest mass.

3.4 Electric, Magnetic and Gravitational Fields

It is clear that neither a static uniform electric field **E** (a time-even polar vector) nor a static uniform magnetic field **B** (a time-odd axial vector) constitutes a chiral system. Likewise time-dependent uniform electric and magnetic fields. Furthermore, contrary to a suggestion first made by Curie,[27] no combination of a static uniform electric and a static uniform magnetic field can constitute a chiral system. Collinear electric and magnetic fields do indeed generate enantiomorphism, but it is time-noninvariant and so corresponds to false chirality. Thus parallel and antiparallel arrangements are interconverted by space inversion and are not superposable:

$$\begin{array}{ccc} \xrightarrow{\mathbf{E}} & & \xleftarrow{\mathbf{E}} \\ & \overset{\hat{P}}{\longleftrightarrow} & \\ \xrightarrow{\mathbf{B}} & & \xrightarrow{\mathbf{B}} \end{array}$$

But they are also interconverted by time reversal combined with a rotation through 180°:

$$\begin{array}{ccccc} \xrightarrow{\mathbf{E}} & & \xrightarrow{\mathbf{E}} & & \xleftarrow{\mathbf{E}} \\ & \overset{\hat{T}}{\longleftrightarrow} & & \overset{\hat{R}_{\pi}}{\longleftrightarrow} & \\ \xrightarrow{\mathbf{B}} & & \xleftarrow{\mathbf{B}} & & \xrightarrow{\mathbf{B}} \end{array}$$

Zocher and Torok[33] also recognized the flaw in Curie's suggestion: they called the collinear arrangement of electric and magnetic fields a time-asymmetric enantiomorphism, and said that it does not permit a time-symmetric optical activity.

In fact the basic requirement for two collinear vectorial influences to generate chirality is that one transforms as a polar vector and the other as an axial vector, with both either time-even or time-odd. The second case is exemplified by the rotating translating cone or sphere discussed above, and by *magneto-chiral* phenomena such as a birefringence and a dichroism induced in a chiral sample by a uniform magnetic field collinear with the propagation vector **k** of a light beam of arbitrary polarization.[53-57] Thus parallel and antiparallel arrangements of **B** and **k** are true chiral enantiomers because they cannot be interconverted by time reversal since **k**, unlike **E**, is time-odd. The magneto-chiral birefringence and dichroism observables transform as time-odd polar vectors.[55]

Analogous to collinear electric and magnetic fields is the case of a rapidly rotating vessel with the axis of rotation perpendicular to the earth's surface.[58] Here we have the time-odd axial angular momentum vector of the spinning vessel either parallel or antiparallel to the earth's gravitational field, itself a time-even polar vector. The physical influence in this case therefore exhibits false chirality.

3.5 True Chirality in Group Theoretical Statistical Mechanics

It is possible to discuss chirality in a statistical context using group-theoretical arguments. The starting point is the following re-formulation of Neumann's principle in the language of group theory (ref. 20 p. 195): any tensor components representing a physical property of a system must transform as the totally symmetric irreducible representation of the system's symmetry group. It is worth quoting *verbatim* the following statement by Landau and Lifshitz (ref. 59 p. 439) since this specifies very clearly the symmetry groups to be used for molecular liquids in both the achiral and chiral cases:

> "Finally, it may be mentioned that in ordinary isotropic liquids also there are two different types of symmetry. If the liquid consists of a substance which does not have stereoisomers, it is completely symmetrical not only under a rotation through any angle about any axis but also under a reflection in any plane, i.e. its symmetry group is the complete group of rotations about a point, together with a centre of symmetry (group K_h). If the substance has two stereoisomeric forms, however, and the liquid contains different numbers of molecules of the two isomers, it will not possess a centre of symmetry and will therefore not allow reflections in planes. Its symmetry group is just the complete group of rotations about a point (group K)."

Hence a molecular property tensor is decomposed into sets of components with respect to either the irreducible representations $D_{g,u}^{(j)}$ of the full rotation group K_h or the irreducible representations $D^{(j)}$ of the rotation group K depending on whether the fluid is composed of achiral or chiral molecules, respectively; and only those components spanning either $D_g^{(0)}$ or $D^{(0)}$ in the two cases are observable.

This approach can be extended to fluids in the presence of an external field of force. The overall symmetry of the system is now that of the force field, which means that molecular property tensor components which are zero in the isotropic fluid can be induced by the field if there are field components spanning the same nontotally symmetric irreducible representations of K_h (or K if the molecules are chiral) as the property tensor components in question.

These considerations have centred on the point group symmetry of the bulk system with property tensor components referred to space-fixed axes. Starting from the axiom that the properties of a molecule subject to intermolecular forces in the liquid, solution or dense gas states have ensemble averages based on *molecular* axes which reduce as fully symmetric representations of the *molecular* point group, Whiffen[60] has generalized these concepts to provide a systematic method for enabling ensemble averages of tensor components, including time correlation functions, on both molecular and space-fixed axes to be identified as zero or else potentially nonzero.

Evans[61,62] has elevated these three basic concepts (Neumann's principle expressed in group-theoretical language, Whiffen's axiom, and the extension to field-induced phenomena) into 'three principles of group-theoretical statistical mechanics', and has applied them to reduce a number of complicated problems in the molecular dynamics of fluids to simple point group symmetry considerations. He has considered chiral influences in some detail, and has pointed out that the above definition of true chirality harmonises with group-theoretical statistical mechanics since it always leads to time-even influences with components spanning the pseudoscalar irreducible representation $D_u^{(0)}$ of K_h which can therefore generate chiral phenomena. The simultaneous influence of a static magnetic field **B** and the propagation vector **k** of an unpolarized light beam provides a good example: $\Gamma(\mathbf{B})$, the irreducible representation spanned by **B,** is the axial vector representation $D_g^{(1)}$; whereas $\Gamma(\mathbf{k})$ pertaining to **k** is the polar vector representation $D_u^{(1)}$ so that

$$\Gamma(\mathbf{B}) \times \Gamma(\mathbf{k}) = D_u^{(0)} + D_u^{(1)} + D_u^{(2)} . \tag{16}$$

It is the scalar product of **B** and **k** that generates the $D_u^{(0)}$ part of the influence, the $D_u^{(1)}$ and $D_u^{(2)}$ parts requiring perpendicular vectors. (Although **k** itself is not a force field, it is associated with a linear photon momentum which implies a force with the same symmetry characteristics as **k** because of momentum transfer during collisions with molecules in the medium).

Notice that Evans' analysis reinforces the conclusions in Section 3.1 above that magnetic optical rotation is not a chiral phenomenon: **B** transforms as the axial vector representation $D_g^{(1)}$ in K_h and so can only induce in achiral isotropic fluids observables transforming as axial vector components, never pseudoscalars. Magneto-chiral birefringence and dichroism also harmonizes with this approach: the associated observables transform as time-odd polar vectors; and although the magnetic field is a time-odd axial vector, both the magnetic field and the magneto-chiral birefringence and dichroism observables transform the same in the point group K of a chiral fluid, namely as $D^{(1)}$ (these effects cannot occur in an achiral fluid because the magnetic field and the magneto-chiral observables now transform differently, namely as $D_g^{(1)}$ and $D_u^{(1)}$, respectively, in K_h).

4 Absolute Asymmetric Synthesis

The use of an external physical influence to produce an enantiomeric excess in what would otherwise be a racemic product of a prochiral chemical reaction is known as an absolute asymmetric synthesis.[4,63,64] The subject still attracts much interest and controversy,[8,9,65] not least because it is an important ingredient in considerations of the prebiotic origins of biological molecules.[66–68] As we shall see, the new concepts of true and false chirality have helped to resolve some of the controversies, and have also exposed some rather subtle and unexpected analogies with $\hat{C}\hat{P}$ violation in elementary particle physics.

4.1 Truly Chiral Influences

If an influence can be classified as truly chiral one can be confident that it has the correct symmetry characteristics to induce absolute asymmetric synthesis, or some associated process such as preferential asymmetric decomposition, in any conceivable situation, although of course the influence might be too weak to produce any observable effect. In this respect it is important to remember Jaeger's dictum[69,70] "The necessary conditions will be that the externally applied forces are a *conditio sine qua non* for the initiation of the reaction which would be impossible without them".

Circularly polarized photons are the obvious choice, and several examples of the use of circularly polarized visible or ultraviolet radiation in asymmetric synthesis or preferential asymmetric photodecomposition are known[4,63–68] and are usually based on electronic circular dichroism.

Much less obvious is the use of an *unpolarized* light beam collinear with a magnetic field. This idea was first mooted by Wagnière and Meier,[71] with a suggested mechanism based on the new phenomenon of magneto-chiral dichroism mentioned in Section 3.4 above. Having seen that this system exhibits true chirality, we can be confident that it can induce absolute asymmetric synthesis regardless of the details of any particular mechanism.[8,9]

Evans[62] has provided a discussion of truly chiral influences in absolute asymmetric synthesis from the standpoint of group theoretical statistical mechanics which, as mentioned in Section 3.5 above, prescribes that the force field from such an influence must provide time-even components transforming as the pseudoscalar irreducible representation $D_u^{(0)}$ of the full rotation group K_h. As well as confirming my earlier analysis, Evans has suggested another influence with the same symmetry characteristics as the magneto-chiral influence, namely a collinear combination of the angular velocity vector $\mathbf{\Omega}$ of the liquid ensemble of molecules with the propagation vector $\mathbf{k}$ of a light wave. The resulting effect, called 'spin chiral dichroism', might be implemented quite effectively using a circularly polarized infrared, microwave or radiofrequency laser beam to induce angular velocity in the molecules by way of the rotating electric vector of the radiation field, together with a collinear unpolarized ultraviolet laser beam to initiate photolysis.

4.2 Falsely Chiral Influences

When considering the possibility or otherwise of absolute asymmetric synthesis being induced by a falsely chiral influence, it is important to realize that a fundamental distinction must be made between reactions that have been allowed to reach thermodynamic equilibrium (*thermodynamic control*) and reactions that have not attained equilibrium (*kinetic control*).

The case of thermodynamic control is quite clear. It is generally accepted that, because a chiral molecule M and its enantiomer $\overline{M}$ are isoenergetic in the presence of, say, collinear electric and magnetic fields, or a spinning vessel with its axis perpendicular to the earth's surface (neglecting the very small differences due to parity violation discussed later), such falsely chiral influences cannot induce absolute asymmetric synthesis in a reaction mixture which is isotropic in the absence of the influence and which has been allowed to reach thermodynamic equilibrium.[72-74] (The energy equivalence of M and $\overline{M}$ follows from a consideration of the invariance properties of the Hamiltonian in the presence of the influence under the combined operations of space inversion and time reversal, or from Neumann's principle).

The situation is less straightforward for reactions under kinetic control[58,75-77] A preliminary insight can be obtained from the fact that Neumann's principle cannot be applied in space-time to a system in which the entropy is changing,[25] which is certainly the case for a reaction mixture away from equilibrium. This gives a hint that false chirality might suffice, for the intrinsically preferred direction of time associated with the changing entropy destroys the time reversal symmetry of the bulk reacting system so that the two time-noninvariant enantiomeric influences remain distinct.[9,78] A general argument based on the principal of detailed balancing appears to negate this last conclusion,[76] and was used to reinforce the criticisms of the claims for absolute asymmetric synthesis in collinear electric and magnetic fields[79] and in spinning vessels.[80] However, as discussed in the next section, detailed balancing itself might not hold.

4.3 The Breakdown of Microscopic Reversibility: Enantiomeric Detailed Balancing

I have suggested that conventional detailed balancing, and the associated kinetic principles, might not be valid for reactions involving chiral molecules in a time-noninvariant enantiomorphous influence.[81] This suggestion was inspired by a remark of Lifshitz and Pitaevskii[82] that, for a system comprising chiral molecules of just one enantiomer, detailed balancing in the literal sense does not obtain because space inversion as well as time reversal is applied to each microscopic process, so that a completely different system is generated which cannot be compared with the original in order to deduce new information as to its properties. This is seen most clearly from the following quantum-mechanical description of the microscopic process.[30] The amplitude for a transition from some initial linear momentum state $\mathbf{p}$ to some final state $\mathbf{p}'$ is written $\langle \mathbf{p}' | \hat{T} | \mathbf{p} \rangle$, where $\hat{T}$ is the operator responsible for the transition. If $\hat{T}$ involves purely electromagnetic interactions it will be invariant under both parity and time reversal, which enables us to write

$$\langle \mathbf{p}' | \hat{T} | \mathbf{p} \rangle \overset{\text{under } \hat{\mathrm{T}}}{=} \langle -\mathbf{p} | \hat{T} | -\mathbf{p}' \rangle \overset{\text{under } \hat{\mathrm{P}}}{=} \langle \bar{\mathbf{p}} | \hat{T} | \bar{\mathbf{p}}' \rangle, \tag{17}$$

where we have allowed the particles to be chiral, the bar denoting the $\hat{\mathrm{P}}$-enantiomer. The first equality in (17), obtained from time reversal alone, is the basis of the conventional principle of microscopic reversibility and, when averaged over the complete system of reacting particles at equilibrium, of the principle of detailed balancing.[83] The second equality, obtained by applying space inversion to the time-reversed transition amplitude, describes the inverse process involving the *enantiomeric* particles.

Conventional detailed balancing is usually adequate for the kinetic analysis of reactions, even those involving chiral molecules, because conventional microscopic reversibility expressed by the first equality in (17) is usually valid. However, in the presence of a time-noninvariant enantiomorphous influence such as collinear electric and magnetic fields, time reversal alone is not a symmetry operation since a different influence is generated: space inversion must also be applied in order to recover the original relative orientations of **E** and **B**. The first equality in (17) is therefore no longer valid, and we must base any kinetic analysis on the relationship

$$\langle \mathbf{p}' | \hat{T} | \mathbf{p} \rangle \overset{\hat{\mathrm{T}}\hat{\mathrm{P}}}{=} \langle \bar{\mathbf{p}} | \hat{T} | \bar{\mathbf{p}}' \rangle. \tag{18}$$

Lifshitz and Pitaevskii considered a system containing just one enantiomer. But if the system contains equal numbers of enantiomeric molecules at equilibrium (a racemic mixture), it does appear to be possible to obtain new information using arguments based on (18). I have suggested that, in this situation, a new *principle of enantiomeric detailed balancing* can be invoked in which the statistical average of all the microscopic processes involving one enantiomer can be balanced by the average of the inverse processes, in the sense of (18), involving the enantiomeric molecules.[8,78] Consider a unimolecular process in which a prochiral molecule R generates a chiral molecule M or its enantiomer $\overline{\mathrm{M}}$:

$$M \underset{k_b}{\overset{k_f}{\rightleftarrows}} R \underset{\bar{k}_f}{\overset{\bar{k}_b}{\rightleftarrows}} \bar{M} . \tag{19}$$

In refs. 8 and 78, enantiomeric detailed balancing was applied to the separate reactions in this scheme (i.e. $[R]k_f = [\bar{M}]\bar{k}_b$ and $[R]\bar{k}_f = [M]k_b$) to show that a time-noninvariant enantiomorphous influence allows a difference in rate constants for enantiomeric processes, i.e $k_f \neq \bar{k}_f$ and $k_b \neq \bar{k}_b$. However, the analysis was incomplete because difficulties arise from the condition that the concentrations $[M]$ and $[\bar{M}]$ of the two enantiomers must be equal at thermodynamic equilibrium. These difficulties were subsequently resolved by realizing that the scheme (19) represents just one racemization pathway: by considering a reaction quadrangle which allows at least one alternative interconversion pathway between the enantiomers, it was shown that the thermodynamic and kinetic requirements can be reconciled.[81] The reason that the conditions $[M] = [\bar{M}]$, $k_f \neq \bar{k}_f$ and $k_b \neq \bar{k}_b$ can hold simultaneously is that, at *true* thermodynamic equilibrium, the different enantiomeric excesses associated with each separate racemization pathway sum to zero.

The conrotatory ring closure of a substituted butadiene to produce a cyclobutene was used as a simple example to show how collinear electric and magnetic fields can bring about a difference in potential energy profiles, and hence rate constants, for enantiomeric reactions.[81] This example exposes the origin of the breakdown of microscopic reversibility in this situation: the transient magnetic dipole moment in the transition state, and hence the interaction with the magnetic field, is velocity-dependent, and reverses in sign for the two directions of motion through the transition state. The function of the electric field is to partially align the molecules in the fluid so that these equal and opposite magnetic interactions for the forward and backward reactions do not average to zero over the bulk sample. In fact the electric field is not required if the molecules are already oriented, which means that a magnetic field alone can induce a breakdown of microscopic reversibility in processes involving chiral molecules aligned in a crystal or on a surface, for example.

In more conventional situations, time reversal invariance and hence microscopic reversibility in the presence of a magnetic field can be recovered by also reversing the moving charged particles that are the source of the magnetic field.[84] This prescription fails in the present situation because the system is only invariant under $\hat{T}\hat{P}$, not $\hat{T}$ alone, so microscopic reversibility is only recovered in the time-reversed *enantiomeric* system. In conventional chemical kinetics, the microscopic reversibility which follows from the assumption of $\hat{T}$ invariance is conceptualized in terms of a potential energy profile that is the same in the forward and backward directions for a given reaction. But for reactions involving chiral molecules in situations where only the combined $\hat{T}\hat{P}$ invariance holds (as in the presence of collinear electric and magnetic fields, or for oriented molecules in the presence of a magnetic field alone), we must extend this picture to show the same potential energy profiles for the forward and backward *enantiomeric* reactions with the forward and backward profiles for the reaction of a given enantiomer in general different, as illustrated in Fig. 2.

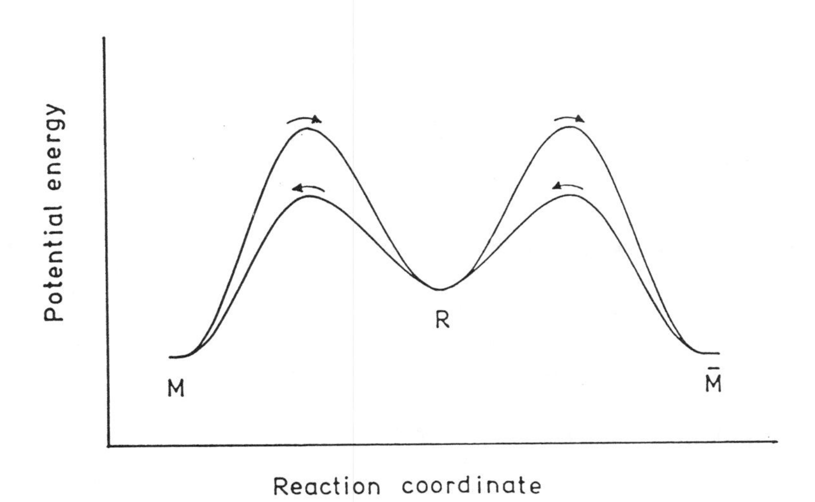

Fig. 2. Potential energy surfaces for enantiomeric reactions in the presence of a time-noninvariant enantiomorphous influence.

Gilat[85] has also suggested that time reversal invariance might not hold for certain processes involving chiral molecules. However, the context of his discussion is different from that given here since it centres on a detailed mechanism he has postulated[86] involving the interaction of chiral molecules with ions in a solvent.

4.4 Unitarity and Thermodynamic Equilibrium

The unitarity of the scattering matrix,[32] together with $\hat{T}\hat{P}$ invariance in the context of a collection of interconverting chiral enantiomers M and $\overline{M}$, can be used to generalize this analysis.[81]

Let T_{ji} be the amplitude for a transition from a state i to a state j. The requirement of unitarity for the scattering matrix (which corresponds to the fact that the sum of the transition probabilities from a given initial state to all final states is unity) leads to the following relationship for the transition amplitude:[32]

$$\sum_j T_{ij} T_{ij}^* = \sum_j T_{ji}^* T_{ji} \tag{20}$$

where the sum over j includes all states j and $\bar{j}$ of both enantiomers. It is important to realize that Hermiticity is not invoked in proving this result: T_{ij} is only Hermitian in a first approximation.[32] If $\hat{T}\hat{P}$ invariance holds, we can use (18) to write

$$T_{ji} = T_{\bar{i}\bar{j}} \tag{21}$$

and this, together with (20), gives

$$\sum_j |T_{ji}|^2 = \sum_j |T_{\bar{i}j}|^2 = \sum_j |T_{ij}|^2. \tag{22}$$

If the system is racemic and in thermal equilibrium, equivalent enantiomeric states are equally populated. The second equality in (22) then shows that transitions out of these states must produce molecules in a given state i and the enantiomeric molecules in the equivalent state $\bar{i}$ in equal numbers. Thus no excess of one enantiomer over the other can develop at thermal equilibrium even when the presence of a time-noninvariant enantiomorphous influence destroys the equality between rates for specific enantiomeric transitions, i.e. when

$$|T_{ji}|^2 \neq |T_{\bar{j}\bar{i}}|^2. \tag{23}$$

Notice that, had we allowed T_{ji} to be Hermitian, rates for specific enantiomeric transitions *would* be equal. From (20) and (21) we can also write

$$\sum_j |T_{ji}|^2 = \sum_j |T_{j\bar{i}}|^2, \tag{24}$$

which shows that the total transition rates out of equivalent enantiomeric states are equal. Since in thermal equilibrium no excess of i over $\bar{i}$ may develop, this implies that any pre-existing excess tends to be diminished. This argument is similar to that used to demonstrate the existence of equal numbers of particles and antiparticles at thermodynamic equilibrium, despite $\hat{C}\hat{P}$ violation, in the big bang model of the early universe.[87]

Thus if an absolute asymmetric synthesis starting with a pure prochiral reagent R could indeed be induced by collinear electric and magnetic fields, say, the enantiomeric excess will ultimately disappear if sufficient time is allowed for the establishment of true thermodynamic equilibrium in which all possible racemization pathways have separately equilibrated. Similarly, the influence could not generate an enantiomeric excess in a racemic mixture. Thus no fundamental thermodynamic principles are violated by the suggestion that a time-noninvariant enantiomorphous influence can, *via* a breakdown of microscopic reversibility, generate a difference in rate constants for enantiomeric processes. It should be mentioned that, at first sight, the proposed difference in enantiomeric rate constants appears to lead to the possibility of constructing a perpetual motion machine of the second kind,[81] which conflicts with the conclusion just reached that no fundamental thermodynamic principles are violated. Exactly the same difficulty arises in the analogous discussion of particle–antiparticle processes in the presence of $\hat{C}\hat{P}$ violation; however, Teller[88] has identified the fallacy present in such schemes for a perpetual motion machine (a violation of unitarity always arises).

There is one last element, not yet demonstrated at the time of writing, required to complete this discussion; namely that the combined influence of collinear electric and magnetic fields on the elementary scattering processes associated with reaction and transport processes involving chiral molecules is able to induce the necessary non-Hermitian contribution to the transition amplitude.

This analysis is important for chemical physics generally because it reinforces the conclusion that it is unitarity, rather than microscopic reversibility, that is necessary for the validity of Boltzmann's H-theorem.[87,89]

5 Symmetry Violation

A symmetry violation (often called symmetry nonconservation) arises when one of the 'non-observables' discussed in Section 2.1 above is actually observed. A consideration of symmetry violation, and how it differs from spontaneous symmetry breaking, provides considerable insight into the phenomenon of molecular chirality.

5.1 The Fall of Parity

Prior to 1957 it had been accepted as self evident that handedness is not built into the world at any level. If two objects exist as non-superposable mirror images of each other, such as the two enantiomers of a chiral molecule, it did not seem reasonable that nature should prefer one over the other. Any difference between enantiomeric systems was thought to be confined to the sign of pseudoscalar observables: the mirror image of any complete experiment involving one enantiomer should be realizable, with any pseudoscalar observable (such as optical rotation angle) changing sign but retaining *precisely* the same magnitude. Then in 1956 Lee and Yang[90] pointed out that, unlike the electromagnetic and strong interactions, there was no evidence for parity conservation in processes involving the weak interaction. Of the experiments they suggested, that carried out by Wu et al.[91] is the most famous.

The Wu experiment studied the β-decay process

$$^{60}Co \rightarrow {}^{60}Ni^{+} + \bar{e} + \tilde{\nu}_e$$

in which, essentially, a neutron n has decayed via the weak interaction into a proton p, an electron $\bar{e}$ and an electron antineutrino $\tilde{\nu}_e$. The nuclear spin magnetic magnetic moment I of each ^{60}Co nucleus in the sample was aligned with an external magnetic field **B**, and the angular distribution of electrons measured. It was found that the electrons are emitted preferentially in the direction *antiparallel* to that of the magnetic field (Fig. 3a). As discussed in

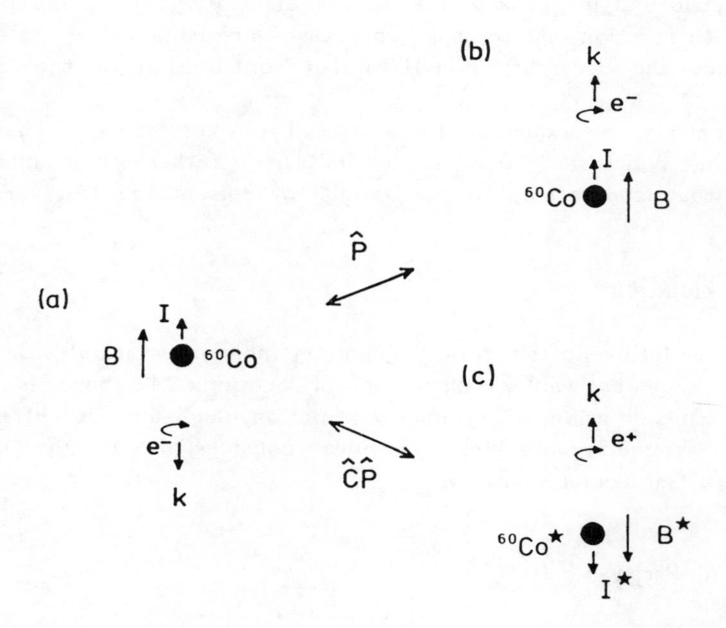

Fig. 3. Parity violation in β-decay. Only experiment (a) is found. The space-inverted version (b) cannot be realized. Symmetry is recovered in experiment (c), which is obtained from (a) by invoking charge conjugation simultaneously with space inversion (Co* is anti-Co, and **B*** and **I*** are reversed relative to **B** and **I** because the charges of the moving source particles have reversed).

Section 2.2, **B** and **I** are axial vectors and so do not change sign under space inversion, whereas the electron propagation vector **k** does because it is a polar vector. Hence in the corresponding space-inverted experiment the electrons should be emitted *parallel* to the magnetic field (Fig. 3b). It is only possible to reconcile Figs. 3a and 3b with parity conservation if there is no preferred direction for electron emission (an isotropic distribution), or if the electrons are emitted preferentially in the plane perpendicular to **B**. The observation of (a) alone provides unequivocal evidence for parity violation. Another important aspect of parity violation in β-decay is that the emitted electrons have a 'left-handed' longitudinal spin polarization, being accompanied by 'right-handed' antineutrinos. The corresponding antiparticles emitted in other β-decay processes, namely positrons and neutrinos, have the opposite handedness. (The projection of the spin angular momentum **s** of a particle along its direction of motion is called the helicity $\lambda = \mathbf{s} \cdot \mathbf{p} / |\mathbf{p}|$. Spin–1/2 particles can have $\lambda = \pm \hbar/2$, the positive and negative states being called right- and left-handed; but this corresponds to the opposite sense of circularity to that used in the usual definition of right- and left-circularly polarized light).

In fact symmetry is recovered by invoking charge conjugation simultaneously with space inversion: the missing experiment is to be found in the antiworld! Thus it can be seen from Fig. 2c that the combined operation of $\hat{C}\hat{P}$ interconverts the two equivalent experiments for which nature appears to have no preference (assuming $\hat{C}\hat{P}$, and hence $\hat{T}$, is not violated: see Section 5.3 below). This result implies that $\hat{P}$ violation is accompanied here by $\hat{C}$ violation: explicitly, absolute charge is distinguished since the charge that we call negative is carried by the electrons, which are emitted with a left-handed spin polarization.

Notice that the Wu experiment provides a nice example of true chirality, as defined in Section 3.2. The two experiments (a) and (b) in Fig. 2 are enantiomeric with respect to space inversion, but cannot be interconverted by time reversal combined with any proper spatial rotation.

5.2 Parity Violation in Atoms and Molecules: the Weak Neutral Current

Since the electromagnetic interaction is formulated in terms of an exchange of virtual photons, it was natural to postulate the existence of a particle, denoted W, that mediated the weak interaction. Like the photon, the W is a boson; but unlike the photon, which is neutral, the W must be charged (W^+ or W^-) since β-decay, for example, involves an exchange of charge between particles. A second difference is that, whereas photons have zero mass, the Ws are massive (this follows from the Yukawa-Wick argument that the range of a force is inversely proportional to the mass of the exchanged quantum: the electromagnetic and weak interactions have infinite and very short ranges, respectively).

Following the Wu experiment, the original Fermi theory of the weak interaction[92] was upgraded in order to take account of parity violation. This was achieved by reformulating the theory in such a way that the interaction takes the form of a left-handed pseudoscalar. However, a number of technical problems remained, which were finally overcome in the 1960s in the celebrated work of Weinberg,[93] Salam[94] and Glashow,[95] which unified the theory of the weak and electromagnetic interactions into a single electroweak interaction theory. The conceptual basis of the theory rests on two pillars: gauge invariance and spontaneous symmetry breaking,[96,97] but the details are beyond the scope of this article. In addition to accommodating the massless photon and the two massive charged W^+ and W^- particles, a new massive neutral particle called Z^0 (the neutral intermediate vector boson) was predicted which can generate a whole new range of *neutral current* phenomena, including parity-violating effects in atoms and molecules. The theory provides a simple relation between the weak and electromagnetic coupling constants ($g\sin\vartheta_w = e$, where g and e are the weak and electromagnetic unit charges, and ϑ_w is the Weinberg angle), and also gives the masses of the W^+, W^- and Z^0 particles. In one of the most important experiments of all time, these three particles were detected in 1983 at CERN in proton-antiproton scattering experiments.[98]

This weak neutral current interaction generates parity-violating interactions between electrons, and between electrons and nucleons. The latter leads to the following electron-nucleus contact interaction in atoms and molecules (in a.u. where $\hbar = e = m_e = 1$):[99,100]

$$\hat{V}^{PV}_{eN} = \frac{G\alpha}{4\sqrt{2}} Q_w \{ \boldsymbol{\sigma}_e \cdot \mathbf{p}_e , \rho_N(\mathbf{r}_e) \}_+ , \tag{25}$$

where G is the Fermi weak coupling constant, α is the fine structure constant, $\boldsymbol{\sigma}_e$ and $\mathbf{p}_e$ are the Pauli spin operator and linear momentum operator of the electron, $\rho_N(\mathbf{r}_e)$ is a normalized nuclear density function and

$$Q_w = Z(1 - 4\sin^2\vartheta_w) - N$$

is an effective weak charge which depends on the proton and neutron numbers Z and N. $\{ \ \}_+$ denotes an anticommutator. The electron-electron interaction is usually neglected, in which case (25) is taken as the parity-violating term to be added to the Hamiltonian of an atom or molecule. Since $\boldsymbol{\sigma}_e$ and $\mathbf{p}_e$ are axial and polar vectors, respectively, and all the other factors are scalars, $\hat{V}^{PV}_{eN}$ transforms as a pseudoscalar, as required, and so can mix even and odd parity electronic states at the nucleus.

Chiral molecules support a unique manifestation of parity violation in the form of a lifting of the exact degeneracy of the energy levels of mirror-image enantiomers.[100-105] Being pseudoscalars, the parity-violating weak neutral current terms in the molecular Hamiltonian are odd under space inversion:

$$\hat{P}\hat{V}^{PV}\hat{P}^{-1} = -\hat{V}^{PV}. \tag{26}$$

As discussed in Section 6 below, the enantiomeric quantum states ψ_L and ψ_R of a chiral molecule are examples of the mixed parity states (7) and so are interconverted by $\hat{P}$. It then follows that $\hat{V}^{PV}$ shifts the energies of the enantiomeric states in opposite directions:

$$\langle \psi_L | \hat{V}^{PV} | \psi_L \rangle = \langle P\psi_R | \hat{V}^{PV} | P\psi_R \rangle = \langle \psi_R | \hat{P}^{+} \hat{V}^{PV} \hat{P} | \psi_R \rangle$$

$$= -\langle \psi_R | \hat{V}^{PV} | \psi_R \rangle = \varepsilon. \tag{27}$$

Attempts to calculate ε are faced with the following difficulty. The electronic coordinate part of $\hat{V}^{PV}$ in (25) is linear in $\mathbf{p}_e$ and is therefore pure imaginary. Since, in the absence of external magnetic fields, the molecular wavefunction can always be chosen to be real, $\hat{V}^{PV}$ has zero expectation values. Also, the presence of $\boldsymbol{\sigma}_e$ means that only matrix elements between different spin states survive. Consequently, it is necessary to invoke a magnetic perturbation of the wavefunction involving spin, the favourite candidate being spin–orbit coupling.[100,103,104] This leads to a tractable method for detailed quantum-chemical calculations of parity-violating energy differences between enantiomers, giving values of the order 10^{-20} a.u.[100,106–108] (The atomic unit of energy, the Hartree, is equivalent to 27.2 eV or to 4.36×10^{-18} J). There are also slight structural differences between enantiomers.[109] It is intriguing that, in all the cases treated so far, the L-amino acids and the D-sugars, which dominate the biochemistry of living organisms, are found to be the more stable enantiomers.[110,111]

Manifestations of parity violation in atoms have now been observed in the form of optical activity phenomena such as tiny optical rotations in vapours of heavy metals;[112–116] and Hegstrom *et al.*[117] have provided an appealing pictorial representation of the associated atomic chirality in terms of a helical electron probability current density. Parity violation has not yet been observed in molecules. *Achiral* molecules are expected to show similar manifestations to atoms, such as tiny optical rotations, but enhanced perhaps by several orders of magnitude.[118–120] However, the most important challenge is the measurement of the parity-violating enantiomeric energy difference in chiral molecules. Quack[121] has reviewed the various ideas and attempts to measure this difference, and has made detailed proposals for some new experiments which could enhance the sensitivity of the measurements to the required level. A rather different aspect is the recent claim, on the basis of *ab-initio* calculations, that the observed 1.4% excess of chiral (-)- quartz crystals over (+)-quartz crystals in terrestrial samples is a manifestation of parity violation.[111,122]

5.3 Violation of Time Reversal and the $\hat{C}\hat{P}\hat{T}$ Theorem

Violation of time reversal was observed in 1964 in the famous experiment of Cronin, Fitch *et al.* involving measurements of rates for different decay modes of the neutral K-meson.[123–125] Despite intensive efforts since then, no other system has shown the effect. As Cronin has said,[124] nature has provided us with just one extraordinarily sensitive system to convey a cryptic message that has still to be deciphered.

Although unequivocal, the effects are very small; certainly nothing like the parity-violating effects in weak processes, which can sometimes be absolute. In fact $\hat{T}$ violation itself is not observed directly: rather, the observations show $\hat{C}\hat{P}$ violation, from which $\hat{T}$ violation is implied by the celebrated $\hat{C}\hat{P}\hat{T}$ theorem (this statement is qualified at the end of this section). This theorem is derived from general considerations of relativistic quantum field theory,[17,32] and states that the Hamiltonian is invariant to the combined operation of $\hat{C}\hat{P}\hat{T}$ even if it is not invariant to one or more of these operations.

One manifestation of $\hat{C}\hat{P}$ violation is the following decay rate asymmetry of the long-lived neutral K-meson, the K_L:[14,17,97,125]

$$\Delta = \frac{\text{Rate}\,(K_L \rightarrow \pi^- e_r^+ \nu_l)}{\text{Rate}\,(K_L \rightarrow \pi^+ e_l^- \tilde{\nu}_r)} \approx 1.00648. \tag{28}$$

As the formula indicates, K_L can decay into either positive pions π^+ plus left helical electrons e_l^- plus right helical antineutrinos $\tilde{\nu}_r$; or into negative antipions π^- plus right helical positrons e_r^+ plus left helical neutrinos ν_l. Since these two sets of decay products are interconverted by $\hat{C}\hat{P}$, this decay rate asymmetry indicates that $\hat{C}\hat{P}$ is violated. If we naively represent this decay process in the form of 'chemical equilibria' as in (19),

$$\pi^+ + e_l^- + \tilde{\nu}_r \underset{k_b}{\overset{k_f}{\leftrightarrows}} K_L \underset{\bar{k}_f}{\overset{\bar{k}_b}{\leftrightarrows}} \pi^- + e_r^+ + \nu_l, \tag{29}$$

a parallel is established with absolute asymmetric synthesis associated with a breakdown in microscopic reversibility discussed in Section 4.2 since in both cases $k_f \neq \bar{k}_f$. Thus the K_L and the two sets of decay products are the equivalents, with respect to $\hat{C}\hat{P}$, of R, M, and $\overline{\text{M}}$ with respect to $\hat{P}$. We can therefore conceptualize the decay rate asymmetry here as arising from a breakdown in microscopic reversibility due to a time-noninvariant $\hat{C}\hat{P}$ enantiomorphism in the forces of nature[81] (the $\hat{C}\hat{P}\hat{T}$ theorem guarantees that the two distinct $\hat{C}\hat{P}$ enantiomorphous influences are interconverted by $\hat{T}$). The analogy is completed by the fact that, as mentioned in Section 4.3, the asymmetries cancel out over all possible channels at true thermodynamic equilibrium.

Another manifestation of $\hat{C}\hat{P}$ violation arises that should be mentioned, but first more needs to be said about the K^o system. Four distinct states are displayed: particle and antiparticle states $|K^o\rangle$ and $|\tilde{K}^o\rangle$, and two combined states

$$|K_1\rangle = \frac{1}{\sqrt{2}}(|K^o\rangle + |\tilde{K}^o\rangle), \tag{30a}$$

$$|K_2\rangle = \frac{1}{\sqrt{2}}(|K^o\rangle - |\tilde{K}^o\rangle), \tag{30b}$$

which have different energies because of coupling between $|K^o\rangle$ and $|\tilde{K}^o\rangle$ via the weak force. Since the particle and antiparticle states are interconverted by $\hat{C}\hat{P}$,

$$\hat{C}\hat{P}|K^o\rangle = |\tilde{K}^o\rangle, \qquad \hat{C}\hat{P}|\tilde{K}^o\rangle = |K^o\rangle, \tag{31}$$

we can appreciate that the combined states are even and odd eigenstates of $\hat{C}\hat{P}$:

$$\hat{C}\hat{P}|K_1\rangle = |K_1\rangle, \qquad \hat{C}\hat{P}|K_2\rangle = -|K_2\rangle. \tag{32}$$

However, these eigenstates are not pure: for example $|K_2\rangle$, which is odd with respect to $\hat{C}\hat{P}$, is occasionally observed to decay into products which are even with respect to $\hat{C}\hat{P}$. This implies that the Hamiltonian contains a small $\hat{C}\hat{P}$-violating term which mixes $|K_1\rangle$ and $|K_2\rangle$. In fact the long-lived state $|K_L\rangle$ introduced above is actually $|K_2\rangle$ with a small admixture of $|K_1\rangle$; and there is an associated short-lived state $|K_S\rangle$ that is $|K_1\rangle$ with a small admixture of $|K_2\rangle$. One example is the decay into the pair of oppositely charged pions $\pi^+\pi^-$ for which the state of zero angular momentum is even with respect to $\hat{C}\hat{P}$ so only the decay of $|K_1\rangle$ should yield such a product state. What is found is a very small but significant amplitude for K_L to decay into $\pi^+\pi^-$:

$$\eta_{+-} = \frac{\text{amplitude}\,(K_L \rightarrow \pi^+\pi^-)}{\text{amplitude}\,(K_S \rightarrow \pi^+\pi^-)} \approx 2.274 \times 10^{-3}. \tag{33}$$

The $\hat{C}\hat{P}$ violation parameters Δ and η_{+-} extracted from the two at first sight rather different experiments summarized by (28) and (33), together with a third parameter η_{oo} derived from a similar experiment to (33), are in fact all related, and the results of all the experiments involving $\hat{C}\hat{P}$ violation can be summarized by a single complex number.[14,124,126] Although it is only $\hat{C}\hat{P}$ violation that has been observed directly, a detailed but rather complicated analysis of the results of these different experiments does appear to show, without invoking the $\hat{C}\hat{P}\hat{T}$ theorem, that $\hat{T}$ itself is violated; however, this analysis does not demonstrate that the $\hat{T}$ violation and the $\hat{C}\hat{P}$ violation compensate each other to the same degree (in other words a contribution from $\hat{C}\hat{P}\hat{T}$ violation cannot be ruled out).[14]

5.4 Time Reversal Violation in Atoms and Molecules: Electric Dipoles and Magnetic Monopoles

The question naturally arises as to whether $\hat{T}$ violation can be detected in atoms and molecules. The most commonly discussed possible manifestation of $\hat{T}$ violation in atoms is the existence of a permanent electric dipole moment μ (in the absence of an accidental degeneracy as in the hydrogen atom) originating in a $\hat{T}$-violating electric dipole moment on the electron or nucleus or a $\hat{T}$-violating electron–nucleus interaction.[14,112,114,116,126,127] In fact the observation of such a dipole in a general atomic state $|J, M\rangle$ requires violation of both $\hat{P}$ and $\hat{T}$ simultaneously:[14,128,129] this can be shown by developing the expectation value $\langle J, M|\hat{\mu}|J, M\rangle$ using (10) and (13)–(15).[20,35,129] To date no such electric dipole has been found, but the very accurate measurements that have been made are important because of the strong limits that they impose on theory.

The central aspect of $\hat{T}$ violation in centrosymmetric systems (which includes nuclei as well as atoms) is that the $\hat{T}$ invariance of the Hamiltonian imposes a reality condition on the wavefunctions. This can be demonstrated by comparing the effect of both the time reversal operator and the rotation operator on a general atomic (or nuclear) state written as some linear combination of angular momentum base states.[14] Consequently, the expectation value of the $\hat{T}$-violating quantity depends on the phase difference between terms in a wavefunction; that is, on deviations from the reality condition. This use of phase difference as a measure of $\hat{T}$ violation is quite general and arises, for instance, in the detailed analysis of the behaviour of the K-meson system outlined in the previous section.[14]

Attempts to measure $\hat{T}$ violation in molecules have centred so far on possible $\hat{T}$-violating electric dipole moments of nucleons, especially the proton because its charge rules out experiments on proton beams analogous to those performed on neutron beams.[130] Much effort has been expended on thallium fluoride[131–133] because this heavy polar molecule is much more easily polarized than an atom thereby enhancing, for example, $\hat{T}$- and $\hat{P}$-violating hyperfine interactions between the spin-1/2 Tl nucleus and the rest of the molecule.[134]

A rather different approach has been suggested involving the measurement of optical rotation induced in an atomic or achiral molecular system by a static uniform electric field collinear with the light beam.[119,135] As I pointed out some time ago,[19,20] such an effect would violate both $\hat{T}$ and $\hat{P}$. Intuitively, molecules with odd numbers of electrons might be expected to show the largest effects because any $\hat{T}$ violation would lift the Kramers degeneracy.[14]

Chiral molecules add an interesting new dimension relative to atomic and achiral molecular systems in experiments to detect $\hat{T}$ violation because $\hat{P}$ violation is no longer required since parity is spontaneously broken in a chiral molecule (see Section 6.3 below). This could boost the corresponding observable by many orders of magnitude if the $\hat{T}$ violation was associated with, say, an electron–nucleon interaction; but molecular chirality would not be expected to help if the $\hat{T}$ violation was associated with a nucleon or electron electric dipole moment because $\hat{P}$ violation is still required on account of the centrosymmetric symmetry of the particle. A good example is the linear electric optical rotation just mentioned: if the sample were chiral, any optical rotation induced by the electric field would be added to the natural optical rotation generated by the molecular chirality. If this additional optical rotation was linear in the electric field strength only $\hat{T}$ would be violated, not $\hat{P}$.[20]

The most interesting aspect of chiral molecules in this context is their potential value in the search for magnetic monopoles[30,136,137] which, as mentioned earlier, transform as time-odd pseudoscalars. The pseudoscalar character is supported by the spontaneously broken parity of the quantum states of a resolved chiral molecule so, unlike the case of an elementary particle which requires violation of both parity and time reversal to show a magnetic monopole, 'only' time reversal violation is required for a chiral molecule to show a magnetic monopole. A particular enantiomer of a given chiral molecule would be expected to carry a particular $\hat{T}$-violating magnetic charge, which would reverse sign in the mirror-image enantiomer. Useful additional ingredients might be Kramers degeneracy, which obtains if the chiral molecule has an odd number of electrons, together with large spin–orbit coupling. I have found a hint of time-odd pseudoscalar properties in second-rank electromagnetic scattering tensors describing spin–flip processes,[35] but they seem to evaporate on close inspection; however, something might survive if $\hat{T}$ violation is incorporated. It is worth quoting Sachs[14] here: "... the fact that the pseudoscalar magnetic charges (monopole moments) change sign under time reversal suggests that, if such monopoles exist, they must have a dynamic origin at some deep level." Serious discussions of magnetic monopoles are always in the context of elementary particle physics, with the required internal motions provided, for example, by colour currents of quantum chromodynamics;[137] however, it is tantalizing that odd-electron chiral molecules seem to satisfy at least some of the requirements.

Unfortunately there appears to be a fundamental aspect of magnetic monopoles which might prevent their observation in mundane objects such as chiral molecules: as well as being odd under $\hat{P}$ and $\hat{T}$, the magnetic monopole is presumably also odd under $\hat{C}$ (like electric charge). This means that the magnetic monopole would be odd under $\hat{C}\hat{P}\hat{T}$. Since the $\hat{C}\hat{P}\hat{T}$ theorem requires invariance of any Hamiltonian under $\hat{C}\hat{P}\hat{T}$, this implies that a magnetic monopole could only exist in a system in which $\hat{C}\hat{P}\hat{T}$ is spontaneously broken (see Section 6.3 below for the distinction between symmetry violation and spontaneous symmetry breaking). A magnetic monopole of precisely the same magnitude but opposite sign would be found in the $\hat{C}\hat{P}\hat{T}$-enantiomeric system. But the only system I can conceive that might be in a spontaneously broken state of $\hat{C}\hat{P}\hat{T}$ is the entire universe. Hence magnetic monopoles, if they exist, might embrace the entire universe.

6 The Mixed Parity States of a Chiral Molecule

It was shown above (Sections 2.3 and 3.1) that, since a chiral molecule can support pseudoscalar observables, it must exist in mixed-parity internal quantum states (vibrational-electronic) associated with a chiral molecular framework. We now explore the nature of these quantum states, and investigate the consequences of a small parity-violating term in the Hamiltonian.

6.1 The Double Well Model

The origin of these mixed-parity states is well known. It is best appreciated by considering vibrational wavefunctions associated with the 'inversion' mode ν_2 of a molecule such as NH_3 which is said to invert between the two equivalent configurations shown in Fig. 4,[138,139] although this motion does not in fact correspond to an inversion through the centre of mass. If the planar configuration were the most stable, the adiabatic potential energy function would have the parabolic form shown on the left with simple harmonic vibrational levels equally spaced. If a potential hill is raised gradually in the middle, the two pyramidal configurations become the most stable and the energy levels approach each other in pairs. For an infinitely high potential hill, the pairs of levels are exactly degenerate, as shown on the right. The rise of the central potential hill modifies the wavefunctions as shown, but does not destroy their parity. The even and odd parity wavefunctions $\psi(+)$ and $\psi(-)$ describe stationary states in all circumstances. On the other hand, the wavefunctions ψ_L and ψ_R, corresponding to the system in its lowest state of oscillation and localized completely in the left

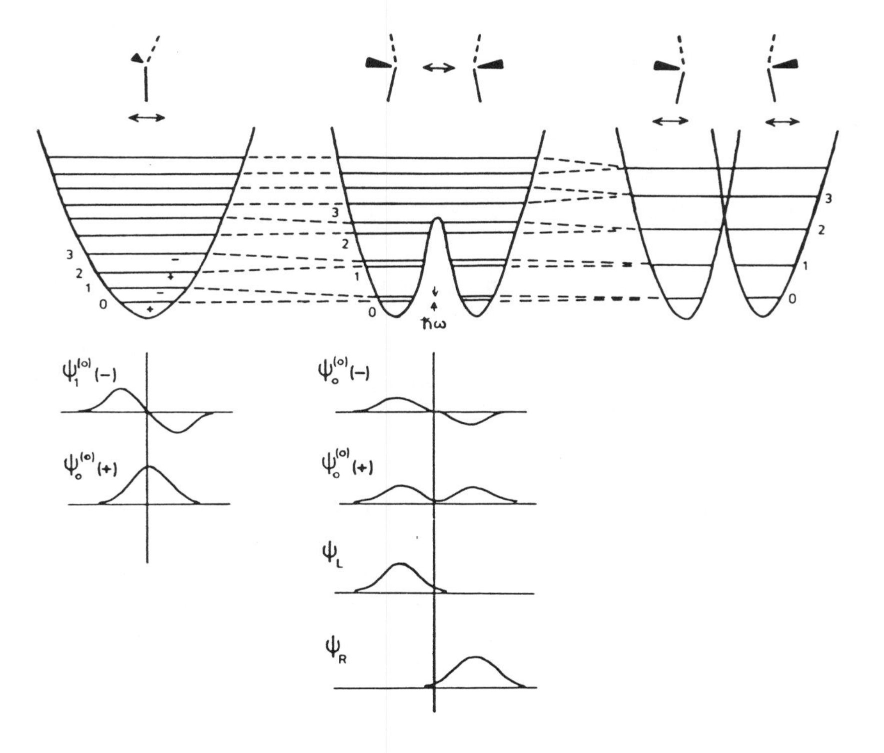

Fig. 4. The vibrational states of a molecule that can invert between two equivalent configurations. $\psi^{(0)}(+)$ and $\psi^{(0)}(-)$ are the amplitudes of the definite parity stationary states with energy W(+) and W(−), and ψ_L and ψ_R are the two mixed-parity non-stationary states at t = 0 and t = π/ω, where $\hbar\omega$ is the tunnelling splitting.

and right wells, respectively, are not true stationary states. They are obtained from the following combinations of the even and odd parity wavefunctions,

$$\psi_L = \frac{1}{\sqrt{2}}[\psi^{(0)}(+) + \psi^{(0)}(-)], \tag{34a}$$

$$\psi_R = \frac{1}{\sqrt{2}}[\psi^{(0)}(+) - \psi^{(0)}(-)], \tag{34b}$$

which are explicit examples of the mixed parity wavefunctions (7).

The wavefunctions (34) are in fact specializations of the general time-dependent wavefunction of a degenerate two-state system (see Section

6.2 below). To be precise, we assume that the system is in the left well at $t = 0$. Then at a later time we have[30]

$$\psi(t) = \frac{1}{\sqrt{2}}\left[\psi^{(0)}(+)\,e^{-iW(+)t/\hbar} + \psi^{(0)}(-)e^{-iW(-)t/\hbar}\right]$$

$$= \frac{1}{\sqrt{2}}\left[\psi^{(0)}(+) + \psi^{(0)}(-)e^{-i\omega t}\right]e^{-iW(+)t/\hbar}, \qquad (35)$$

where $\hbar\omega = W(-) - W(+)$ is the energy separation of the two opposite parity states, which in this context is interpreted as a splitting arising from tunneling through the potential energy barrier separating the two wells. Thus at $t = 0$ (35) reduces to (34a) corresponding to the molecule being found in the left well, as required; and at $t = \pi/\omega$ (35) reduces to (34b) corresponding to the molecule being found in the right well. The angular frequency ω is interpreted as the frequency of a complete inversion cycle. The tunnelling splitting $\hbar\omega$ is determined by the height and width of the barrier, and is zero if the barrier is infinite.

One source of confusion in this model is that the parity of the vibrational wavefunctions is defined with respect to a *reflection* σ across the plane of the nuclei,[140]

$$\sigma\psi_v = (-1)^v\psi_v, \qquad (36)$$

where v is the vibrational quantum number (the normal vibrational coordinate for ν_2 changes sign under σ); whereas the basic definition of the parity operation is an *inversion* with respect to space-fixed axes. In the conventional treatment of inverting nonplanar symmetric tops,[138,140] the rotational wavefunction of a planar symmetric top such as BF_3 is multiplied by the time-dependent wavefunction (35) corresponding to the 'inversion' vibration. The parity operation corresponds to an inversion of all the particle positions (nuclei plus electrons), and is achieved by rotating the complete BF_3 molecule through π about the threefold axis, followed by a reflection across the plane containing the nuclei. Since the rotation is an external matter, it affects only the rotational wavefunctions and is used to classify their parity. The reflection is a purely internal matter, so the parity of the vibrational—electronic parts of the quantum state is determined by their behaviour under reflection across the plane of the nuclei. This type of consideration has been placed on a more sophisticated footing by the use of permutation—inversion groups to specify the parity of the complete wavefunction of a general nonrigid molecule in the gas phase.[141–143]

Since an analogous potential energy diagram can be drawn for any chiral molecule with a high barrier separating left and right wells corresponding to the two enantiomeric states, we now have a model for the source of the mixed parity internal (vibrational-electronic) states of a resolved enantiomer.

The horizontal axis might represent the position of an atom above a plane containing three different atoms, the torsion coordinate of a chiral biphenyl, or some more complicated collective coordinate of the molecule. If such a state is prepared, but the tunnelling splitting is finite, its energy will be indefinite because it is a superposition of two opposite parity states of different energy. The splitting of the two definite parity states, and hence the uncertainty in the energy of an enantiomer, is inversely proportional to the left–right conversion time π/ω: this is an example of the general result that the width of an energy level corresponding to a quasi-stationary state with average lifetime t is $\Delta W = \hbar/t$.[144]

A central point is therefore the relation between the time scale of the optical activity (or any time-even pseudoscalar observable) measurement and the lifetime of the resolved enantiomer. A new aspect of the uncertainty principle appears to arise here, which I have stated loosely as follows:[20,144,145]

> *If, for the duration of the measurement, there is complete certainty about the enantiomer, there is complete uncertainty about the parity of its quantum state; whereas if there is complete certainty about the parity of its quantum state, there is complete uncertainty about the enantiomer.*

Thus experimental resolution of the definite parity states of tartaric acid, for instance, an enantiomer of which has a lifetime probably greater than the age of the universe, is impossible unless the duration of the experiment is virtually infinite; whereas for a nonresolvable chiral molecule such as H_2O_2 spectroscopic transitions between definite parity states are observed routinely.

6.2 Two-State Systems and Parity Violation

It was shown in the previous section how the mixed parity states of a resolved chiral molecule can be visualized in terms of a double well potential. We now develop this aspect further by considering the quantum mechanics of a degenerate two-state system. This provides insight into the apparent paradox of the stability of optical enantiomers, which was recognized at the beginning of the quantum era when it was found that the existence of optical enantiomers was difficult to reconcile with basic quantum mechanics.[146–148]

The essence of the 'paradox' is that, because the potential energy term in the Hamiltonian of a molecule originates in Coulomb interactions between point charges, the complete molecular Hamiltonian is always invariant under space inversion and so the energy eigenstates (the stationary states) must be parity eigenstates. However, a resolved chiral molecule cannot be in such an eigenstate because the parity operation generates the mirror image enantiomer, which is a different system. Yet typical chiral molecules, such as alanine, appear to be no less stable than typical achiral molecules. Hund[146] suggested a resolution of the paradox using arguments of the type given in the previous section, namely that typical chiral molecules have such large barriers to inversion that the lifetime of a prepared enantiomer is virtually infinite. Hund's approach has been brought up to date by injecting a small parity-violating term into the Hamiltonian which, as demonstrated in the rest of this section, can result in the two enantiomeric states becoming the true stationary states.[104,105]

For a general two-state system in the orthonormal basis (ψ_1, ψ_2), not necessarily degenerate, the exact energy eigenvalues and eigenfunctions corresponding to the true stationary states are[13,149]

$$W_{\pm} = \frac{1}{2}(H_{11} + H_{22}) \pm \frac{1}{2}[(H_{11} - H_{22})^2 + 4|H_{12}|^2]^{\frac{1}{2}}\ , \tag{37a}$$

$$\psi_+^{(0)} = \cos\vartheta\, e^{-i\varphi/2}\psi_1 + \sin\vartheta\, e^{i\varphi/2}\psi_2, \tag{37b}$$

$$\psi_-^{(0)} = -\sin\vartheta\, e^{-i\varphi/2}\psi_1 + \cos\vartheta\, e^{i\varphi/2}\psi_2, \tag{37c}$$

where

$$\tan 2\vartheta = \frac{2|H_{12}|}{(H_{11} - H_{22})} \quad \text{with} \quad 0 \le 2\vartheta < \pi, \tag{37d}$$

$$H_{21} = |H_{21}|e^{i\varphi}. \tag{37e}$$

The superscripts (0) denote the amplitudes of the corresponding time-dependent wavefunctions, and $H_{ab} = \langle \psi_a^{(0)} | \hat{H} | \psi_b^{(0)} \rangle$ are matrix elements of the total Hamiltonian of the system. The subscripts ± here denote higher and lower energy levels, not the parity.

Clearly ψ_1 and ψ_2 are not the stationary states (eigenstates) of the Hamiltonian of the system and so couple with each other through H_{21}; whereas the stationary states ψ_+ and ψ_- do not. So if a two-state system is prepared in a non-stationary state ψ_1 or ψ_2, it might appear falsely to be influenced by a time-dependent perturbation lacking some fundamental symmetry of the internal Hamiltonian of the system. In general, ψ_1 and ψ_2 will be interconverted by a particular symmetry operation of the Hamiltonian, whereas $\psi_+^{(0)}$ and $\psi_-^{(0)}$ must transform according to one or other of the irreducible representations of the symmetry group comprising the identity and the operation in question.

By restricting attention to the ground and first excited state of the normal mode of vibration that interconverts the enantiomers in Fig. 4, we can identify ψ_1 and ψ_2 with ψ_R and ψ_L. If initially we neglect the small parity-violating terms, the Hamiltonian has inversion symmetry so that, since $\hat{P}\psi_R = \psi_L$ and $\hat{P}\psi_L = \psi_R$, the enantiomeric states ψ_R and ψ_L are degenerate. The stationary state amplitudes (37b) and (37c) now specialize to

$$\psi_+^{(0)} = \frac{1}{\sqrt{2}} e^{-i\varphi/2}(\psi_R + e^{i\varphi}\psi_L), \tag{38a}$$

$$\psi_-^{(0)} = \frac{1}{\sqrt{2}} e^{-i\varphi/2}(-\psi_R + e^{i\varphi}\psi_L), \tag{38b}$$

and so transform according to one or other of the irreducible representations of the inversion group comprising $\hat{P}$ plus the identity. Which has even and which has odd parity depends on the choice of φ: for example, if $\varphi = \pi$ (so that H_{LR} is real and negative), $\psi_+^{(0)}$ is odd and $\psi_-^{(0)}$ is even. The separation of the two stationary states is simply twice the coupling energy of the two enantiomeric states,

$$W_+ - W_- = 2|\langle\psi_L|\hat{H}|\psi_R\rangle| = 2\delta, \tag{39}$$

and is interpreted as a splitting caused by tunneling through the potential energy barrier separating the two enantiomers (Fig. 4).

We now allow the Hamiltonian to contain a small parity-violating term $\hat{V}^{PV}$ such as the electron–nucleus weak neutral current interaction (25). According to (27) this shifts the energies of the two enantiomeric states in opposite directions by an amount ε. The enantiomeric states are now no longer degenerate, so using the general two-state results (37) we have

$$W_+ - W_- = 2(\varepsilon^2 + \delta^2)^{\frac{1}{2}}, \tag{40a}$$

$$\tan 2\vartheta = \delta/\varepsilon. \tag{40b}$$

The general time-dependent wavefunction is given by the sum of each stationary state amplitude mulitiplied by its exponential time factor:

$$\psi(t) = \frac{1}{\sqrt{2}} e^{-i\varphi/2}[(\cos\vartheta\,\psi_R + \sin\vartheta\, e^{i\varphi}\psi_L)e^{-iW_+t/\hbar} + (-\sin\vartheta\,\psi_R + \cos\vartheta\, e^{i\varphi}\psi_L)e^{-iW_-t/\hbar}]. \tag{41}$$

This only has a simple interpretation in the two limits of $\varepsilon = 0$ and $\delta = 0$. When $\varepsilon = 0$ (zero parity violation)

$$\psi(t) = \frac{1}{2} e^{-i\varphi/2}[(\psi_R + e^{i\varphi}\psi_L)e^{-i\delta t/\hbar} + (-\psi_R + e^{i\varphi}\psi_L)e^{i\delta t/\hbar}]e^{-i(W_+ + W_-)t/2\hbar} \tag{42}$$

which reduces, within a phase factor, to (35) (do not confuse the notation $\psi_\pm$ and $W_\pm$ for higher and lower energy states with $\psi(\pm)$ and $W(\pm)$ for even- and odd-parity states). Thus at $t = 0$ the system is entirely in ψ_L and at $t = \pi\hbar/2\delta$ it is entirely in ψ_R: the system oscillates between the handed states ψ_L and ψ_R with $2\delta/\hbar$ being the frequency of a complete inversion cycle and $\psi_+ = \psi(-)$ and $\psi_- = \psi(+)$ the stationary states which in this case have definite parity. But when $\delta = 0$ (zero tunnelling splitting)

$$\psi(t) = \frac{1}{\sqrt{2}} e^{-i\varphi/2}(\psi_R e^{-i\varepsilon t/h} + e^{i\varphi}\psi_L e^{i\varepsilon t/h}) e^{-i(W_+ + W_-)t/2\hbar}, \tag{43}$$

so at $t = 0$ the system is entirely in $\psi(+)$ and at $t = \pi\hbar/2\varepsilon$ it is entirely in $\psi(-)$: now the system oscillates between the definite parity states $\psi(+)$ and $\psi(-)$ with $2\varepsilon/\hbar$ being the frequency of a complete cycle and ψ_R and ψ_L the stationary states. In fact Quack[150] has proposed an experiment, based on (43), to detect parity violation by preparing excited rovibrational molecular states (within the ground electronic level) with well defined parity from chiral molecules initially in their mixed parity ground states, followed by spectroscopic observation of transitions from 'forbidden' excited rovibrational states of opposite parity as a function of time.

The time-dependence of the optical activity observable depends on the nature of the state in which the molecule is prepared initially. Consider first the molecule prepared in a handed state ψ_L and ψ_R, which means that at $t = 0$ the state is given by (34a) or (34b), respectively. At some later time t the corresponding states will be

$$\psi_L(t) = e^{-i\varphi/2}(\sin\vartheta\, \psi_+^{(0)} e^{-iW_+t/\hbar} + \cos\vartheta\, \psi_-^{(0)} e^{-iW_-t/\hbar}), \tag{44a}$$

$$\psi_R(t) = e^{i\varphi/2}(\cos\vartheta\, \psi_+^{(0)} e^{-iW_+t/\hbar} - \sin\vartheta\, \psi_-^{(0)} e^{-iW_-t/\hbar}), \tag{44b}$$

which are obtained by inverting (37b) and (37c) and multiplying each stationary state amplitude by its exponential time factor. Thus for a molecule prepared in ψ_L, the time dependence of the optical rotation angle is given by[20,104,105]

$$\alpha(t) = \alpha_L\{\varepsilon^2 + \delta^2\cos[2(\delta^2 + \varepsilon^2)^{\frac{1}{2}} t/\hbar]\} \big/ (\delta^2 + \varepsilon^2), \tag{45}$$

where α_L is the optical rotation angle of the left-handed enantiomer. So if $\varepsilon \neq 0$, the optical rotation oscillates asymmetrically (but if $\varepsilon = 0$ it oscillates between equal and opposite values associated with the two enantiomers). Taking the time average, we find

$$\overline{\alpha}/\alpha_{max} = \varepsilon^2/(\delta^2 + \varepsilon^2). \tag{46}$$

Thus parity violation causes a shift away from zero of $\bar{\alpha}$. This is the basis of an experiment suggested by Harris and Stodolsky[104] to detect the parity-violating energy shift between enantiomers.

But if the molecule is prepared in one of the stationary states

$$\psi_+(t) = (\cos\vartheta\, e^{-i\varphi/2}\psi_R + \sin\vartheta\, e^{i\varphi/2}\psi_L)\, e^{-iW_+t/\hbar}, \tag{47a}$$

$$\psi_-(t) = (-\sin\vartheta\, e^{-i\varphi/2}\psi_R + \cos\vartheta\, e^{i\varphi/2}\psi_L)\, e^{-iW_-t/\hbar}, \tag{47b}$$

the optical rotation will be given by

$$\alpha_+(t) = -\alpha_-(t) = -\alpha_L\varepsilon \Big/ (\varepsilon^2 + \delta^2)^{\frac{1}{2}}. \tag{48}$$

Thus if $\varepsilon = 0$, the optical rotation will be zero, as required, since the stationary states will have definite parity; but if $\varepsilon \neq 0$, the stationary states will acquire equal and opposite parity-violating optical rotation *that does not change with time.*

It follows from (41) and (37) that, as $\delta/\varepsilon \to 0$, ψ_L and ψ_R become the true stationary states. In fact for typical chiral molecules, δ corresponds to tunnelling times of the order of millions of years: Harris and Stodolsky[104] have estimated ε to correspond to times of the order of seconds to days, so at low temperature and in a vacuum, a prepared enantiomer will retain its handedness effectively forever. So the ultimate answer to the paradox of the stability of optical enantiomers might lie in the weak interactions.

6.3 Parity Violation and Parity Breaking

The appearance of parity-violating phenomena is interpreted in quantum mechanics by saying that, contrary to what had been previously supposed, the Hamiltonian lacks inversion symmetry (the weak interaction potential being a pseudoscalar). This means that $\hat{P}$ and $\hat{H}$ no longer commute, so the associated law of conservation of parity no longer holds. Such *symmetry violation* (non-conservation) must be clearly distinguished from *spontaneous symmetry breaking*: current usage in the physics literature applies the latter term to describe the situation that arises when a system displays a lower symmetry than expected from its Hamiltonian.[151] Natural optical activity is therefore a phenomenon arising from spontaneous parity breaking since, as we have seen, a resolved chiral molecule displays a lower symmetry than its associated Hamiltonian: if the small parity-violating term in the Hamiltonian is neglected, the symmetry operation that the Hamiltonian possesses but the chiral molecule lacks is parity, and it is this parity operation that interconverts the two enantiomeric parity-broken states.

It should be mentioned that the term 'spontaneous symmetry breaking' is often reserved exclusively for quantum mechanical systems with an infinite number of degrees of freedom for which there can be special dynamic instabilities which inevitably generate an unsymmetric state.[152] Quack[121] adheres to this usage, but in order to accommodate the unsymmetric states that occur in finite systems such as molecules and which are so important in the discussion of molecular chirality, he has introduced the term 'symmetry breaking *de facto*' to describe the situation where the unsymmetric states arise through the choice of initial conditions: in this case the existence of symmetric states is still possible. Quack also introduced the term 'symmetry breaking *de lege*' to describe what I have called symmetry violation (non-conservation) above for the case where the asymmetry resides in the Hamiltonian. Although Quack's new terminology has great virtue, in what follows I have used spontaneous symmetry breaking to cover the generation of unsymmetric states in both finite and infinite systems. One reason for this is that, as explained in the next section, there is sometimes no clear distinction between the symmetry breaking in the macroscopic (infinite) system and the symmetry breaking in the microscopic (finite) constituents of that macroscopic system; also the 'inevitability' of an unsymmetric state of an infinite system needs to be qualified since the temperature is a crucial factor.

The conventional view, formulated in terms of the double well model as in Section 6.1, is that parity violation plays no part in the stabilization of chiral molecules. The optical activity remains observable only so long as the observation time is short compared with the interconversion time between enantiomers, which is proportional to the inverse of the tunnelling splitting. Such spontaneous parity-breaking optical activity therefore averages to zero over a sufficiently long observation time. Hence statements such as[153] "... processes involving pseudoscalar quantities will not obey the law of parity" exposes a common misconception: the law of parity is saved in systems displaying spontaneous parity breaking because their pseudoscalar properties average to zero over a sufficiently long observation period on account of tunnelling or, equivalently, the space-inverted experiment is realizable. In either interpretation absolute chirality is not observable.

There is a similar misconception in connection with the correlation of entities that have opposite parities, such as the coupling of rotational and translational motion in collections of chiral molecules. For example, it has been stated that[154] "... in a system such that the Hamiltonian has inversion symmetry, properties of different parity are totally uncorrelated for all time"; also "... if the system contains optically active molecules, the Hamiltonian does not have even parity and none of these theorems apply". Any Hamiltonian involving only electromagnetic interactions always has even parity: it is the spontaneously broken parity states of the chiral molecules that mediate the coupling of opposite parity entities, and any associated pseudoscalar properties will consequently average to zero over a sufficiently long period on account of tunnelling.

These considerations lead us to an important criterion for distinguishing spontaneous parity breaking and parity violating natural optical activity phenomena: the former are time-dependent and average to zero; the latter are constant in time (recall the stationary states acquiring time-independent optical activity in the previous section when $\varepsilon \neq 0$). Hence if a small chiral molecule could be isolated sufficiently from the environment, a parity-violating element is indicated if the optical activity remains observable for longer than the expected interconversion time between the enantiomers.[155]

6.4 Spontaneous Symmetry Breaking in Isolated Chiral Molecules and in Condensed Media

Spontaneous symmetry breaking has attracted much attention in recent years in both elementary particle and condensed matter physics, but with rather different emphasis on the various aspects. Anderson has written at length on broken symmetry in condensed matter, with some valuable asides on the molecular aspects.[156,157]

Ferromagnetism provides an important example. The Hamiltonian for an iron crystal is invariant under spatial rotations. However, the ground state of a magnetized sample is not invariant: it distinguishes a specific direction, the direction of magnetization. This non-zero magnetization in zero applied field also breaks time reversal symmetry. When the temperature is raised above the Curie point, the magnetization disappears and the rotational and time reversal symmetries become manifest. In fact the term 'spontaneous symmetry breaking' itself derives from the term 'spontaneous magnetization'. Notice that a vestige of the rotational symmetry still survives in the ferromagnetic phase in that the sense of magnetization is arbitrary; but this would be hidden from an observer living inside the crystal.

Temperature is a central feature here, because behaviour reflecting the full symmetry of the Hamiltonian can be recovered at sufficiently high temperature. Molecules behave rather differently from macroscopic systems in that there are no sharp transitions between symmetric and asymmetric states in molecules.[157] For example, in a molecule described by a double well model, thermal agitation will cause the 'inversion' transition to take place (unless the height and width of the barrier are effectively infinite) so that there is no absolute 'one-sidedness' at any temperature: in other words its spontaneous parity-broken characteristics decrease continuously with increasing temperature. In condensed media, on the other hand, large numbers of particles can cooperate to produce sudden rather than gradual thermal transitions between symmetric and asymmetric states of the complete macroscopic sample. In a ferroelectric crystal below its phase transition temperature, for example, the field associated with each molecular dipole (or the dipole associated with each unit cell) acts to hold the others in the same direction: the reversal of any one will not reverse the others but simply represents a local fluctuation. At the phase transition temperature, however, a sufficient number of dipoles are reversed that the system suddenly transforms into a state where opposite signs for each dipole occur with equal probability. Thus a distinction must be made between spontaneous symmetry breaking in a macroscopic system and in the individual microscopic constituents, although in many cases, including the example of ferroelectricity considered here, broken symmetry in the microscopic constituents (so that they are dipolar) is a prerequisite for the macroscopic spontaneous symmetry breaking associated with ferroelectricity. This distinction between microscopic and macroscopic spontaneous symmetry breaking can sometimes become rather blurred, as in superconductivity.[157]

The relationship between the microscopic and macroscopic aspects of the spontaneous parity-broken states of chiral systems is still an open question, with much discussion concerning the rôle of the environment.[121,155,158-169] While most would agree that the environment must have some influence, Woolley has suggested that the environment is everything so that "optical activity has to be understood in a macroscopic context as a loss of inversion symmetry of the whole material medium, and that chirality is not a property that can be related to isolated molecules (the rotational strength for an isolated molecule vanishes identically if one ignores the weak neutral current)".[162] Here an isolated molecule means a closed system of electrons and nuclei that evolve in time in vacuum. According to Woolley, the essential ingredient is that the chiral molecules are coupled to an environment that is described by a quantum field theory.[164] This environment may be given a traditional representation such as a reaction field,[167] or all other molecules in a substance apart from a reference molecule;[163] but it may also be the quantized electromagnetic field[158,159] or other boson fields such as phonons.[169] In my view Woolley is suggesting that the dynamical mechanism causing optical activity is similar to that producing spontaneous symmetry breaking in condensed matter; furthermore that it corresponds to the situation where the distinction between microscopic and macroscopic broken symmetry (the chirality of individual molecules and that of the bulk sample) is lost. I am not yet entirely convinced about this interesting concept.[170,171]

7 Chirality and Relativity

It was shown in Section 3.3 that a spinning cone or sphere translating along the axis of spin possesses true chirality. This is an interesting concept because it exposes a link between chirality and special relativity. Consider a particle moving away from an observer with a right-handed helicity. If the observer accelerates to a sufficiently high velocity that he starts to catch up with the particle, it will then appear to be moving towards the observer and so takes on a left-handed helicity. In its rest frame the helicity of the particle is undefined and its chirality vanishes. Only for massless particles such as photons and neutrinos is the chirality conserved since they always move at the velocity of light in any reference frame.

This relativistic aspect of chirality is in fact a central feature of modern elementary particle theory, especially in connection with the weak interaction where the parity-violating aspects are velocity-dependent. A good illustration is provided by the interaction of electrons with neutrinos: neutrinos are quintessential chiral objects since only left-handed neutrinos and right-handed antineutrinos exist.[17,32,96,97] Consider first the extreme case of electrons moving close to the velocity of light. Only left-handed relativistic electrons interact with left-handed neutrinos via the weak force; right-handed relativistic electrons do not interact at all with neutrinos. But right-handed relativistic positrons interact with right-handed antineutrinos. For nonrelativistic electron momenta, the weak interaction still violates parity, but the amplitude of the violation is reduced to order v/c.[97] This is used to explain the interesting fact that the $\pi^- \rightarrow e^- \tilde{\nu}_e$ decay is a factor of 10^4 smaller than the $\pi^- \rightarrow \mu^- \tilde{\nu}_\mu$ decay, even though the available energy is much larger in the first decay.[96] Thus in the rest frame of the pion, the lepton (electron or muon) and the antineutrino are emitted in opposite directions so that their linear momenta cancel. Also, since the pion is spinless, the lepton must have a right-handed helicity in order to cancel the right-handed helicity of the anti-neutrino. Thus both decays would be forbidden if e and μ had the velocity c because the associated maximal parity violation dictates that both be pure left-handed. However, on account of its much greater mass, the muon is emitted much more slowly than the electron, so there is a much greater amplitude for it to be emitted with a right-handed helicity.

It should be mentioned that the discussion in the previous paragraph applies only to charge-changing weak processes, mediated by W^+ or W^- particles. Weak neutral current processes, mediated by Z^0 particles, are rather different since, even in the relativistic limit, both left- and right-handed electrons participate, but with slightly different amplitudes.[95,96]

The word *chirality* has been used up to this point in the article in its qualitative chemical sense. But in elementary particle physics, chirality is given a precise quantitative meaning: it is the eigenvalue of the Dirac matrix operator $\hat{\gamma}_5$, with values +1 and −1 associated with pure right-handed and pure left-handed leptons.[6] But only massless leptons (such as neutrinos), which always move at the velocity of light, are in eigenstates of γ_5 and so have precise chirality. Leptons with mass (such as electrons) always move more slowly than c and so do not have well-defined chirality. In fact the very existence of mass is associated with 'chiral symmetry breaking'.[17] On the other hand, helicity (defined in Section 5.1 above) can be defined for both massless and massive particles. but only for the former is it completely invariant to the frame of the observer. For massless particles the helicity is actually equivalent to the chirality (for an antiparticle the helicity and chirality have the opposite sign). The interesting suggestion has been made recently that, if the physical problem singles out a preferred spatial origin, such as the source of an electromagnetic field, then chirality becomes sharply defined even for a particle with mass.[172]

Another rather different connection between chirality and relativity should be mentioned. It was shown in Section 6.3 that spontaneous parity-breaking and parity-violating optical activity are distinguished by the fact that the first is time-dependent while the second is independent of time. Because a clock on a moving object slows down relative to a stationary observer, a molecule exhibiting spontaneous parity-breaking optical activity will become increasingly stable with increasing velocity relative to a stationary observer, and as it approaches the speed of light it will become infinitely stable. This means that spontaneous parity-breaking optical activity in a chiral object moving at the speed of light becomes indistinguishable from parity-violating optical activity.

8 Molecule-Antimolecule Pairs: True Enantiomers

We are now in a position to appreciate that parity violation has great conceptual significance in the discussion of molecular chirality, because only the space-inverted enantiomers of *truly* chiral systems show a parity-violating energy difference.[7] This follows from the fact that, although the parity-violating weak neutral current Hamiltonian (25) is odd under space inversion, it is invariant under both time reversal and any proper spatial rotation: since the last two operations together interconvert the two space-inverted enantiomers of a system displaying false chirality, it follows from a development analogous to (27) that the energy difference is zero.

Since the space-inverted enantiomers of a truly chiral object are not strictly degenerate, they are not true enantiomers (since the concept of enantiomer implies the *exact* opposite). So where is the true enantiomer of a chiral object to be found? In the antiworld, of course! The molecule with the opposite absolute configuration but composed of antiparticles will have exactly the same energy as the original.[20,35,173] This follows from the $\hat{C}\hat{P}\hat{T}$ theorem and the assumption that $\hat{T}$ is not violated. So true enantiomers are interconverted by $\hat{C}\hat{P}$. Since $\hat{P}$ violation automatically implies $\hat{C}$ violation here, it also follows that there is a small energy difference between a chiral molecule in the real world and the corresponding chiral molecule with the same absolute configuration in the antiworld (i.e. between $\hat{C}$—enantiomers), with a magnitude equal to that of the parity-violating energy difference between space-inverted enantiomers in the real world ($\hat{P}$—enantiomers). Jungwirth et al. have provided a quantum-mechanical development, analogous to (27), that supports this conclusion.[171]

This more general definition of the enantiomers of a truly chiral system is consistent with the chirality that free atoms display on account of parity violation.[20,170] The weak neutral current interaction generates only one type of chiral atom in the real world: the conventional enantiomer of a chiral atom obtained by space inversion alone does not exist. Clearly the enantiomer of a chiral atom is generated by the combined $\hat{C}\hat{P}$ operation. Thus the corresponding atom composed of antiparticles will of necessity have the opposite 'absolute configuration' and will show an opposite sense of optical rotation.

The space-inverted enantiomers of objects such as translating spinning cones or spheres that only exhibit chirality on account of their motion also show parity-violating differences. One manifestation is that, as mentioned in Section 7, left-handed and right-handed particles (or antiparticles) have different weak interactions. Again, *true* enantiomers are interconverted by $\hat{C}\hat{P}$: for example, a left-handed electron and a right-handed positron.

9 Concluding Remarks: Lessons for Chemistry and Physics

This discussion of the symmetry aspects of molecular chirality has drawn on many concepts from modern physics, especially the physics of elementary particles. It is particularly striking how the pursuit of analogies between the quantum states of a chiral molecule and those of various elementary particles reinforces Heisenberg's perception of a kinship between molecules and elementary particles.[174,175] This insight ought to encourage theoretical chemists to keep abreast of developments in elementary particle physics in order to introduce concepts that could form the basis of a new quantum chemistry.

Elementary particle physicists can also learn something from chemistry through such analogies. For example, Bohm[176] has reviewed the relationship between the collective motions of a diatomic molecule and the charge-monopole system in order to expose some useful insights into certain aspects of the structure of hadrons (particles which undergo strong interactions–baryons and mesons). Bohm et al.[177] have extended such molecular and also nuclear analogies into the relativistic domain to develop a model of collective motions of extended relativistic objects in order to calculate the mass spectrum and radiative transitions of hadrons.

Also Barut[178] has developed a composite model of elementary particles in which some particles, due to their internal structure, are not in eigenstates of $\hat{P}$ or $\hat{C}\hat{P}$, analogous to the spontaneously parity-broken states of chiral molecules. Then, just as a chiral molecule can support pseudoscalar observables, so these broken-symmetry particle states are responsible for the observation of 'symmetry-violating' quantities. The appearance or otherwise of symmetry violation in elementary particle processes then depends on the timescale of the interaction between particles: for example parity violation arises in this model because a left-handed neutrino, say, can escape *via* tunnelling from some long-lived intermediate resonance state of the interacting particles.

Another example is Wigner's comparison[179] of the four distinct states $|K^o\rangle$, $|\tilde{K}^o\rangle$, $|K_1\rangle$ and $|K_2\rangle$ of the neutral K-meson to the four possible states ψ_L, ψ_R, $\psi(+)$ and $\psi(-)$ of a chiral molecule: just as $|K^o\rangle$ and $|\tilde{K}^o\rangle$ are interconverted by $\hat{C}\hat{P}$ and $|K_1\rangle$ and $|K_2\rangle$ are even and odd eigenstates with respect to $\hat{C}\hat{P}$, so ψ_L and ψ_R are interconverted by $\hat{P}$ and $\psi(+)$ and $\psi(-)$ have even and odd parity. However, Wigner's analogy falters when we introduce $\hat{C}\hat{P}$ violation into the K^o system and $\hat{P}$ violation into the chiral molecule. Although $\hat{C}\hat{P}$ violation mixes $|K_1\rangle$ and $|K_2\rangle$ just as $\hat{P}$ violation mixes $\psi(+)$ and $\psi(-)$, there does not appear to be a $\hat{C}\hat{P}$ analogue of the lifting of the degeneracy of ψ_L and ψ_R through $\hat{P}$ violation because one of the consequences of the $\hat{C}\hat{P}\hat{T}$ theorem is that a particle and its associated antiparticle have the same mass and lifetime.[14,17,32] A better molecular analogy would be with the four states of a chiral molecule in a time-noninvariant $\hat{P}$-enantiomorphous influence such as collinear electric and magnetic fields. Indeed, a detailed quantum-mechanical analysis of a chemical reaction system such as the butadiene—cyclobutene interconversion in collinear electric and magnetic fields, in which expressions for analogues of the $\hat{C}\hat{P}$-violation parameters mentioned in Section 5.3 were derived, might help to remove some of the mystery surrounding $\hat{C}\hat{P}$ violation. Barut's composite particle model[178] could also be useful in this context since it provides an intuitive picture of $\hat{C}\hat{P}$ violation in the K^o system involving the breakdown of the perfect symmetry of a double potential well.

Molecular analogies might assist in discussions of the possible characteristics of manifestations of $\hat{T}$ violation that are being sought in atoms and molecules such as the permanent electric dipole moments mentioned in Section 5.4. It is natural to suppose that such $\hat{T}$ violation would be akin to that observed in the K meson system in which direct observation of $\hat{C}\hat{P}$ violation implies, *via* the $\hat{C}\hat{P}\hat{T}$ theorem, $\hat{T}$ violation to the same degree. However, as mentioned in Section 5.3 above, the analogy with absolute asymmetric synthesis induced by collinear electric and magnetic fields leads to the conceptualization of this breakdown in microscopic reversibility as arising from a time-noninvariant $\hat{C}\hat{P}$-enantiomorphous influence in the forces of nature: of the two possible influences, only one is found in our world (analogous to, say, parallel rather than antiparallel electric and magnetic fields). This implies that manifestations of $\hat{T}$ violation must be associated with a process, not with a state, and can only appear as a breakdown in microscopic reversibility in particle-antiparticle processes (since these involve $\hat{C}\hat{P}$ enantiomers). *The implication of this interpretation of $\hat{C}\hat{P}$ violation is that searches for electric dipole moments, etc., in stationary states of atoms and molecules as manifestations of $\hat{T}$ violation might be futile!*

Finally, there are tantalizing similarities between some new theories of high-temperature superconductivity[180, 181] and the ideas presented here on the breakdown of microscopic reversibility induced in reaction and transport processes involving chiral particles by a time-noninvariant enantiomorphous influence. These new theories contain two essential ingredients: the spontaneous breaking of $\hat{P}$ and $\hat{T}$ simultaneously (which parallels the breaking of $\hat{P}$ and $\hat{T}$ by the time-noninvariant enantiomorphous influence) and the existence of vortices in spin liquids (which parallels the chiral molecules). It is easy to see that $\hat{P}$ and $\hat{T}$ must be broken simultaneously in a superconductor because once a current is established in, say, a superconducting ring, it will persist forever: this means that the bulk sample supports a property, the electric current, which transforms as a time-odd polar vector.

Acknowledgements

I thank O. Costa de Beauregard, M. W. Evans, R. G. Moorhouse, M. Quack, D. H. Whiffen and R. G. Woolley for helpful discussions, and the Wolfson Foundation for assistance with the preparation of this manuscript.

References

1 FRESNEL, A., 1824, *Bull. Soc. Philomath.*, p. 147.
2 PASTEUR, L., 1848, *Ann. Chim.* **24,** 457.
3 PASTEUR, L., 1884, *Rev. Scientifique,* **7**, 2.
4 MASON, S. F., 1982, *Molecular Optical Activity and the Chiral Discriminations* (Cambridge University Press, Cambridge).

5 LORD KELVIN, 1904, *Baltimore Lectures* (C. J. Clay, London).
6 SAKURAI, J. J., 1964, *Invariance Principles and Elementary Particles* (Princeton University Press, Princeton).
7 BARRON, L. D., 1986, *Chem. Phys. Lett.*, **123**, 423.
8 BARRON, L. D., 1986, *J. Am. Chem. Soc.*, **108**, 5539.
9 BARRON, L. D., 1986, *Chem. Soc. Rev.*, **17**, 189.
10 BARRON, L. D., *Theoretical Models of Chemical Bonding*, Vol. 1, edited by Z. B. Maksiċ (Springer, Berlin), in press.
11 HALDANE, J. B. S., 1960, *Nature*, **185**, 87.
12 MASON, S. F., 1987, *Nouveau Journal de Chimie*, **10**, 739.
13 FEYNMAN, R. P., LEIGHTON, R. B., and SANDS, M., 1964, *The Feynman Lectures on Physics* (Addison–Wesley, Reading, Massachusetts).
14 SACHS, R. G., 1987, *The Physics of Time Reversal* (University of Chicago Press, Chicago).
15 COSTA de BEAUREGARD, O., 1987, *Time, the Physical Magnitude* (Reidel, Dordrecht).
16 WIGNER, E. P., 1959, *Group Theory* (Academic Press, New York).
17 LEE, T. D., 1984, *Particle Physics and Introduction to Field Theory* (Harwood, Chur).
18 WIGNER, E. P., 1927, *Z. Phys.*, **43**, 624.
19 BARRON, L. D., 1972, *Nature*, **238**, 17.
20 BARRON, L. D., 1982, *Molecular Light Scattering and Optical Activity* (Cambridge University Press, Cambridge).
21 de FIGUEIREDO, I. M. B., and RAAB, R. E., 1980, *Proc. Roy. Soc.*, **A369**, 501.
22 GRAHAM, C., 1980, *Proc. Roy. Soc.*, **A369**, 517.
23 STEDMAN, G. E., 1983, *Am. J. Phys.*, **51**, 753.
24 STEDMAN, G. E., 1985, *Adv. Phys.*, **34**, 513.
25 BIRSS, R. R., 1966, *Symmetry and Magnetism* (North–Holland, Amsterdam).
26 SHUBNIKOV, A. V., and KOPTSIK, V. A., 1974, *Symmetry in Science and Art* (Plenum Press, New York).
27 CURIE, P., 1884, *J. Phys. (Paris) (3)*, **3**, 393.
28 LANDAU, L. D., and LIFSHITZ, E. M., 1977, *Quantum Mechanics* (Pergamon Press, Oxford).
29 KAEMPFFER, F. A., 1965, *Concepts in Quantum Mechanics* (Academic Press, New York).
30 SAKURAI, J. J., 1985, *Modern Quantum Mechanics* (Benjamin/Cummings, Menlo Park, California).
31 HEINE, V., 1960, *Group Theory in Quantum Mechanics* (Pergamon Press, Oxford).
32 BERESTETSKII, V. B., LIFSHITZ, E. M., and PITAEVSKII, L. P., 1982, *Quantum Electrodynamics* (Pergamon Press, Oxford).
33 ZOCHER, H., and TOROK, C., 1953, *Proc. Natl. Acad. Sci. USA*, **39**, 681.

34 ROSS, H. J., SHERBORNE, B. S., and STEDMAN, G. E., 1989, *J. Phys.*, **B22**, 459.
35 BARRON, L. D., 1981, *Mol. Phys.*, **43**, 1395.
36 BUCKINGHAM, A. D., GRAHAM, C., and RAAB, R. E., 1971, *Chem. Phys. Lett.*, **8**, 622.
37 BUCKINGHAM, A. D., 1979, *Philos. Trans. Roy. Soc.*, **A293**, 239.
38 ATKINS, P. W., and GOMES, J. A. N. F., 1976, *Chem. Phys. Lett.*, **39**, 519.
39 MISLOW, K., and BICKART, P., 1976/77, *Isr. J. Chem.*, **15**, 1.
40 ATKINS, P. W., 1980, *Chem. Phys. Lett.*, **74**, 358.
41 JONES, R. V., 1976, *Proc. Roy. Soc.*, **A349**, 423.
42 BOHM, A., 1979, *Quantum Mechanics* (Springer—Verlag, New York).
43 BEERLAGE, M. J. M., FARAGO, P. S., and Van der WIEL, M. J., 1981, *J. Phys.*, **B14**, 3245.
44 FARAGO, P. S., 1981, *J. Phys.*, **B14**, L743.
45 RICH, A., Van HOUSE, J., and HEGSTROM, R. A., 1982, *Phys. Rev. Lett.*, **48**, 1341.
46 CAMPBELL, D. M., and FARAGO, P. S., 1985, *Nature*, **318**, 52.
47 ZEL'DOVICH, Ya B., and SAAKYAN, D. B., 1980, *Sov. Phys. JETP*, **51**, 1118.
48 STEPHEN, T. M., XUEYING SHI, and BURROW, P. D., 1988, *J. Phys.*, **B21**, L169.
49 KABIR, P. K., KARL, G., and OBRYK, E., 1974, *Phys. Rev.*, **D10**, 1471.
50 COX, J. N., and RICHARDSON, F. S., 1977, *J. Chem. Phys.*, **67**, 5702.
51 HARRIS, R. A., and STODOLSKY, L., 1979, *J. Chem. Phys.*, **70**, 2789.
52 GAZDY, B., and LADIK, J., 1982, *Chem. Phys. Lett.*, **91**, 158.
53 BARANOVA, N. B., and Zel'DOVICH, B. Ya, 1979, *Mol. Phys.*, **38**, 1085.
54 WAGNIÈRE, G., and MEIER, A., 1982, *Chem. Phys. Lett.*, **93**, 78.
55 BARRON, L. D., and VRBANCICH, J., 1984, *Mol. Phys.*, **51**, 715.
56 WAGNIÈRE, G., 1984, *Z. Naturforsch., Teil* **A39**, 254.
57 WAGNIÈRE, G., 1988, *Z. Phys.*, **D8**, 229.
58 DOUGHERTY, R. C., 1980, *J. Am. Chem. Soc.*, **102**, 380.
59 LIFSHITZ, E. M., and PITAEVSKII, L. P., 1980, *Statistical Physics*, Part 1 (Pergamon Press, Oxford).
60 WHIFFEN, D. H., 1988, *Mol. Phys.*, **63**, 1053.
61 EVANS, M. W., 1989, *Phys. Rev.*, **A39**, 6041.
62 EVANS, M. W., 1988, *Chem. Phys. Lett.*, **152**, 33.
63 MORRISON, J. D., and MOSHER, H. S., 1976, *Asymmetric Organic Reactions* (American Chemical Society, Washington D. C.).
64 IZUMI, Y., and TAI, A., 1977, *Stereo-Differentiating Reactions* (Academic Press, New York).
65 MASON, S. F., 1983, *Int. Rev. Phys. Chem.*, **3**, 217.
66 MASON, S. F., 1988, *Chem. Soc. Rev.*, **17**, 347.
67 BONNER, W. A., 1988, *Topics in Stereochemistry*, **18**, 1.
68 de MIN, M., LEVY, G., and MICHEAU, J. C., 1988, *J. Chim. Physique*, **85**, 603.

69 JAEGER, F. M., 1930, *Optical Activity and High Temperature Measurements* (McGraw-Hill, New York).
70 BONNER, W. A., to be published.
71 WAGNIÈRE, G., and MEIER, A., 1983, *Experientia,* **39**, 1090.
72 de GENNES, P. G., 1970, *C. R. Hebd. Seances Acad. Sci., Ser. B,* **270**, 891.
73 MEAD, C. A., MOSCOWITZ, A., WYNBERG, H., and MEUWESE, F., 1977, *Tetrahedron Lett.,* p. 1063.
74 PERES, A., 1980, *J. Am. Chem. Soc.,* **102**, 7390.
75 RHODES, W., and DOUGHERTY, R. C., 1978, *J. Am. Chem. Soc.,* **100**, 6247.
76 MEAD, C. A., and MOSCOWITZ, A., 1980, *J. Am. Chem. Soc.,* **102**, 7301.
77 de GENNES, P. G., 1981, *Symmetry and Broken Symmetry,* edited by N. Boccara (Editions IDSET, Paris).
78 BARRON, L. D., 1987, *Biosystems,* **20**, 7.
79 GERIKE, P., 1975, *Naturwissenschaften,* **62**, 38.
80 EDWARDS, D., COOPER, K., and DOUGHERTY, R. C., 1980, *J. Am. Chem. Soc.,* **102**, 381.
81 BARRON, L. D., 1987, *Chem. Phys. Lett.,* **135**, 1.
82 LIFSHITZ, E. M., and PITAEVSKII, L. P., 1981, *Physical Kinetics* (Pergamon Press, Oxford).
83 TOLMAN, R. C., 1938, *The Principles of Statistical Mechanics* (Oxford University Press, Oxford).
84 ONSAGER, L., 1931, *Phys. Rev.,* **37**, 405.
85 GILAT, G., 1987, *Chem. Phys. Lett.,* **137**, 492.
86 GILAT, G., 1986, *Chem. Phys. Lett.,* **125**, 129.
87 KOLB, E. W., and WOLFRAM, S., 1980, *Nucl. Phys.,* **B172**, 224.
88 TELLER, E., 1968, *Symmetry Principles at High Energy,* edited by A. Perlmutter, C. A. Hurst, and B. Kursunoglu (Benjamin, New York), p. 1.
89 AHARONY, A., 1973, *Modern Developments in Thermodynamics,* edited by B. Gal-Or (Wiley, New York), p. 95.
90 LEE, T. D., and YANG, C. N., 1956, *Phys. Rev.,* **104**, 254.
91 WU, C. S., AMBLER, E., HAYWARD, R. W., HOPPES, D. D., and HUDSON, R. P., 1957, *Phys. Rev.,* **105**, 1413.
92 FERMI, E., 1934, *Z. Phys.,* **88**, 161.
93 WEINBERG, S., 1980, *Rev. Mod. Phys.,* **52**, 515.
94 SALAM, A., 1980, *Rev. Mod. Phys.,* **52**, 525.
95 GLASHOW, S. E., 1980, *Rev. Mod. Phys.,* **52**, 539.
96 AITCHISON, I. J. R., and HEY, A. J., 1989, *Gauge Theories in Particle Physics* (Adam Hilger, Bristol).
97 GOTTFRIED, K., and WEISSKOPF, V. F., 1984, *Concepts of Particle Physics,* Vol. 1 (Clarendon Press, Oxford).
98 RUBBIA, C., 1985, *Rev. Mod. Phys.,* **57**, 699.
99 BOUCHIAT, M. A., and BOUCHIAT, C., 1974, *J. Phys. (Paris),* **35**, 899.

100 HEGSTROM, R. A., REIN, D. W., and SANDARS, P. G. H., 1980, *J. Chem. Phys.,* **73**, 2329.
101 REIN, D. W., 1974, *J. Mol. Evol.,* **4**, 15.
102 LETOKHOV, V., 1975, *Phys. Lett.,* **53A**, 275.
103 ZEL'DOVICH, B. Ya, SAAKYAN, D. B., and SOBEL'MAN, I. I., 1977, *Sov. Phys. JETP Lett.,* **25**, 95.
104 HARRIS, R. A., and STODOLSKY, L., 1978, *Phys. Lett.,* **78B**, 313.
105 HARRIS, R. A., 1980, *Quantum Dynamics of Molecules,* edited by R. G. Woolley (Plenum Press, New York), p. 357.
106 MASON, S. F., and TRANTER, G. E., 1984, *Mol. Phys.,* **53**, 1091.
107 MASON, S. F., and TRANTER, G. E., 1985, *Proc. Roy. Soc.,* **A397**, 45.
108 WIESENFELD, L., 1988, *Mol. Phys.,* **64**, 739.
109 TRANTER, G. E., 1985, *Chem. Phys. Lett.,* **121**, 339.
110 TRANTER, G. E., 1986, *J. Theor. Biol.,* **119**, 467.
111 MacDERMOTT, A. J., and TRANTER, G. E., 1989, *Croat. Chem. Acta,* **62** (2A), 165.
112 FORTSON, E. N., and LEWIS, L. L., 1984, *Phys. Rep.,* **113**, 289.
113 BOUCHIAT, M. A., and POTTIER, L., 1986, *Science,* **234**, 1203.
114 COMMINS, E. D., 1988, *Sov. J. Quantum Electron.,* **18**, 703.
115 STACEY, D. N., 1984, *Acta Phys. Polonica,* **A66**, 377.
116 BARKOV, L. M., ZOLOTOREV, M. S., and MELIK-PASHAEV, D. A., 1988, *Sov. J. Quantum Electron.,* **18**, 710.
117 HEGSTROM, R. A., CHAMBERLAIN, J. P., SETO, K., and WATSON, R. G., 1988, *Am. J. Phys.,* **56**, 1086.
118 LABZOVSKII, L. N., 1978, *Sov. Phys. JETP,* **48**, 434.
119 SUSHKOV, O. P., and FLAMBAUM, V. V., 1978, *Sov. Phys. JETP,* **48**, 608.
120 KOZLOV, M. G., 1988, *Sov. J. Quantum Electron.,* **18**, 713.
121 QUACK, M., 1989, *Angew. Chem. Int. Ed. Engl.,* **28**, 571.
122 MacDERMOTT, A. J., and TRANTER, G. E., 1989, *Chem. Phys. Lett.,* **163**, 1.
123 CHRISTENSON, J. H., CRONIN, J. W., FITCH, V. L., and TURLAY, R., 1964, *Phys. Rev. Lett.,* **13**, 138.
124 CRONIN, J. W., 1981, *Rev. Mod. Phys.,* **53**, 373.
125 FITCH, V. L., 1981, *Rev. Mod. Phys.,* **53**, 367.
126 WOLFENSTEIN, L., 1986, *Ann. Rev. Nucl. Part. Sci.,* **36**, 137.
127 SANDARS, P. G. H., 1975, *Atomic Physics,* Vol.4, edited by G. zu Putlitz, E. W. Weber and A. Winnacker (Plenum Press, New York), p. 71.
128 LANDAU, L., 1957, *Sov. Phys. JETP,* **5**, 336.
129 SANDARS, P. G. H., 1968, *J. Phys.,* **B1**, 499; *ibid,* 511.
130 RAMSAY, N. F., 1982, *Ann. Rev. Nucl. Part. Sci.,* **32**, 211.
131 HINDS, E. A., and SANDARS, P. G. H., 1980, *Phys. Rev.,* **A21**, 480.
132 WILKENING, D. A., RAMSAY, N. F., and LARSON, D. J., 1983, *Phys. Rev.,* **A29**, 425.

133 SCHROPP, D., CHO, D., VOLD, T., and HINDS, E. A., 1987, *Phys. Rev. Lett.,* **59**, 991.
134 SANDARS, P. G. H., 1967, *Phys. Rev. Lett.,* **19**, 1396.
135 BARKOV, L. M., ZOLOTAREV, M. S., and MELIK-PASHAEV, D. A., 1988, *Sov. Phys. JETP Lett.,* **48**, 144.
136 SANDARS, P. G. H., 1966, *Contemp. Phys.,* **7**, 419.
137 GODDARD, P., and OLIVE, D., 1978, *Rep. Prog. Phys.,* **41**, 1357.
138 TOWNES, C. H., and SCHAWLOW, A. L., 1955, *Microwave Spectroscopy* (McGraw-Hill, New York).
139 HERZBERG, G., 1945, *Infrared and Raman Spectra of Polyatomic Molecules* (Van Nostrand, New York).
140 ALLEN, H. C., and CROSS, P. C., 1963, *Molecular Vib-Rotors* (Wiley, New York).
141 OKA, T., 1973, *J. Mol. Spectrosc.,* **48**, 503.
142 QUACK. M., 1977, *Mol. Phys.,* **34**, 477.
143 BUNKER, P. R., 1979, *Molecular Symmetry and Spectroscopy* (Academic Press, New York).
144 DAVYDOV, A. S., 1976, *Quantum Mechanics* (Pergamon Press, Oxford).
145 BARRON, L. D., 1979, *J. Am. Chem. Soc.,* **101**, 269.
146 HUND, F., 1927, *Z. Phys.,* **43**, 805.
147 ROSENFELD, L., 1928, *Z. Phys.,* **52**, 161.
148 BORN, M., and JORDAN, P., 1930, *Elementare Quantenmechanik* (Springer, Berlin).
149 COHEN-TANNOUDJI, C., DIU, B., and LALOË, F., 1977, *Quantum Mechanics* (Wiley, New York).
150 QUACK, M., 1986, *Chem. Phys. Lett.,* **132**, 147.
151 MICHEL, L., 1980, *Rev. Mod. Phys.,* **52**, 617.
152 STROCCHI, F., 1985, *Elements of Quantum Mechanics of Finite Systems* (World Scientific, Singapore).
153 ULBRICHT, T. L. V., 1959, *Quart. Rev. Chem. Soc.,* **13**, 48.
154 BERNE, B. J., and PECORA, R., 1976, *Dynamic Light Scattering* (Wiley, New York).
155 HARRIS, R. A., and STODOLSKY, L., 1981, *J. Chem. Phys.,* **74**, 2145.
156 ANDERSON, P. W., 1972, *Science,* **177**, 393.
157 ANDERSON, P. W., 1983, *Basic Notions of Condensed Matter Physics* (Benjamin/Cummings, Menlo Park, California).
158 PFEIFER, P., 1980, *Chiral Molecules—a Superselection Rule Induced by the Radiation Field.* Thesis, Swiss Federal Institute of Technology, Zurich.
159 PFEIFER, P., 1983, *Energy Storage and Redistribution in Molecules* (Plenum Press, New York), p. 315.
160 WOOLLEY, R. G., 1975, *Adv. Phys.,* **25**, 27.
161 WOOLLEY, R. G., 1978, *J. Am. Chem. Soc.,* **100**, 1073.

162 WOOLLEY, R. G., 1980, *Isr. J. Chem.*, **19**, 30.
163 WOOLLEY, R. G., 1982, *Struct. Bonding*, **52**, 1.
164 WOOLLEY, R. G., 1988, *Molecules in Physics, Chemistry and Biology*, Vol. 1, edited by J. Maruani (Kluwer, Dordrecht), p. 45.
165 SIMONIUS, M., 1978, *Phys. Rev. Lett.*, **40**, 980.
166 JOOS, E., and ZEH, H. D., 1985, *Z. Phys.*, **B59**, 223.
167 CLAVERIE, P., and JONA-LASINIO, G., 1986, *Phys. Rev.*, **A33**, 2245.
168 LEGGETT, A. J., CHAKRAVARTY, S., DORSEY, A. T., FISHER, M. P. A., GARG, A., and ZWERGER, W., 1987, *Rev. Mod. Phys.*, **59**, 1.
169 DAVIES, E. B., 1979, *Comm. Math. Phys.*, **64**, 191; *ibid*, 1980, **75**, 263.
170 BARRON, L. D., 1981, *Chem. Phys. Lett.*, **79**, 392.
171 WOOLLEY, R. G., 1981, *Chem. Phys. Lett.*, **79**, 395.
172 BIEDENHARN, L. C., and HORWITZ, L. P., 1984, *Found. Phys.*, **14**, 953.
173 JUNGWIRTH, P., SKÁLA, L., and ZAHRADNÍK, R., 1989, *Chem. Phys. Lett.*, **161**, 502.
174 HEISENBERG, W., 1966, *Introduction to the Unified Field Theory of Elementary Particles* (Wiley, New York).
175 HEISENBERG, W., 1976, *Physics Today*, **29**, No. 3, 32.
176 BOHM, A., 1989, *Symmetries in Science III*, edited by B. Gruber and F. Iachello (Plenum Press, New York), p. 85.
177 BOHM, A., BOYA, L. J., KIELANOWSKI, P., KMIECIK, M., LOEWE, M., and MAGNOLLAY, P., 1988, *Int. J. Mod. Phys.*, **A3**, 1103.
178 BARUT, A. O., 1983, *Found. Phys.*, **13**, 7.
179 WIGNER, E. P., 1965, *Sci. Am.*, **213**, No. 6, 28.
180 MARCH-RUSSELL, J., and WILCZEK, F., 1988, *Phys. Rev. Lett.*, **61**, 2066.
181 WEN, X. G., and ZEE, A., 1989, *Phys. Rev. Lett.*, **62**, 2873.

Vibrational Circular Dichroism Intensities: *Ab Initio* Calculations

by

Arvi Rauk

Department of Chemistry,
University of Calgary,
Calgary, AB, Canada T2N 1N4

Introduction

Of the various chiroptical techniques, one of the newest, vibrational circular dichroism (VCD) promises to yield the greatest amount of structural and stereochemical information. Since the earliest theoretical models [1] and the first successful experimental measurements [2,3], VCD has seen rapid advances in instrumentation and theory. Reliable measurements over most of the vibrational spectrum can be carried out, and reliable theoretical computations of signs and intensities of the fundamental transitions in small to medium sized molecules can be performed. It is not the intention of this article to present a comprehensive review of the development of VCD. This has been accomplished in numerous excellent review articles by the leaders in the field [4–11]. Instead, we will concentrate rather narrowly on the theoretical developments which have led to successful computer implementation, at an *ab initio* quantum mechanical level and to review the results of such calculations insofar as they have been reported in the literature to date.

Theory of VCD

Under Beer's law conditions, for a transition, $o \rightarrow n$, the integrated absorbance, ϵ, and differential absorbance (CD), $\Delta\epsilon$, of a substance are proportional to the dipole, and optical rotatory strengths, D_{on}, and R_{on}, respectively:

$$D_{on} = < \Psi_o | \mu_{elec} | \Psi_n > \cdot < \Psi_n | \mu_{elec} | \Psi_o > \tag{1}$$

$$R_{on} = -i < \Psi_o | \mu_{elec} | \Psi_n > \cdot < \Psi_n | \mu_{mag} | \Psi_o > \tag{2}$$

where the electric and magnetic dipole operators, μ_{elec} and μ_{mag}, are the sums of electron and nuclear operators,

$$\mu_{elec} = \mu^e_{elec} + \mu^n_{elec} = -\sum_i e\vec{r}_i + \sum_I Z_I e\vec{R}_I \tag{3}$$

P. G. Mezey (ed.), New Developments in Molecular Chirality, 57–92.

$$\mu_{mag} = \mu^e_{mag} + \mu^n_{mag} = -\sum_i \frac{e}{2mc}\vec{r}_i \times \vec{p}_i + \sum_I \frac{Z_I e}{2M_I c}\vec{R}_I \times \vec{P}_I \tag{4}$$

The quantities, $e, m, \vec{r}_i$ and $\vec{p}_i$, are the charge, mass, position and momentum of the i'th electron, $Z_I e$, M_I, $\vec{R}_I$ and $\vec{P}_I$ are the charge, mass, position and momentum of the I'th nucleus, and c is the speed of light. Let us assume that the transition involved is a vibronic transition between the lowest and the first vibrational levels on the ground electronic state adiabatic Born–Oppenheimer (BO) potential energy surface. Thus the states, o and n, may be labelled (0,0) and (0,1), respectively, and the wavefunctions in (1) and (2) can be written in BO form, differing in the nuclear component:

$$\Psi_o = \Psi(r,Q)_{0,0} = \psi(r,Q)_0\Phi(Q)_{0,0} \tag{5}$$

$$\Psi_n = \Psi(r,Q)_{0,1} = \psi(r,Q)_0\Phi(Q)_{0,1} \tag{6}$$

where the implicit dependence of the wavefunctions on the electronic and nuclear coordinates is given, the latter in terms of displacements along normal modes, $Q = \{Q_1,..,Q_{3N-6}\}$ (for the special case of a linear molecule, there are 3N–5 normal modes). In this notation the equilibrium geometry would correspond to all $Q_i = 0$. The electric dipole transition moment with wavefunctions (5) and (6) is:

$$< \psi_0\Phi_{0,0}|\mu_{elec}|\psi_0\Phi_{0,1} > = < \Phi_{0,0}| < \psi_0|\mu^e_{elec}|\psi_0 > + \mu^n_{elec}|\Phi_{0,1} > \tag{7}$$

The quantity, $< \psi_0|\mu^e_{elec}|\psi_0 > + \mu^n_{elec} = \mu^0_{elec}$, is the adiabatic electric dipole moment of the ground state. Its nuclear coordinate dependence, in terms of a particular normal mode, may be expressed as a Taylor expansion about the equilibrium geometry, truncated after the linear term,

$$\mu(Q)^0_{elec} \approx \left(\mu^0_{elec}\right)_0 + \left(\frac{\partial\mu^0_{elec}}{\partial Q}\right)_0 Q \tag{8}$$

Substitution of (8) into (7) yields the familiar result that the magnitude of the electric dipole transition moment is proportional to the change in the molecular dipole moment with respect to the particular normal mode, namely

$$< \psi_0\Phi_{0,0}|\mu_{elec}|\psi_0\Phi_{0,1} > = \left(\frac{\partial\mu^0_{elec}}{\partial Q}\right)_0 < \Phi_{0,0}|Q|\Phi_{0,1} > \tag{9}$$

With BO wavefunctions, (5) and (6), the outcome for the magnetic dipole transition moment is less satisfactory, namely,

$$< \psi_0\Phi_{0,0}|\mu_{mag}|\psi_0\Phi_{0,1} > = < \Phi_{0,0}| < \psi_0|\mu^e_{mag}|\psi_0 > + \mu^n_{mag}|\Phi_{0,1} >$$

$$= < \Phi_{0,0}|\mu^n_{mag}|\Phi_{0,1} > \tag{10}$$

The electronic contribution to the transition moment vanishes since the magnetic dipole operator is Hermitian, but purely imaginary, and for molecules with non-degenerate ground electronic states the wavefunctions may be taken as real. In order to introduce the electronic contribution to the magnetic dipole transition moment, many empirical models have evolved which explicitly incorporate nuclear displacements and the electronic–nuclear interaction in some fashion. These have been discussed in a number of reviews on VCD and will not be further discussed here. Alternatively, one is forced to a non–BO treatment to arrive at a description for VCD intensities[12–15]. Let the wave function be expanded in terms of adiabatic BO wavefunctions. For an arbitrary electronic–vibrational state (e',v') one may write

$$\Psi(r,Q)_{e',v'} \approx \psi(r,Q)_{e'}\Phi(Q)_{e',v'} - \sum_{e\,(\neq e')} a(Q)_{e'v',ev}\psi(r,Q)_e\Phi(Q)_{e,v} \tag{11}$$

where the coefficient, $a(Q)_{e'v',ev}$, is approximated by first order Rayleigh–Schroedinger perturbation theory using the part of the nuclear kinetic energy which was neglected in the BO approximation as the perturbation. Thus,

$$a(Q)_{e'v',ev} = \frac{< \psi(r,Q)_e\Phi(Q)_{e,v}|\hat{T}|\psi(r,Q)_{e'}\Phi(Q)_{e',v'} >}{E(Q)_{e,v} - E(Q)_{e',v'}} \tag{12}$$

where

$$\hat{T} = -\frac{\hbar^2}{2}\left[\left(\frac{\partial^2}{\partial Q^2}\right)_e + 2\left(\frac{\partial}{\partial Q}\right)_e\left(\frac{\partial}{\partial Q}\right)_n\right] \tag{13}$$

In (13), the subscripts indicate whether the operator acts on the electronic or the nuclear part of the wavefunction. The nuclear dependence of the *electronic* wavefunction may be expressed as a Taylor expansion about the equilibrium geometry of the ground electronic state, truncating after the linear term. Thus,

$$\psi(Q)_{e'} \approx (\psi_{e'})_0 + \left(\frac{\partial\psi_{e'}}{\partial Q}\right)_0 Q \tag{14}$$

The approximation (14) is most valid for the ground electronic state. The first term of (13) will be omitted from further consideration since it is a second order term. Substitution of (14) into (11) and (12), keeping only the second term of (13), yields, for the wavefunctions of interest for VCD,

$$\Psi(r,Q)_{0,v'} \approx (\psi_0)_0\Phi_{0,v'} + \left(\frac{\partial\psi_0}{\partial Q}\right)_0 Q\Phi_{0,v'} + \hbar^2 \sum_{e,v} \frac{\left\langle \psi_e \middle| \frac{\partial\psi_0}{\partial Q}\right\rangle_0 \left\langle \Phi_{e,v} \middle| \frac{\partial\Phi_{0,v'}}{\partial Q}\right\rangle_0}{(E_{e,v} - E_{0,v'})_0} (\psi_e)_0\Phi_{e,v} \tag{15}$$

The subscript *0* indicates that the quantities in parentheses are to be evaluated at the equilibrium geometry. Transition moments may be evaluated with (15), setting $v' = 0$ or 1, as appropriate, and using $\theta = \theta^e + \theta^n$ as a general operator of the type (3) or (4), keeping only terms to first order. Thus,

$$< \Psi_{0,0}|\theta|\Psi_{0,1} > \approx < \psi_0\Phi_{0,0}|\theta|\psi_0\Phi_{0,1} >_0 \tag{16a}$$

$$+ < \psi_0\Phi_{0,0}|\theta|\frac{\partial\psi_0}{\partial Q}Q\Phi_{0,1} >_0 + < \frac{\partial\psi_0}{\partial Q}Q\Phi_{0,0}|\theta|\psi_0\Phi_{0,1} >_0 \tag{16b}$$

$$+ \hbar^2 \sum_{e,v} < \psi_0\Phi_{0,0}|\theta|\psi_e\Phi_{e,v} > \frac{\left\langle \psi_e \middle| \frac{\partial\psi_0}{\partial Q}\right\rangle_0 \left\langle \Phi_{e,v} \middle| \frac{\partial\Phi_{0,1}}{\partial Q}\right\rangle_0}{(E_{e,v} - E_{0,1})_0} \tag{16c}$$

$$+ \hbar^2 \sum_{e,v} \frac{\left\langle \psi_e \middle| \frac{\partial\psi_0}{\partial Q}\right\rangle_0 \left\langle \Phi_{e,v} \middle| \frac{\partial\Phi_{0,0}}{\partial Q}\right\rangle_0}{(E_{e,v} - E_{0,0})_0} < \psi_e\Phi_{e,v}|\theta|\psi_0\Phi_{0,0} > \tag{16d}$$

If one assumes that the energy difference between specific vibronic levels on different electronic energy surfaces is approximately equivalent to the separation of the BO surfaces at the equilibrium geometry, namely

$$\frac{1}{(E_{e,v} - E_{0,v'})_0} \approx \frac{1}{(E_e - E_0)_0} \tag{17}$$

then one may formally close the sums over vibronic wavefunctions [13] in the third and fourth terms of (16). After some further manipulation, recognizing the antisymmetry of the operator, $\frac{\partial}{\partial Q}$, one arrives at

$$< \Psi_{0,0}|\theta|\Psi_{0,1} > \approx < \Phi_{0,0}|\theta^n|\Phi_{0,1} > \tag{18a}$$

$$+ \left[\left\langle \psi_0|\theta^e|\frac{\partial\psi_0}{\partial Q}\right\rangle + \left\langle \frac{\partial\psi_0}{\partial Q}|\theta^e|\psi_0\right\rangle\right]_0 < \Phi_{0,0}|Q|\Phi_{0,1} > \tag{18b}$$

$$+ \hbar^2 \sum_e \frac{< \psi_0|\theta^e|\psi_e >_0 \left\langle \psi_e|\frac{\partial\psi_0}{\partial Q}\right\rangle_0}{(E_e - E_0)_0} \left\langle \Phi_{0,0}|\frac{\partial\Phi_{0,1}}{\partial Q}\right\rangle \tag{18c}$$

$$+ \hbar^2 \sum_e \frac{\left\langle \psi_0|\frac{\partial\psi_e}{\partial Q}\right\rangle_0 < \psi_e|\theta^e|\psi_0 >_0}{(E_e - E_0)_0} \left\langle \Phi_{0,1}|\frac{\partial\Phi_{0,0}}{\partial Q}\right\rangle \tag{18d}$$

If θ is the electric dipole operator (3), then (18c) and (18d) cancel, the two terms of (18b) are equal, and one has the result

$$< \Psi_{0,0}|\mu_{elec}|\Psi_{0,1} > = < \Phi_{0,0}|\mu^n_{elec}|\Phi_{0,1} > + 2\left\langle \psi_0|\mu^e_{elec}|\frac{\partial\psi_0}{\partial Q}\right\rangle_0 < \Phi_{0,0}|Q|\Phi_{0,1} > \tag{19}$$

Equation (19) and (9) are formally equivalent. If θ is the magnetic dipole operator (4), then the two terms in (18b) cancel, (18c) and (18d) are equivalent, and one has the result:

$$< \Psi_{0,0}|\mu_{mag}|\Psi_{0,1} > = < \Phi_{0,0}|\mu^n_{mag}|\Phi_{0,1} >$$

$$+ 2\hbar^2 \sum_e \frac{< \psi_0|\mu^e_{mag}|\psi_e >_0 \left\langle \psi_e|\frac{\partial\psi_0}{\partial Q}\right\rangle_0}{(E_e - E_0)_0} \left\langle \Phi_{0,0}|\frac{\partial\Phi_{0,1}}{\partial Q}\right\rangle \tag{20}$$

Equations (19) and (20) constitute the Vibronic Coupling (VC) theory of Nafie and Freedman [13]. Evaluation of the magnetic dipole transition moment as expressed in (20) has been implemented at the *ab initio* level by Rauk and Dutler [16,17]. The sum over excited states poses special problems but can be successfully carried out as discussed below.

Stephens [14] has noted that the perturbation of the electronic wavefunction by the external magnetic field (of the incident radiation), which manifests itself in (20) as a sum over excited electronic states, may be treated in an explicit manner. The field dependence of the electronic ground state wave function may be elicited by expansion in Taylor and perturbation series, and identifying the first order corrections to each. Thus, Taylor expansion, truncating after the linear term, yields

$$\psi(\vec{B})_0 \approx (\psi_0)_{\vec{B}=0} + \left(\frac{\partial \psi_0}{\partial \vec{B}}\right)_{\vec{B}=0} \cdot \vec{B} \tag{21}$$

Alternatively, expansion as a Rayleigh–Schroedinger perturbation series, with

$$\mathcal{H}(\vec{B}) = \mathcal{H}^o + \mu^e_{mag} \cdot \vec{B} \tag{22}$$

yields, to first order,

$$\psi(\vec{B})_0 \approx (\psi_0)_{\vec{B}=0} - \sum_{e(\neq 0)} \frac{< \psi_e | \mu^e_{mag} | \psi_0 >}{E_e - E_0} \cdot \vec{B} \psi_e \tag{23}$$

From comparison of (21) and (23), one may identify the relationship

$$\left(\frac{\partial \psi_0}{\partial \vec{B}}\right)_{\vec{B}=0} = - \sum_{e(\neq 0)} \frac{< \psi_e | \mu^e_{mag} | \psi_0 >}{E_e - E_0} \psi_e \tag{24}$$

Insertion of the complex conjugate of (24) into (20) yields an expression in which the sum over states does not appear, namely the interesting result that the electronic contribution to the magnetic dipole transition moment is proportional to the overlap of two derivatives of the ground state electronic wave function:

$$< \Psi_{0,0} | \mu_{mag} | \Psi_{0,1} > = < \Phi_{0,0} | \mu^n_{mag} | \Phi_{0,1} > + 2\hbar^2 \left\langle \frac{\partial \psi_0}{\partial \vec{B}} \middle| \frac{\partial \psi_0}{\partial Q} \right\rangle_o \left\langle \Phi_{0,0} \middle| \frac{\partial \Phi_{0,1}}{\partial Q} \right\rangle \tag{25}$$

where the subscript o indicates that the derivatives are to be evaluated at zero magnetic field and at the equilibrium geometry, respectively. Stephens and coworkers have implemented (25) at the *ab initio* level, obtaining $\frac{\partial \psi_0}{\partial \vec{B}}$ initially numerically and subsequently by means of a second coupled perturbed Hartree–Fock calculation (the first, required to perform a normal coordinate analysis, yields $\frac{\partial \psi_0}{\partial Q}$ as a bonus). An alternative derivation of the result of (25) has recently been given [18].

In practise, the transition dipole matrix elements are evaluated from Hartree–Fock SCF (single determinantal electronic wavefunctions), harmonic oscillator nuclear wave functions, and in terms of cartesian or internal coordinates. A number of simplifications ensue. The matrix elements over harmonic oscillator nuclear wavefunctions, with associated frequency, ω_i, are

$$< \Phi_{0,0}|Q_i|\Phi_{0,1} > = \left(\frac{\hbar}{2\omega_i}\right)^{\frac{1}{2}} \tag{26}$$

$$\left\langle \Phi_{0,0}| \frac{\partial \Phi_{0,1}}{\partial Q_i} \right\rangle = \left(\frac{\omega_i}{2\hbar}\right)^{\frac{1}{2}} \tag{27}$$

and displacements of atomic coordinates from the equilibrium values, R^o_{Ia} , are related to normal coordinates, Q_i, by the relations

$$R_{Ia} - R^o_{Ia} = \sum_i S_{Ia,i}Q_i \ , \qquad \frac{\partial R_{Ia}}{\partial Q_i} = S_{Ia,i} \ , \qquad \frac{\partial}{\partial Q_i} = \sum_{Ia} \frac{\partial}{\partial R_{Ia}} S_{Ia,i} \tag{28}$$

Using relations (26) and (28), equation (9) is readily rewritten in familiar atomic polar tensor notation,

$$< \psi_0\Phi_{0,0}|(\mu_{elec})_\beta|\psi_0\Phi_{0,1} > = \left(\frac{\hbar}{2\omega_i}\right)^{\frac{1}{2}} \sum_{Ia} \left(\frac{\partial(\mu^0_{elec})_\beta}{\partial R_{Ia}}\right)_0 S_{Ia,i}$$

$$= \left(\frac{\hbar}{2\omega_i}\right)^{\frac{1}{2}} \sum_{Ia} \mathbf{P}^I_{\alpha\beta} S_{Ia,i} \tag{29}$$

The atomic polar tensor $\mathbf{P}^I_{\alpha\beta}$, is a measure of the change in the β'th component of the adiabatic dipole moment of the ground electronic state with respect to a displacement from equilibrium $R_{Ia} - R^o_{Ia}$ of the I'th nucleus in direction α, and is readily available by numerical or analytical methods from quantum chemistry codes such as GAUSSIAN 82 [19] (and later versions) as well as the CADPAC programs [20].

The magnetic dipole transition moment may be written in a parallel fashion in terms of atomic axial tensors $\mathbf{M}^I_{\alpha\beta}$, although the derivation is somewhat more involved since there is a nuclear contribution within the BO approximation (equation (10)). Expressing the nuclear position in terms of a displacement from equilibrium $\vec{R}_I = \vec{R}^o_I + (\vec{R}_I - \vec{R}^o_I)$ for a particular normal mode, one has for the nuclear contribution

$$< \Phi_{0,0}|(\mu^n_{mag})_\beta|\Phi_{0,1} > = < \Phi_{0,0}| \sum_I \frac{Z_I e}{2M_I c}(\vec{R}^o_I \times \vec{P}_I)_\beta|\Phi_{0,1} > \tag{30a}$$

$$= < \Phi_{0,0}| \sum_I \frac{-i\hbar Z_I e}{2M_I c}(\vec{R}^o_I \times \frac{\partial}{\partial \vec{R}_I})_\beta|\Phi_{0,1} > \tag{30b}$$

$$= \sum_{\alpha\gamma} \sum_{I} \frac{Z_I e}{2ic} \left(\frac{\omega_i}{2\hbar} \right)^{\frac{1}{2}} \epsilon_{\alpha\beta\gamma} \vec{R}^o_{I\gamma} S_{Ia,i} \tag{30c}$$

Equation (30a) results from the fact that $< \Phi_{0,0} | (\vec{R}_I - \vec{R}^o_I) \times \vec{P}_I | \Phi_{0,1} > = 0$ for a normal mode. In deriving (30b) we have used $\vec{P}_I = -i\hbar(\frac{\partial}{\partial \vec{R}_I})$. The final result, (30c), arises from use of

$$< \Phi_{0,0} | \frac{\partial}{\partial \vec{R}_I} | \Phi_{0,1} > = \frac{M_I}{\hbar^2} \omega_i < \Phi_{0,0} | (\vec{R}_I - \vec{R}^o_I) | \Phi_{0,1} > \tag{31}$$

as well as relations (26) and (28). Expansion of the cross product requires, $\epsilon_{\alpha\beta\gamma}$, the antisymmetric unit third-rank tensor. The second term of (25) may be converted to nuclear position dependence by use of (27) and (28). Thus,

$$2\hbar^2 \left\langle \frac{\partial \psi_0}{\partial \vec{B}} \, | \, \frac{\partial \psi_0}{\partial Q} \right\rangle_o \left\langle \Phi_{0,0} | \frac{\partial \Phi_{0,1}}{\partial Q} \right\rangle = 2 \left(\frac{\hbar^3 \omega_i}{2} \right)^{\frac{1}{2}} \sum_{Ia} \left\langle \frac{\partial \psi_0}{\partial B_a} \, | \, \frac{\partial \psi_0}{\partial R_{Ia}} \right\rangle_o S_{Ia,i} \tag{32}$$

Inserting (30c) and (32) into (25) yields

$$< \Psi_{0,0} | (\mu_{mag})_\beta | \Psi_{0,1} >_i = -(2\hbar^3 \omega_i)^{\frac{1}{2}} \sum_{Ia} \left[\sum_{\gamma} \epsilon_{\alpha\beta\gamma} R^o_{Ia} \frac{Z_I e i}{4\hbar c} + \left\langle \frac{\partial \psi_0}{\partial B_a} \, | \, \frac{\partial \psi_0}{\partial R_{Ia}} \right\rangle_o \right] S_{Ia,i} \tag{33a}$$

$$= -(2\hbar^3 \omega_i)^{\frac{1}{2}} \sum_{Ia} \mathbf{M}^I_{a\beta} S_{Ia,i} \tag{33b}$$

where the atomic axial tensor $\mathbf{M}^I_{a\beta}$ is defined as the quantity in the square brackets in (33a).

The atomic polar and axial tensors, $\mathbf{P}^I_{a\beta}$ and $\mathbf{M}^I_{a\beta}$, may be separated into electronic and nuclear parts,

$$\mathbf{P}^I_{a\beta} = \mathbf{E}^I_{a\beta} + \mathbf{N}^I_{a\beta} \tag{34}$$

where $\mathbf{N}^I_{a\beta} = Z_I e \delta_{\alpha\beta}$, and

$$\mathbf{E}^I_{a\beta} = 2 \left\langle \psi_0 | (\mu^e_{elec})_\beta | \frac{\partial \psi_0}{\partial R_{Ia}} \right\rangle_0 \tag{35}$$

Similarly,

$$\mathbf{M}^I_{a\beta} = \mathbf{I}^I_{a\beta} + \mathbf{J}^I_{a\beta} \tag{36}$$

where the two parts, $\mathbf{I}^I_{\alpha\beta}$ and $\mathbf{J}^I_{\alpha\beta}$, may readily be identified with the *second* and *first* terms, respectively, in the square brackets of (33a);

$$\mathbf{J}^I_{\alpha\beta} = \sum_{\gamma} \epsilon_{\alpha\beta\gamma} R^o_{I\alpha} \frac{Z_I e i}{4\hbar c} \tag{37}$$

and

$$\mathbf{I}^I_{\alpha\beta} = \left\langle \frac{\partial \psi_0}{\partial B_\alpha} \,|\, \frac{\partial \psi_0}{\partial R_{I\alpha}} \right\rangle_o \tag{38}$$

Stephens has recently derived a relationship between the electronic components, $\mathbf{E}^I_{\alpha\beta}$ and $\mathbf{I}^I_{\alpha\beta}$, with nuclear shielding tensors, $\gamma^I_{\alpha\beta}$ and $\xi^I_{\alpha\beta}$, of Lazzeretti, Zanasi, and coworkers [21,22], namely

$$\mathbf{E}^I_{\alpha\beta} = -(Z_I e)\gamma^I_{\alpha\beta}(0) \tag{39}$$

$$\mathbf{I}^I_{\alpha\beta} = -\frac{i}{2\hbar}(Z_I e)\xi^I_{\alpha\beta}(0) \tag{40}$$

where the zero in parentheses indicates a static (zero frequency) magnetic field. Expressions for $\gamma^I_{\alpha\beta}(0)$ and $\xi^I_{\alpha\beta}(0)$ [22–24] involve sums over excited states but have been implemented at the SCF level using the RPA approximation to evaluate the sums over states [21], and therefore provide a third alternative method for calculating dipole and optical rotatory strengths [24].

Two other reformulations not explicitly discussed above have appeared. Stephens has recast the atomic polar tensor component $\mathbf{E}^I_{\alpha\beta}$ in a form parallel to that of $\mathbf{I}^I_{\alpha\beta}$ (above) by considering a momentum representation, and $\mathbf{I}^I_{\alpha\beta}$ itself has been recast in the same spirit [25]. The expressions have been implemented in CADPAC [20]. Freedman and Nafie have reformulated the vibronic coupling theory in terms of atomic orbital basis sets which float with nuclear positions and velocities[26]. The resulting expressions have not yet been implemented for *ab initio* wavefunctions.

Results

The results of published *ab initio* calculations of vibrational dipole and optical rotatory strengths of a small but rapidly growing list of compounds are discussed below, more or less in chronological order.

Computational Details. The following steps must be carried out in order to use *ab initio* methods to calculate VCD intensities:

- Equilibrium geometry. All quantities are evaluated at the equilibrium geometry of the molecule. Ideally, this would be the experimental geometry where available, although for mathematical reasons should be the "equilibrium geometry" obtained by the computational scheme. All VCD calculations to date have used single–determinantal (Hartree–Fock) SCF wavefunctions expanded in terms of a moderate to large basis set of gaussian–type atomic orbitals. Geometry optimization at this level of theory by analytical gradient procedures normally yields geometries close to experimental, so either the experimental or theoretical geometry may be used. Where systematic deviations are known to occur, the *ab initio* geometry may be "corrected" [27].
- Force Field to obtain displacement vectors $S_{I\alpha,i}$ (28). The experimental force field may be used if available, although this is rarely the case. Usually, a harmonic frequency analysis is carried out at the theoretical equilibrium geometry using RHF–SCF wavefunctions expanded in the same basis set as was used to obtain the geometry. The procedure may be carried out by numerical or analytical second differentiation of the energy. The quantum mechanical force field thus obtained has force constants which are systematically too large [28]. These may be scaled uniformly [28], or non–uniformly [29] to improve agreement of the calculated harmonic frequencies with experimentally observed values. Uniform scaling does not alter the theoretical $S_{I\alpha,i}$ values. It is generally agreed that some non–uniform scaling, judiciously carried out, is necessary to maximize agreement of computed intensities with experimental values.
- Derivatives of the ground state wavefunction. Evaluation of electric and magnetic dipole transition moments require derivatives of the electronic ground state wavefunction with respect to nuclear displacements, $\frac{\partial \psi_0}{\partial R_{I\alpha}}$. These are obtained as a by–product of the solution of the coupled perturbed Hartree–Fock equations required for the harmonic frequency analysis. In addition, Stephens method [14] (equation (25)) requires derivatives with respect to an applied magnetic field. Efficient analytical procedures exist for the evaluation of both kinds of derivatives.
- Sums over excited states. The vibronic coupling theory of Nafie and Freedman [13] (equa-

tion (20)) and the nuclear shielding tensor reformulation [23,24] (equations (37) and (38)), require the explicit evaluation of a sum over excited states. In each case, only single particle operators connect states so that no more than singly excited configurations will mix with the RHF–SCF ground state. At least for the vibronic coupling theory, "completeness" of the sum over excited states requires that at some stage derivatives of the usual set of basis functions with respect to nuclear or electron displacements be added to the basis set [30]. It is not certain that the use of derivative functions can be avoided entirely if the sum over states is avoided, as in Stephens' method [14].

***trans*–1,2–Dideuteriocyclopropane *1*.** The calculated VCD properties of (*S,S*)–1,2–d_2–*1* are the first *ab initio* results to be reported [31]. The theoretical method is that of Stephens [14]. Several geometries of *1*, including the 4–31G optimized geometry were employed. Three different force fields were employed, an empirical force field due to Duncan and Burns [32], a scaled *ab initio* 4–31G force field by Blom and Altona [33], and a scaled 6–31G** force field due to Komornicki and coworkers [34]. The magnetic field dependence of the SCF 4–31G wavefunction and the scalar products associated with equation (25) were calculated by numerical differentiation using a modified version of the GAUSSIAN 80 system of programs [35]. Calculated VCD intensities were relatively insensitive to the choice of force field in both the CH and CD stretching regions, as well as the mid–infrared region. Frequencies, rotatory strengths, and dipole strengths obtained with the Komornicki, et al. force field are listed in Table 1.

***trans*–1,2–Dideuteriocyclobutane *2*.** The calculated VCD properties [36] of (*R,R*)–1,2–d_2–*2* are the first to be compared with experimental data [37]. The theoretical method is that of Stephens [14]. The equilibrium geometry and SCF wavefunction were determined using a modified GAUSSIAN 80 system of programs [35] and the 4–31G basis set. The force field was that of Annamalai and Keiderling [38]. The geometry dependence of the ground state and the magnetic field dependence of the equilibrium ground state wavefunction were determined by numerical differentiation. The calculated VCD intensities were represented as simulated spectra of an 1:1 equilibrium mixture of axial and equatorial isomers. Comparison with experimental spectra available for the CH and CD stretching region, and in the mid–IR range, 700 cm^{-1} – 1500 cm^{-1}. was very good. The authors note that *ab initio* VCD results did "not replicate results obtained from alternate theories" [36], except for modes of frequency above 1200 cm^{-1} where the sign pattern obtained with the fixed partial charge (FPC) model was reproduced.

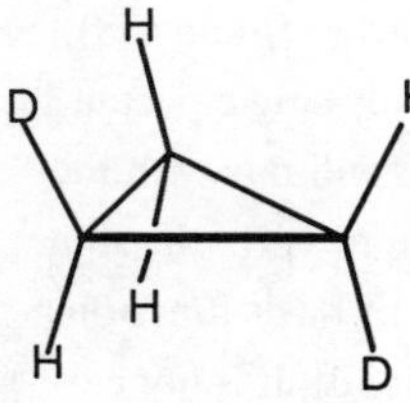

(S,S)–1,2–d_2–**1** 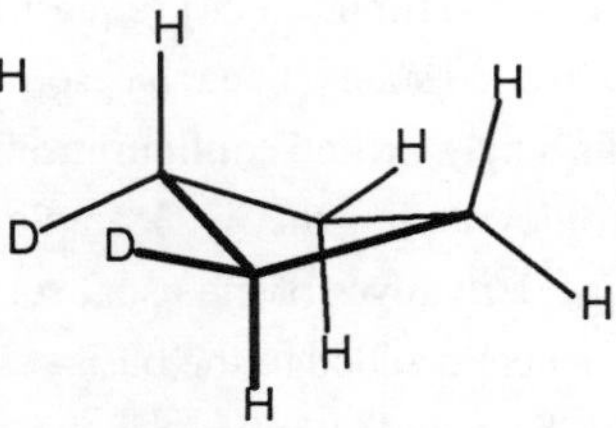 (R,R)–1,2–d_2–**2** 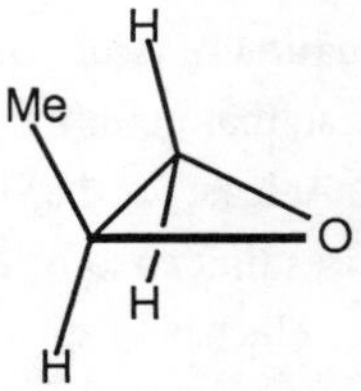 (S)–**3** 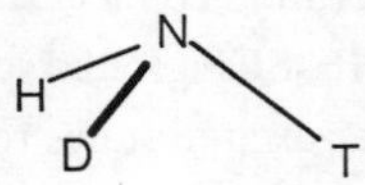 (S)–NHDT

2–Methyloxirane (propylene oxide) ***3*****.** Together with those of **2**, the VCD properties [36] of ***3*** are the first to be compared with experiment [39]. The theoretical method is that of Stephens [14]. The equilibrium geometry and SCF wavefunction were determined using a modified GAUSSIAN 80 system of programs [35] and the 4–31G basis set. The authors derived an *ab initio* force field by numerical differentiation of the 4–31G SCF energy and scaled it to match experimentally observed frequencies [40]. The geometry dependence of the ground state and the magnetic field dependence of the equilibrium ground state wavefunction were determined by numerical differentiation. The calculated VCD intensities were represented as simulated spectra. Comparison to the VCD spectrum of neat (–)–(*S*)–2–methyloxirane [39] measured in the 850 cm^{-1} – 1550 cm^{-1} range, was not as good as for the parallel study of **2**, but was encouraging, nevertheless. According to the authors, "FPC model calculations...bear no resemblance (to the *ab initio* results) in any frequency range [36]. Stephens and coworkers have recently recalculated the VCD spectra of **3** using analytical methods [41,42] as implemented in CADPAC [20] rather than finite difference methods. They also employed the distributed origin gauge (DOG), and substantially larger basis sets (6–31G**, 6–311G**, DZ/1P) [43]. Atomic polar and axial tensors were recalculated with the 4–31G basis set for comparison. Some results of this study on (*R*)–***3*** are reproduced in Table 2 together with their experimental observations. For the most part, the experimentally observed pattern of signs and intensities is reproduced, although the wrong sign is predicted for several transitions and several others differ substantially in intensity in the 1100 cm^{-1} to 1170 cm^{-1} region. The two basis sets of Table 2, and two others (6–311G** and DZ/1P) which were also tested give essentially the same results, a somewhat discouraging finding.

The vibronic coupling theory of Nafie and Freedman [13] has been applied to *(R)–**3*** by Rauk and Yang [44], using the 6–31G$^{*(0.3)}$ basis set, a modified 6–31G* set. The modification, which simply involves changing the exponent of all d–functions to 0.30, allows the VC theory based computations to reproduce reasonably well results for *trans*–dideuteriooxirane, hydrazine, hydrogen peroxide, and α–deuteroethanol (see below) obtained from a fully derivatized 6–31G$^{\sim}$ basis set. The results with the 6–31G$^{*(0.3)}$ basis set for (*R*)–*3* are compared to those of Stephens in Table 2. All except one of the calculated optical rotatory strengths agree in sign with the observed transitons.

NHDT . Stephens and coworkers have carried out several extensive studies centred on the hypothetical molecule, NHDT: basis set effects [24,45], guage dependence [24,46], and methodology [24]. Stephens and coworkers conclude [24] that the optimum results are obtained if large polarized basis sets are used, atomic polar tensors are calculated in the position formulation (eqn (35)), atomic axial tensors are calculated using the distributed origin gauge with origins at nuclei and in the angular momentum or torque representation. The most recent study [24] represents the first application of the use of nuclear shielding tensors and the SCF–RPA method to the calculation of VCD intensities. The results deemed to be most accurate are reproduced in Table 3 for reference.

***trans*–1,2–Dicyanocyclopropane *4*.** Stephens and coworkers [47] have reported results of VCD calculations on (*S,S*)–*4* using Stephens' method [14], obtaining atomic polar and axial tensors analytically by CADPAC [20,42] and the 6–31G** basis set. The force field was derived from scaling the *ab initio* 6–31G* force field at the 6–31G** equilibrium geometry. The results for uniform and non–uniform (6 scale factors) scaling were compared to dipole and rotatory strengths estimated from the experimental spectra of Heintz and Keiderling [48]. The VCD originating from the skeletal deformations, including the cyano groups were reasonably well reproduced, difficulties being experienced only with the CH (as opposed to CH_2) stretches.

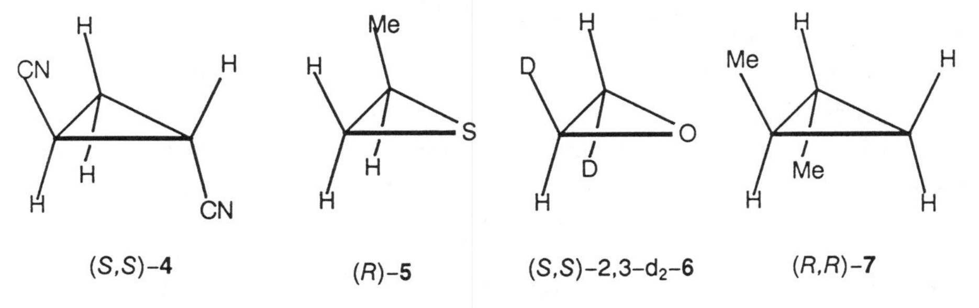

Methylthiirane *5*. Dothe, Lowe and Alper (DLA) have calculated the IR and VCD intensities for (*R*)–*5* [49] using the method of Stephens [14]. A "corrected" 6–31G* geometry, and scaled *ab initio* force field were used to determine the cartesian displacement vectors for numerical evaluation of the atomic polar and axial tensors, the dependence of the latter on the external magnetic field also being determined numerically. The atomic axial tensor was evaluated in the common origin gauge (COG) (center-of-mass and origin at S) and the distributed origin gauge (DOG). The three simulated IR and VCD spectra obtained from the calculated intensities were compared with the experimental IR and VCD spectra of Polavarapu and coworkers [50]. DLA concluded that none of the three simulated VCD spectra agreed well with the experimental spectum, and surprisingly found that the worst agreement was obtained with the origin independent DOG.

***trans*-2,3-Dideuteriooxirane *6*.** Both Stephens' method and the VC theory of Nafie and Freedman have been applied to the calculation of the VCD intensities of (*S,S*)-2,3-d_2-*6*. Jalkanen, *et al.* [51] reported the theoretical IR and VCD intensities of the CH and CD stretching region, comparing simulated spectra with experimental solution phase VCD spectra [52]. The basis set used for all aspects of the calculation was the 6-31G(extended) polarized basis set ((11s5p2d/5s2p) contracted to (4s3p2d/3s2p)). The *ab initio* force field obtained analytically at the experimental geometry of Hirose [53] was scaled against experimental frequencies of oxirane and oxirane-d_4. Atomic polar and axial tensors were calculated analytically by the CADPAC program [20] with the distributed origin gauge [54].

Dutler and Rauk [16,17] have published IR and VCD intensities for (*S,S*)-2,3-d_2-*6*, as well as its mono- and trideuterio isotopomers, (*S*)-2-d-*6* and (*S*)-2,2,3-d_3-*6*. The 6-31G$^{\sim}$ basis set was used in all aspects of the calculations. The 6-31G$^{\sim}$ basis set is constructed by addition to the 6-31G basis set all non-redundant derivatives of the 6-31G basis functions with respect to nuclear displacements [16,17]. Comparison was made to VC theoretical results obtained with the 6-31G** basis set. The *ab initio* force field was derived analytically at the 6-31G$^{\sim}$ geometry and scaled uniformly, or with up to seven scale factors. Spectral data for all three isotopomers of *6* were tabulated and displayed as simulated spectra. The latter are reproduced in Figure 1. Agreement with experimental IR intensities of Nakanaga [55] was very good, marginally better for the uniformly scaled force field. As in the case of Jalkanen, *et al.* [51], visual agreement with the experimental VCD spectrum in the CH and CD stretching region is excellent. In the mid-IR, complete agreement was obtained with the observed pattern of VCD signs and estimated intensities [56].

***trans*-1,2-Dimethylcyclopropane *7*.** The IR and VCD spectra of (*R*, *R*)-*7* have been measured and theoretically calculated by Amos, Handy, Drake and Palmieri (AHDP) [18]. AHDP used 6-31G* equilibrium geometry and unscaled force field but reported rotatory strengths only with a 3-21G basis set. AHDP employed Stephens' method [14] and Amos' implementation

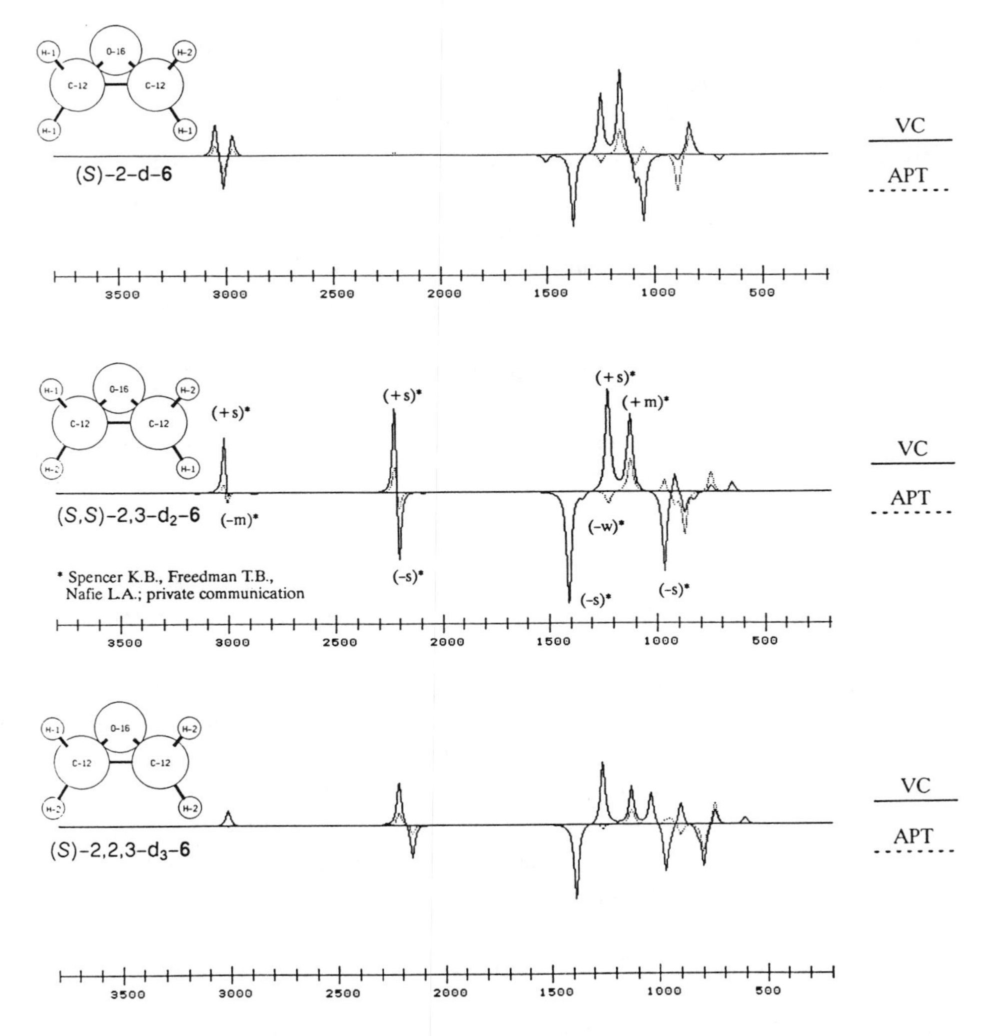

Figure 1. Simulated APT and VC VCD Spectra of (S)-2-deuterooxirane, (S,S)-2,3-dideuterooxirane, and (S)-2,2,3-trideuterooxirane: 6-31g~ (derivative) basis set at fully optimized geometry and uniform scaling (0.814).

in CADPAC [20]. They claimed substantial agreement between the theoretical VCD spectrum simulated with 3–21G rotatory strengths, and their experimental spectrum which covers the C–H stretching region.

Hydrazine *8*. The vibrational optical activity of hydrazine has been investigated by Rauk, Dutler and Yang (RDY) [57]. The geometry was determined by complete optimization at the restricted Hartree–Fock level of theory using the analytical procedures of the GAUSSIAN 82 system of programs [58] and the internal 6–31G** basis set. Normal coordinate analysis at the 6–31G** equilibrium geometry was carried out by analytical second differentiation of the Born–Oppenheimer RHF energy using the basis set designated as 6–31G$^{\sim}$ (see above). Harmonic frequencies for isotopically substituted species, as well as IR and VCD intensities were accomplished by the program system Freq85 [17], which implements the vibronic coupling theory of Nafie and Freedman [13] at the *ab initio* level. For improved agreement with experiment, frequencies were uniformly scaled by 0.9 [28].

The RHF/6–31G** geometry of hydrazine agrees reasonably well with experiment. The molecular point group is C_2 and the absolute configuration is *P*. Raw and scaled harmonic frequencies calculated from the 6–31G$^{\sim}$ force field are presented in Table 5. Hydrazine has been the subject of numerous IR and Raman investigations [59], as well as a thorough theoretical study by Morokuma and coworkers [60] who carried out a complete assignment of the vibrational spectrum based on *ab initio* force fields. RDY's calculated frequencies (after uniform scaling) and assignments agree entirely with theirs except in the NH stretching modes, $\nu 1$ and $\nu 8$, which both calculate to be nearly degenerate and which are not resolved experimentally. RDY's IR and VCD intensities are tabulated in Table 5 and simulated spectra are displayed in Figure 2. The most intense band in the IR spectrum, $\nu 12$ at 1001 cm^{-1} (scaled) corresponds to the out–of–phase umbrella motion of the two nitrogen atoms. The in–phase umbrella motion, $^{\nu}6$, at 873 cm^{-1} is also intense. These motions give rise to an intense conservative CD couplet in the VCD spectrum (Figure 2b) of N_2H_4, as well its tetradeuterio analogue, N_2D_4 (Figure 2a). A symmetrical CD couplet is also predicted for the NH_2 rocking modes, $\nu 4$ and $\nu 11$, near 1300 cm^{-1}. The effect of deuteration on IR and VCD spectra, besides the obvious frequency shift due to the increased mass, is to reduce the intensities of those modes where the motion of H/D dominates. Thus the VCD spectrum of N_2D_4 shown in Figure 2 is only about one quarter as intense as the VCD spectrum of N_2H_4. The rocking modes of the ND_2 groups do not give rise to a CD couplet as was observed in the case of N_2H_4 since these modes are shifted closer to the N–N stretch, $\nu 5$, and the increased coupling of $\nu 4$ and $\nu 5$ redistributes their negative rotatory intensity to both sides of the positive CD band due to $\nu 11$.

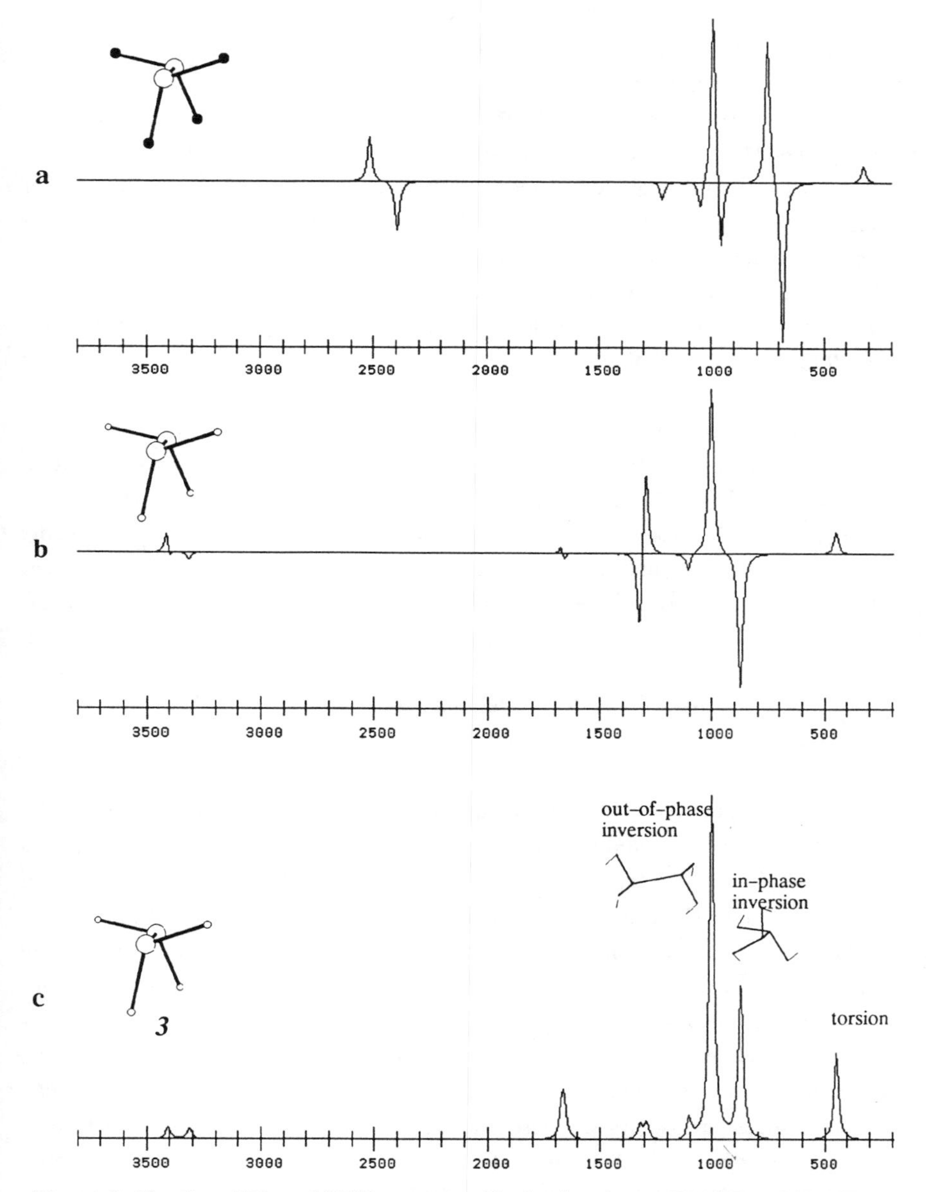

Figure 2. Simulated IR and VCD spectra of hydrazine: **b, c** NH_2NH_2; **a** ND_2ND_2. The spectrum **a** is magnified 4x relative to **b**.

Hydrogen peroxide *9*. The VCD and IR intensities of H_2O_2 have been calculated by Rauk, Dutler and Yang (RDY) [57] using the vibronic coupling formalism [13]. The geometry of hydrogen peroxide was determined by complete optimization at the restricted Hartree–Fock level of theory using the analytical procedures of the GAUSSIAN 82 system of programs [58] and the internal 6–31G** basis set. Normal coordinate analysis at the 6–31G** equilibrium geometry was carried out by analytical second differentiation of the Born–Oppenheimer RHF energy [14] using the 6–31G$^{\sim}$ basis set. Harmonic frequencies for isotopically substituted species, as well as IR and VCD intensities were accomplished by the program system Freq85 [17], which implements the vibronic coupling theory of Nafie and Freedman [13] at the *ab initio* level. For improved agreement with experiment, frequencies were uniformly scaled by 0.9 [28].

The RHF/6–31G** geometry of hydrogen peroxide agrees reasonably well with experiment. Systematic deviations at the RHF level could be corrected by inclusion of electron correlation in the calculations [61] but this was not carried out. The molecular point group is C_2 and the absolute configuration is *P*. Raw and scaled (by 0.9) harmonic frequencies calculated from the 6–31G$^{\sim}$ force field are presented in Table 6 together with calculated IR and VCD intensities. Rogers and Hillman [62] have previously calculated absolute intensities for hydrogen peroxide by APT theory, using a combination of theoretical and empirical APTs for H_2O, CH_3OH, and HOCl. Their values, shown in Table 6, are in qualitative agreement with the 6–31G$^{\sim}$ values. The most intense band in the IR is calculated to be the torsional mode, $\nu 4$. Its intensity may be exaggerated somewhat since the optimized torsional angle is too small by 6 deg.

The predicted VCD spectrum of *P* –H_2O_2 is shown in Figure 3 where it is compared to the predicted VCD spectra of the mono– and dideuterio isotopomers. The torsional mode, $\nu 4$, is calculated to give rise to the most intense VCD band (Figure 3). The most striking feature of the VCD spectrum is the CD couplet between 1300 cm^{-1} and 1500 cm^{-1}. Polavarapu [63] has analysed the expected intensity distribution among symmetric and antisymmetric bending vibrational modes in A_2B_2 systems and has predicted that for H_2O_2 of *P* chirality, the higher frequency symmetric O–O–H bend, should be considerably weaker in the IR than the antisymmetric O–O–H bend, and have negative rotatory strength of about the same magnitude as the antisymmetric mode if the fixed–partial–charge (FPC) mechanism dominates. RDY's calculated VCD spectrum displays just these features, but with a two–fold asymmetry favouring the antisymmetric bending CD band, $\nu 6$. From Table 6, it is apparent that the symmetric and antisymmetric O–H stretching modes, $\nu 1$ and $\nu 5$, respectively, are not consistent with features expected from a simple coupled oscillator mechanism from which one would predict

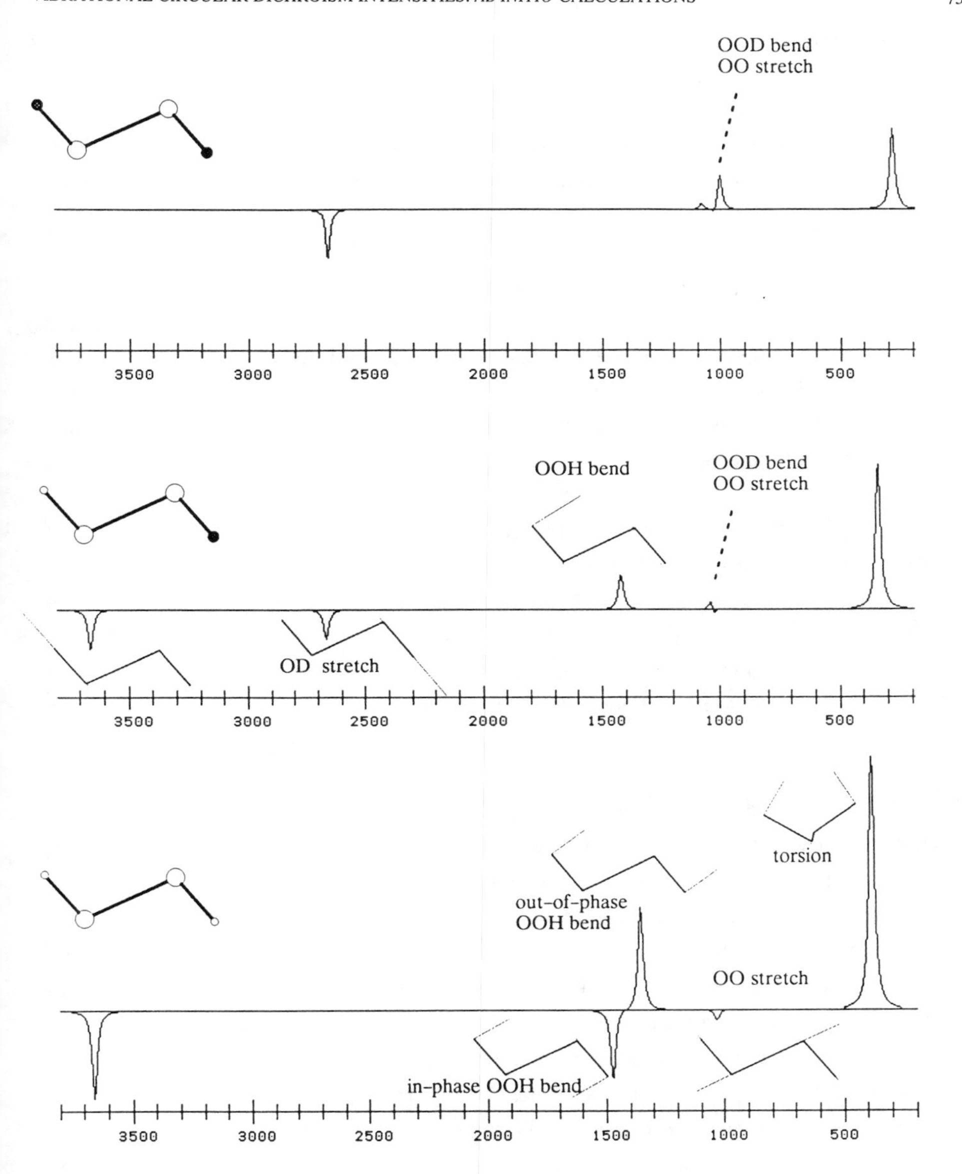

Figure 3. Simulated VCD spectra of HOOH, HOOD, and DOOD, plotted on the same scale.

a negative rotatory strength for the symmetric O–H mode, $\nu 1$. The signs are reversed and there is a five–fold asymmetry favouring the antisymmetric band. The near degeneracy calculated and observed [64] of the frequencies of the two stretching modes has the effect of completely cancelling the weaker positive CD band. An explanation for the origin of the "anomalous" rotatory strength is offered in connection with the VCD spectrum of DOOH discussed below.

The simulated VCD spectrum of DOOD is shown in Figure 3. The VCD intensities are greatly attenuated by deuterium substitution but the same features as were found in H_2O_2 are reproduced in the O–D stretching and torsional regions. The mid–IR region is predicted to be markedly different, however. The isotopic shift places the O–O–D bending frequencies into the vicinity of the O–O stretch and the strong mixture of the modes results in the disappearance of the CD couplet found for H_2O_2 itself.

The simulated VCD spectrum of DOOH is also shown in Figure 3. In this system, the O–H and O–D stretches and the O–O–H and O–O–D bends are well separated and essentially decoupled. Each of the isotopes, H or D, resides in a chiral environment of the same handedness. The O–O–H bend results in a positively signed band of similar intensity to the two stretches. The O–O–D bend and O–O stretch are almost degenerate and, as was the case for DOOD, the mixture of the two modes results in almost a complete cancellation of VCD intensity.

In DOOH, both stretching modes, O–H and O–D, yield a negative VCD signal, the origin of which may be explained by Nafie and Freedman's '3–bond electronic current model' [65] which is applied to DOOH in Figure 4. The O–H bond is strongly polarized in the direction expected from electronegativity considerations. The bond dipole is positive and a stretch of the bond results in a positive change in the bond dipole (and the permanent molecular dipole moment) in the direction of the O–H(D) bond. The remainder of the molecule, the O–O–D(H) fragment, forms an arc along which negative charge is drawn toward the O–H(D) bond being stretched. The induced flow of charge around the arc is associated with an induced magnetic dipole moment perpendicular to the O–O–D(H) plane on the side opposite to the O–H(D) dipole. Thus the scalar product of the change in the electric dipole moment and the induced magnetic dipole moment is negative. The calculated angle between the electric and magnetic dipole transition moments is $|128|$ deg.. In HOOH and DOOD where the stretching modes are coupled, the induced flow of charge along the bonds is cooperative in the case of the antisymmetric stretch, resulting in greater intensity for this mode.

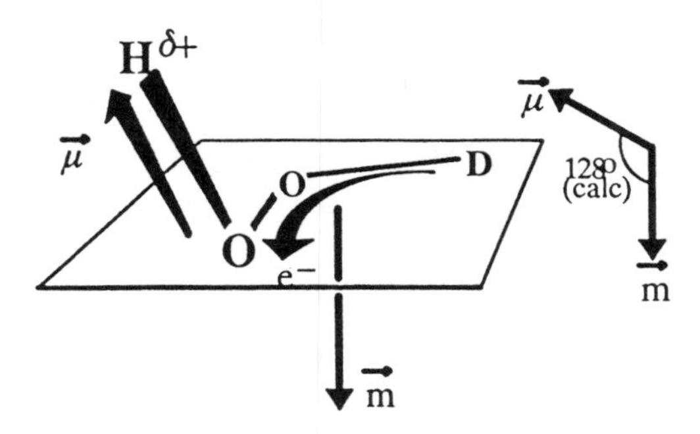

Figure 4. Model for rotatory strength of O-H stretching vibration of HOOD of *P*- chirality. Motion of the positive H away from the O-O-D group induces a flow of electrons toward it. The associated electric (μ) and magnetic (m) dipole transition moments are directed as shown. The calculated angle between the moments is -128 deg.

Ethanol–α–d *10.* Optically active α–deuterioethanol provides an opportunity for comparison of results calculated by Stephens' formalism and the vibronic coupling theory to each other and to experiment. Dothe, Lowe and Alper (DLA)[66] have reported VCD and IR intensities for ethanol and all deuterated isotopomers, calculated using Stephens' method, scaled quantum mechanical force fields evaluated by numerical differentiation and the 6–31G* basis set. Polar and axial tensors were evaluated at the distributed origin gauge [46] and a common origin (centre of mass) gauge. The distributed origin results, which were deemed to be more reliable), are reproduced in Table 7 for both *gauche* rotamers and for the *anti* form. The data in Table 7 has the signs reversed to correspond to the (*S*) enantiomer. For comparison with experiment, DLA had available the VCD results of Pultz [67] who measured the VCD of (R)–1–d–ethanol and (R)–1,O–d_2–ethanol in the C–H stretching region, and of the latter in the mid–IR region. By attributing the dominant features of the experimental spectrum entirely to an equimixture of the two *gauche* conformers, DLA found substantial agreement with the observed pattern of signs except for the band at 2930 cm^{-1} attributed to the symmetric methyl deformation for which they obtained the opposite sign.

Dutler, Rauk, Shaw, and Wieser (DRSW)[68] have applied the vibronic coupling formalism and the 6–31G$^{\sim}$ basis set to calculate the VCD spectrum of the deuterated ethanols. DRSW scaled their *ab initio* force field in the same way as DLA, namely to match the published IR spectra of Perchard and Josien [69]. DRSW determined the relative energies of the *gauche* and *anti* forms at a high level of theory and included zero point vibrational energy differences to arrive at a Boltzmann distribution of the components of the equilibrium mixture. Their results are also reported in Table 7 for the (*S*) enantiomer. The resultant VCD spectrum of (*S*)–1–d–ethanol is simulated in Figure 5 and compared with the experimental spectrum [68]

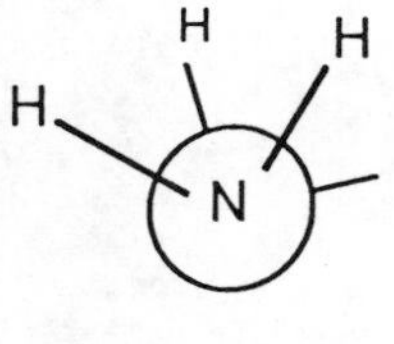

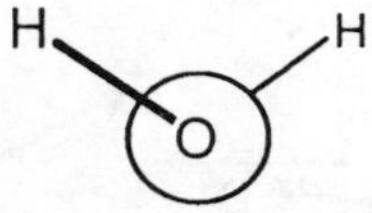

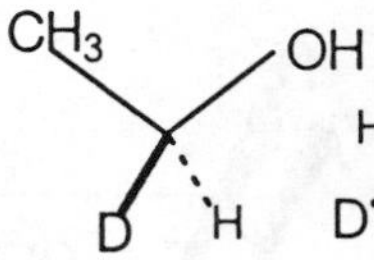

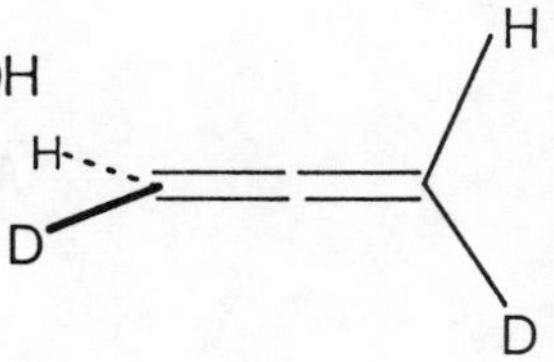

(*P*)–8 (*P*)–**9** (*S*)–1–d–**10** (*S*)–**11**

in the 800 cm^{-1} – 1500 cm^{-1} region. The main features of the experimental spectrum are very well reproduced in the simulation.

An analysis of the theoretical VCD of the individual components of the equilibrium mixture is presented in Figure 6. The strong negative feature at 1120 cm^{-1} in the experimental spectrum originates from each component. The mode is very complex, being a mixture of Me rocking, CO and CC stretching, and CHD twist–wag motions. Other features of the experimental spectrum are clearly identifiable as originating from separate components. The strong positive Cotton effect at 1250 cm^{-1} arises from the predominantly COH bending mode of the *anti* conformer, there being only weak CD from the two *gauche* forms in this region. The positive CD between 1030 cm^{-1} and 1100 cm^{-1} is predicted to originate in the main from the (*P*)–*gauche* conformer with perhaps a small component from the *anti* form. Both *gauche* forms contribute to the positive CD at 1310 cm^{-1} but not the *anti* which is predicted to have a negative CD in this region.

The results of DLA (Table 7) agree in the main with those of DRSW at least in sign. Absolute intensities are very different, in some cases by more than an order of magnitude. The origin of the differences is not certain. In the Stephens' method results of DLA, origin dependence of the magnetic dipole transition moment has been eliminated by use of the distributed origin gauge. However, the basis set, though large, does not have the extensive polarization which Stephens' investigations on NHDT [24,45,46] suggest is necessary for converged results (see above) . On the other hand, the results of DRSW were obtained with a very large (equivalent in size to 6–311G**), properly derivatized, basis set, but were evaluated in the common origin gauge (centre of mass) and are origin dependent. The agreement between calculated and experimental spectra suggests that the final results of DRSW are not far from the correct values. More comparisons are necessary before one could conclude that the centre of mass is a reasonable choice of origin and/or that the origin dependence is not severe with the 6–31G~ basis set.

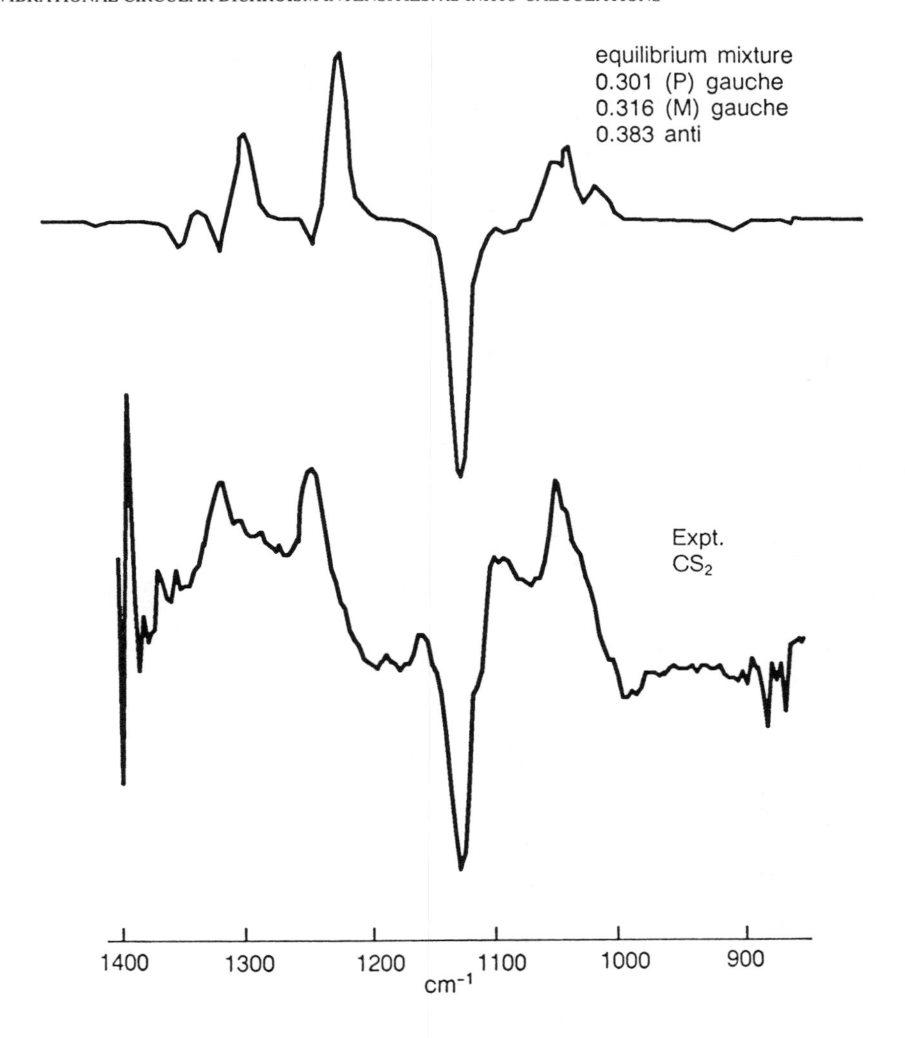

Figure 5. Comparison of the simulated theoretical VCD spectrum of (*S*)-α-d-ethanol with the experimental VCD spectrum (0.13 molar in CS_2.).

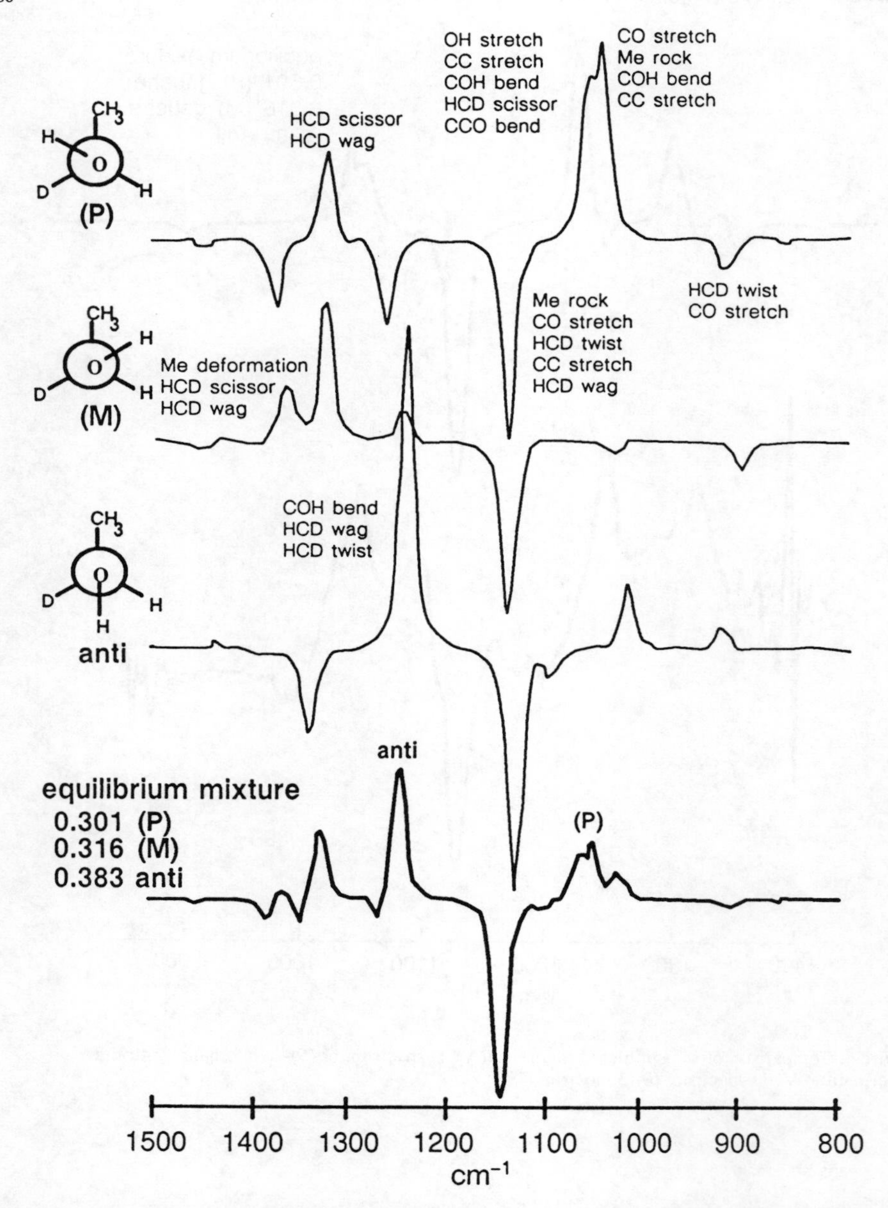

Figure 6. Simulated VCD spectra of the individual conformers of (*S*)–α–d–ethanol, and of the equilibrium mixture

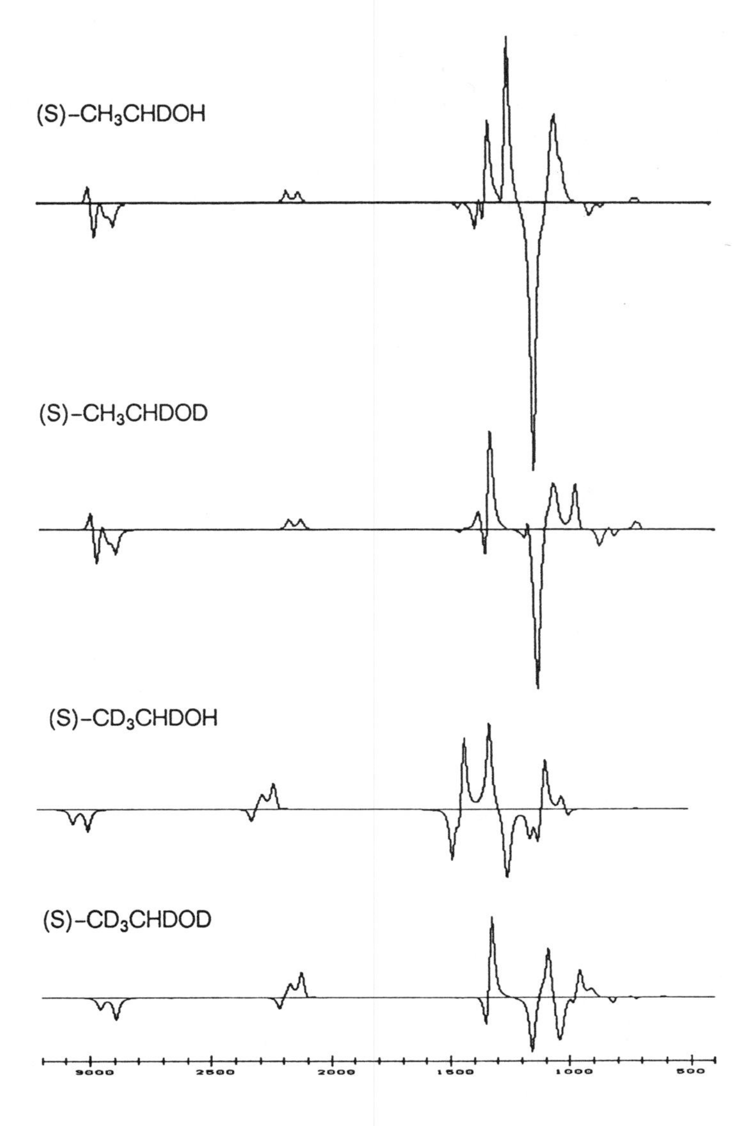

Figure 7. Simulated VCD spectra by the vibronic coupling method (6–31G$^{\sim}$ basis set) of the equilibrium mixtures of deuterated ethanols. All spectra are on the same scale.

1,3-Dideuterioallene *11*. Stephens and coworkers [74] have reported a rigorous study of (*S*)-*11* using the Stephens formalism [14] and calculation of the required atomic polar and axial tensors over two large basis sets, TZ/2P and 6–31G(ext) [74]. The results were compared to various empirical and approximate models. The *ab initio* results are reproduced in Table 8. The results for the rotatory strengths obtained with both basis sets agree in sign and are quite close in absolute magnitude, the greatest deviation being a factor of 2 for the CH stretching vibrations. Experimental optical rotatory strengths were not available but overall agreement with experimentally determined dipole strengths was also satisfactory.

References

1. (a) Deutsche, C. W.; Moscowitz, A. *J. Chem. Phys.* **1968,** *49,* 3257. (b) Deutsche, C. W.; Moscowitz, A. *J. Chem. Phys.* **1970**, *53*, 2630.
2. Hsu, E.C.; Holzwarth, *J. Chem. Phys.*, **1973**, *59*, 4678–4685.
3. Holzwarth, G.; Hsu, E. C.; Mosher, H. S.; Faulkner, T. R.; Moscowitz, A. *J. Am. Chem. Soc.*, **1974**, *96*, 251-253.
4. Stephens, P. J., *Proc. Soc. Photo. Instr., Eng.*, **1976**, *88*, 75–77.
5. Nafie, L. A.; Diem, M. *Acc. Chem. Res.*, **1979**, *12*, 296–302.
6. Stephens, P. J.; Clark, R., *NATO Adv. Study Inst. Ser. C,* **1978**, *48*, 263–287.
7. Keiderling, T. A. *Appl. Spectrosc. Rev.*, **1981**, *17*, 189–226.
8. Nafie, L. A. *Appl. Spectrosc.* **1982,** *36*, 489–495.
9. Stephens, P. J.; Lowe, M. A. *Ann. Rev. Phys. Chem.*, **1985**, *36*, 213–241.
10. Freedman, T. B.; Nafie, L. A. *Topics in Stereochemistry*, **1987**, *17,* 113–206.
11. Polavarapu, P. L. in *Vibrational Spectra and Structure*; H. D. Bist, J. R. Durig, and J. S. Sullivan, Eds.; Elsevier, New York (1989), vol 17B, p319.
12. .(a) Craig, D. P.; Thirunamachandran, T. *Mol. Phys.* **1978,** *35*, 825. (b) Craig, D. P.; Thirunamachandran, T. *Can. J. Chem.* **1985**, *63*, 1773.
13. Nafie, L. A.; Freedman, T. B. *J. Chem. Phys.*, **1983**, *78*, 7108.
14. Stephens, P. J. *J. Phys. Chem.* **1985**, *89*, 748–752.
15. Buckingham, A. D.; Fowler, P. W.; Galwas, P. A. *Chem. Phys.*, **1987**, *112*, 1.
16. Dutler, R.; Rauk, A. *J. Am. Chem. Soc.*, **1989**, *111*, 6957–6966.
17. R. Dutler, Ph. D. Dissertation, The University of Calgary, (1988).
18. Amos, R. D.; Handy, N. C.; Drake, A. F.; Palmieri, P. *J. Chem. Phys.* **1988**, *89*, 7287–7297.
19. Binkley, J. S.; Frisch, M. J.; Defrees, D. J.; Raghavachari, K.;. Whitesides, R. A; Schlegel, H. B.; Fluder, E. M.; Pople, J. A., Department Of Chemistry, Carnegie–Mellon University, Pittsburgh, PA.
20. Amos, R. D., The Cambridge Analytical Derivatives Package CCP 1/84/4, SERC, Daresbury Laboratory, Daresbury, Warrington (1984).
21. (a) Lazzeretti, P.; Zanasi, R. *Phys. Rev. A* **1986,** *33*, 3727. (b) Lazzeretti, P.; Rossi, E.; Zanasi, R. *Phys. Rev. A* **1983**, *27*, 1301. (c) Lazzeretti, P.; Rossi, E.; Zanasi, R. *J. Chem. Phys.* **1983**, *79*, 889. (d) Lazzeretti, P.; Zanasi, R. *J. Chem. Phys.* **1985**, *83*, 1218. (e) Lazzeretti, P.; Zanasi, R. *J. Chem. Phys.* **1986**, *84*, 3916. (f) Lazzeretti, P.; Zanasi, R. *J. Chem. Phys.* **1987**, *87*, 472. (g) Zanasi, R.; Lazzeretti, P. *J. Chem. Phys.* **1986**, *85*, 5932. (h) Lazzeretti, P.; Zanasi, R.; Bursi, R. *J. Chem. Phys.* **1988**, *89*, 987.
22. Lazzeretti, P.; Zanasi, R.; Stephens, P. J. *J. Phys. Chem.*, **1986**, *90*, 6761.
23. Lazzeretti, P.; Zanasi, R. *Chem. Phys. Lett*, **1984**, *112*, 103.
24. Jalkanen, K. J.; Stephens, P. J.; Lazzeretti, P.; Zanasi, R. *J. Chem. Phys.*, **1989**, *90*, 3204–3213.

25. Amos, R. D.; Jalkanen, K. J.; Stephens, P. J. *J. Phys. Chem.*, **1988**, *92*, 5571–5575.

26. Freedman, T. B.; Nafie, L. A. *J. Chem. Phys.*, **1988**, *89*, 374–384.

27. Blom, C. E.; Slingerland, P. J.; Altona, C. *Mol. Phys.* **1976**, *31*, 1359.

28. Pople, J. A.; Schlegel, H. B.; Krishnan, R.; De Frees, D. J.; Binkley, J. S.; Frisch, M. J.; Whiteside, R. A.; Hout, R. F.; Hehre, W. J. *Int. J. Quantum Chem. Symp.*,**1981**, *15,* 269.bi

29. (a) Pulay, P.; Lee, J. G.; Boggs, J. E. *J. Chem. Phys.*, **1983**, *79*, 3382–3391. (b) Pulay, P. In *Modern Theoretical Chemistry*; Schaefer III, H. F., Ed.; Plenum Press: New York, 1977; Vol. 4.

30. (a) Sadlej, A. J. *Chem. Phys. Lett.* **1977**, *47*, 50. (b) Nakatsuji, H.; Kanda, K.; Yonezawa, T. *Chem. Phys. Lett.* **1980**, *75*, 340.

31. Lowe, M. A.; Segal, G. A.; Stephens, P. J. *J. Am. Chem. Soc.* **1986**, *108*, 248–256.

32. Duncan, J. L.; Burns, G. R. *J. Mol. Spectrosc.* **1976**, *30*, 253.

33. Blom, C. E.; Altona, C. *Mol. Phys.* **1976**, *31*, 1377.

34. Komornicki, A.; Pauzat, F.; Ellinger, Y. *J. Phys. Chem.* **1983**, *87*, 3847.

35. Binkley, J. S.; Whiteside, R. A.; Krishnan, R.; Seeger, R.; Defrees, D. J.; Schlegel, J. B.; Topiol, S.; Kahn, L. R.; Pople, J. A. *GAUSSIAN–80, Quantum Chemistry Program Exchange*, Indiana University, Bloomington, IN.

36. Lowe, M. A.; Stephens, P. J.; Segal, G. A. *Chem. Phys. Lett.* **1986**, *123*, 108–116.

37. (a) Annamalai, A.; Keiderling, T. A.; Chickos, J. S. *J. Am. Chem. Soc.* **1984**, *106*, 6254. (b) Annamalai, A.; Keiderling, T. A.; Chickos, J. S. *J. Am. Chem. Soc.* **1985**, *107*, 2285.

38. Annamalai, A.; Keiderling, T. A. *J. Mol. Spectry.* **1985**, *109*, 46.

39. (a) Polavarapu, P. L.; Michalska, D. F. *J. Am. Chem. Soc.* **1983**, *105*, 6190. (b) Polavarapu, P. L.; Michalska, D. F. *Mol. Phys.* **1984**, *52*, 1225. (c) Polavarapu, P. L.; Michalska, D. F. *Mol. Phys.* **1985**, *55*, 723. (d) Polavarapu, P. L.; Hess, B. A.; Schaad, L. J. *J. Chem. Phys.* **1985**, *82*, 1705.

40. Lowe, M. A.; Alper, J. S; Kawiecki, R. A.; Stephens, P. J. *J. Phys. Chem.*

41. Amos, R. D.; Handy, N. C.; Jalkanen, K. J.; Stephens, P. J. *Chem. Phys. Lett.*, **1986**, *133*, 21.

42. Amos, R. D. *Advan. Chem. Phys.* **1987**, *67*, 99.

43. Kawiecki, R. W.; Devlin, F.; Stephens, P. J.; Amos, R. D.; Handy, N. C. *Chem. Phys. Lett.* **1988**, *145*, 411–417.

44. A. Rauk and D. Yang, *to be published.*

45. Jalkanen, K. J.; Stephens, P. J.; Amos, R. D.; Handy, N. C. *Chem. Phys. Lett.*, **1987**, *142*, 153–158.

46. Jalkanen, K. J.; Stephens, P. J.; Amos, R. D.; Handy, N. C. *J. Phys. Chem.*, **1988**, *92*, 1781.

47. Jalkanen, K. J.; Stephens, P. J.; Amos, R. D.; Handy, N. C. *J. Am. Chem. Soc.*, **1987**, *109*, 7193–7194.

48. Heintz, V. J.; Keiderling, T. A. *J. Am. Chem. Soc.* **1981**, *103*, 2395.

49. Dothe H.; Lowe, M. A.; Alper, J. S. *J. Phys. Chem.* **1988**, *92*, 6246–6249.

50. Polavarapu, P. L.; Hess, Jr., B. A.; Schaad, L. J.; Henderson, D. O.; Fontana, L. P.; Smith, H. E.; Nafie, L. A.; Freedman, T. B.; Zuk, W. M. *J. Chem. Phys.* **1987**, *86*, 1140.

51. (a) Jalkanen, K. J.; Stephens, P. J.; Amos, R. D.; Handy, N. C. *J. Am. Chem. Soc.*, **1988**, *110*, 2012–2013. (b) Jalkanen, K. J.; Stephens, P. J.; Amos, R. D.; Handy, N. C. *J. Am. Chem. Soc.*, **1988**, *110*, 5598.

52. Freedman, T. B.; Palerlini, M. G.; Lee, N.; Nafie, L. A.; Schwab, J. M.; Ray, T. *J. Am. Chem. Soc.*, **1987**, *109*, 4727..

53. Hirose, C. *Bull. Chem. Soc. Jpn.* **1974**, *47*, 1311.

54. Stephens, P. J. *J. Phys. Chem.* **1987**, *91*, 1712–1715.

55. (a) Nakanaga, T.; *J. Chem. Phys.*, **1980**, *73*, 5451–5458. (b) Nakanaga, T.; *J. Chem. Phys.*, **1981**, *74*, 5384–5392.

56. Spencer, K. M.; Freedman, T. B.; Nafie, L. A. , personal communication.

57. Rauk, A.; Dutler, R.; Yang, D. *Can. J. Chem.* **1990**, *68*, 258.

58. Rauk, A.; Dutler, R. *J. Comput. Chem.*, **1987**, *8*, 324.

59. (a) Giguere, P. A.; Liu, I. D. *J. Chem. Phys.* **1952**, *20,* 136. (b) Catalano, E.; Sanborn, R. H.; Frazer, J. W. *J. Chem. Phys.* **1963**, *38*, 2265. (c) Ziomek, J. S.; Zeidler, M. D. *J. Mol. Spectrosc.* **1963**, *11*, 163. (d) Durig, J.; Bash, S. F.; Mercer, E. E. *J. Chem. Phys.* **1966**, *44,* 4328. (e). Yamaguchi, A; Ichishima, I.; Shimanouchi, T.; Mizushima, S. *Spectrochim. Acta*, **1960**, *16*, 1471. (f) Hamada, Y.; Hirakawa, A. Y.; Tamagake, K.; Tsuboi, M. *J. Mol. Spectrosc.* **1970**, *35*, 420.

60. Tanaka, N.; Hamada, Y.;. Sugawara, Y; Tsuboi, M.; Kato, S.; Morokuma, K. *J. Mol. Spectrosc.* **1983**, *99,* 245.

61. Defrees, D. J.; Raghavachari, K.; Schlegel, H. B.; Pople, J. A. *J. Am. Chem. Soc.* **1982**, *104,* 5576. (b) Blair R. A.; Goddard III, W. A. *J. Am. Chem. Soc.* **1982**, *104,* 2719.

62. (a) Rogers, J. D.; Hillman, J. J. *J. Chem. Phys.* **1981**, *75,* 1085. (a) Rogers, J. D.; Hillman, J. J. *J. Chem. Phys.* **1982**, *76,* 4046.

63. Polavarapu, P. L. *J. Chem. Phys.* **1987**, 87*,* 4419.

64. Giguere, P. A.; Srinivasan, T. K. K. *J. Raman Spectrosc.* **1974,** *2*, 125.

65. Nafie, L. A.; Freedman, T. B. *J. Mol. Struct.*, in press.

66. Dothe H.; Lowe, M. A.; Alper, J. S. *J. Phys. Chem.* **1989**, *93*, 6632–6637.

67. Pultz, V. M. Vibrational Circular Dichroism Studies of Some Small Chiral Molecules. Ph.D. Dissertation, University of Minnesota, 1983.

68. (a) Dutler, R.; Rauk, A.; Shaw, R. A.; Wieser, H. *J. Am. Chem. Soc.* (submitted for publication). (b) Dutler, R.; Rauk, A.; Shaw, R. A.; Wieser, H. presented at 10th Canadian Symposium on Theoretical Chemistry, Banff, Canada; August 24–30, 1989.

69. Perchard, J.–P.; Josien, M.–L. *J. Chim. Phys. Phys.–Chim. Biol.* **1968**, *65*, 1834–1875.

70. Duncan, J. L.; Mills, I. M. *Spectrochim. Acta* **1964**, *20*, 523.

71. F. P. J. Valero, D. Goorvitch, F. S. Bonomo, R. W. Boese, *Applied Optics* **1981**, *20*, 4097.

72. Giguere, P. A.; Liu, D. I. *Can. J. Chem.* **1952**, *30*, 948

73. Niki, H.; Maker, P. D.; Savage, C. M.; Breitenbach, L. P. *Chem. Phys. Lett.* **1980**, *73,* 43.

74. Annamalai, A.; Jalkanen, K. J.; Narayanan, U.; Tissot, M. –C.; Keiderling, T. A.; Stephens, P. J. *J. Phys. Chem.* **1990**, *94,* 194–199.

75. Narayanan, U.; Annamalai, A.; Keiderling, T. A. *Spectrochim. Acta* **1988**, *44A*, 785.

Table 1. Rotational and Dipole Strengths of (*S,S*)-1,2-dideuteriocyclopropane *1*[a]

Symmetry	Frequency[b]	R[c]	D[d]	Symmetry	Frequency[b]	R[c]	D[d]
A	3159	344	4	B	3202	159	29
	3118	90	27		3163	-588	28
	2328	-258	4		2318	245	24
	1536	33	2		1361	214	6
	1417	-184	3		1207	-372	6
	1240	68	0		1130	-67	26
	1170	0	15		1023	20	1
	1111	215	10		898	75	123
	970	20	18		756	-64	8
	823	85	109		621	-16	1
	634	-19	4				

[a] Ref 31.

[b] In cm^{-1}, force field from ref 34.

[c] In 10^{-45} $esu^2\,cm^2$

[d] In 10^{-40} $esu^2\,cm^2$

Table 2. Rotational and Dipole Strengths of (-)-(*S*)-propylene oxide *3*

Frequency	Frequency[a]	D[c]				R[d]			
6-31G*(0.3)b	obs	4-31G[a,e]	6-31G**[a,e]	6-31G*(0.3)b	obs[a]	4-31G[a,e]	6-31G**[a,e]	6-31G*(0.3)b	obs[a]
1490	1499	10.6	20.4	17.2	20.9	-2.7	-7.1	-3.5	-5.2
1461	1458	20.8	11.9	14.6	20.5	3.5	0.4	4.6	
1449	1446	14.5	7.2	10.3	18.3	-1.0	-1.1	-2.0	
1407	1407	38.8	40.8	34.2	63.7	-10.3	-11.5	-13.0	-20.8
1361	1369	12.0	16.5	8.7	18.3	-2.0	-3.0	-5.7	-8.9
1243	1265	16.3	18.3	15.8	32.6	10.7	10.5	7.9	17.5
1159	1168	7.4	16.3	4.6	4.9	5.8	6.6	4.5	-5.3
1132	1146	3.8	2.0	16.0	18.2	2.9	0.8	19.0	47.8
1130	1132	22.3	26.6	2.8	20.5	4.1	11.7	-7.3	-40.6
1089	1104	40.9	34.9	28.1	32.8	-8.6	-0.4	5.7	15.2
1013	1023	37.9	26.1	24.6	67.4	7.9	-0.6	-4.7	-11.2
938	950	73.0	105.2	64.8	76.8	5.4	10.9	48.0	90.8
878	895	20.5	24.7	9.0	19.2	-20.3	-26.1	-26.7	-50.3
812	829	180.2	216.4	226.8	303.0	7.3	4.5	-17.3	
739	747	19.5	28.2	37.6	65.4	-4.3	-4.4	-13.9	-28.2

[a] Data extracted from Table 1 of ref 43.

[b] Vibronic Coupling Theory, ref 44: frequencies in cm^{-1}, uniformly scaled by 0.9.

[c] In 10^{-40} $esu^2\,cm^2$

[d] In 10^{-44} $esu^2\,cm^2$

[e] Basis set used for atomic polar and axial tensors in Stephens' formalism.

Table 3. Rotational and Dipole Strengths of (*S*)-NHDT[a]

Frequency[b]	D[c]	R[d]
3562	4.1	0.2
2601	4.6	-1.2
2176	5.0	1.2
1484	64.7	-5.1
1195	35.1	9.9
827	518.2	-4.9

[a] From Table 3 of ref 24; VD/3P basis set, SCF-RPA wavefunction.
[b] In cm^{-1}, force field from ref 70.
[c] In 10^{-40} esu^2 cm^2
[d] In 10^{-44} esu^2 cm^2

Table 4. Vibronic Coupling Theory Rotational Strengths of Oxiranes ***6***: (*S*)-1-d-***6***, (*S,S*)-2,3-d_2-***6***, and (*S*)-2,2,3-d_3-***6***.[a]

(*S*)-1-d-***5***		(*S,S*)-2,3-d_2-***5***					(*S*)-2,2,3-d_3-***5***	
Frequency[b]	R[c]	Symm.	Freq[b]	R[c]	R[c]	R[c]	Frequency[b]	R[c]
699	-13.4	B	653	22.2			605	16.8
820	9.7	A	750	15.7			744	36.4
838	70.3	B	828	-17.5			796	-93.1
890	-17.5	A	872	-48.1			823	-17.3
1050	-145.5	B	917	52.3			902	54.5
1086	-55.0	A	966	-181.8 (-s)			949	-19.4
1154	67.5	B	1120	157.9 (+m)			971	-103.8
1156	127.0	A	1137	37.2			1039	71.9
1243	135.2	A	1221	234.9 (+s)			1128	82.1
1374	-162.0	B	1339	-11.8 (-w)			1258	141.8
1503	-17.2	A	1407	-261.7 (-s)			1384	168.9
2214	1.7	B	2206	-265.7 (-s)	-302[d]		2154	-74.5
2968	51.0	A	2221	295.4 (+s)	296[d]		2216	100.0
3011	-84.5	A	3010	-355.2 (-m)	-743[d]		2277	1.9
3050	78.5	B	3013	434.1 (+s)	855[d]		3012	34.0

[a] Ref 16, 6-31G~ basis set, common origin at centre of mass.
[b] Uniformly scaled *ab initio* force field; scale factor = 0.814; units cm^{-1}.
[c] Units cm/mol
[d] Ref 51; converted to units of cm/mol using frequencies in this table.

Table 5. Frequencies, dipole strengths, and rotational strengths for hydrazine NH_2NH_2.

Normal Mode			Frequency cm^{-1} calc[b]	scal[c]	exp[d]	calc[d]	Dipole Strength[a] calc[b] esu^2cm^2	calc[b] km/mol	Rotational Strength esu^2cm^2[e]
1	a	$\nu7$	498	448	377[f]	401	319.1	39.8	459.5
2	a	$\nu6$	970	873	780[g]	898	301.9	73.4	-1495.8
3	b	$\nu12$	1112	1001	937[h]	1045	562.1	156.7	1559.9
4	a	$\nu5$	1226	1103	1098[g]	1167	30.4	9.3	-165.9
5	b	$\nu11$	1437	1293	1283[i]	1320	22.1	8.0	685.8
6	a	$\nu4$	1469	1322	1324[i]	1376	18.3	6.7	-627.0
7	b	$\nu10$	1843	1659	1587[g]	1645	32.3	14.9	-68.4
8	a	$\nu3$	1856	1670	1628[g]	1705	26.9	12.5	71.4
9	b	$\nu9$	3682	3314	3297[j]	3287	4.9	4.5	-18.8
10	a	$\nu2$	3691	3322		3300	0.4	0.4	-5.6
11	a	$\nu1$	3781	3403		3413	1.7	1.6	-80.0
12	b	$\nu8$	3788	3409	3390[j]	3402	4.5	4.3	111.2

[a] The dipole strength, D(km/mol), is derived from the calculated quantity, D(esu^2cm^2 x 10^{40}) by D(km/mol) = 0.0002507xD(esu^2cm^2 x 10^{40}) x $\nu(cm^{-1})$ where ν is the calculated (unscaled) frequency

[b] Ref 56, frequencies and intensities using 6–31G~ basis set at 6–31G** geometry

[c] Calculated frequencies multiplied by 0.9 (ref 28)

[d] Ref 60; calculated frequencies by CI - 6–31G* force field

[e] x 10^{45}

[f] Ref 59(e)

[g] Ref 59(a) and ref 59(c)

[h] Ref 59(f)

[i] Ref 59(d)

[j] Ref 59(b)

Table 6. Frequencies, dipole strengths, and rotational strengths for H_2O_2

Normal Mode			Frequency cm^{-1}			Dipole Strength[a]				Rotational Strength
			calc	scal[d]	exp[e]	calc[b] esu^2cm^2	calc[b] km/mol	calc[c] km/mol	(exp) km/mol	esu^2cm^{2}[f]
1	$\nu 4$	a	426	383	317	1870.8	199.8	123.		1904.6
2	$\nu 3$	a	1146	1031	864	2.8	0.8	0.1		-27.1
3	$\nu 6$	b	1504	1354	1266	301.3	113.6 .	101.	84.[g] 113.[h]	202.6
4	$\nu 2$	a	1636	1472	1394	0.7	0.3	0.9		-127.7
5	$\nu 1$	a	4063	3657	3607	21.5	21.9	14.5		14.3
6	$\nu 5$	b	4064	3658	3608	93.2	94.9	46.2		-77.5

[a] The dipole strength, D(km/mol), is derived from the calculated quantity, $D(esu^2cm^2 \times 10^{40})$ by $D(km/mol) = 0.0002507 \times D(esu^2cm^2 \times 10^{40}) \times \nu(cm^{-1})$ where ν is the calculated (unscaled) frequency

[b] Ref 57 frequencies and intensities using 6–31G~ basis set at 6–31G** geometry

[c] Ref 62.

[d] Calculated frequencies multiplied by 0.9, ref 28.

[e] Ref 64

[f] $\times 10^{45}$

[g] Ref 71

[h] Value quoted by ref 72 as modified from ref 73 by P. D. Maker

Table 7. Comparison of VCD Intensities of (*S*)-1-d-Ethanol by the Vibronic Coupling Theory and Stephens' Formalism.

anti				*P-gauche*				*M-gauche*			
Frequency[a]	D[b]	R(VC)[b]	R(PJS)[c]	Frequency[a]	D[b]	R(VC)[b]	R(PJS)[c]	Frequency[a]	D[b]	R(VC)[b]	R(PJS)[c]
225.7	603.2	-14.5		242.3	312.9	-649.2		242.3	314.7	653.9	
264.0	1402.6	-8.4		265.9	1607.7	581.9		266.4	1616.1	-586.5	
415.4	106.5	-27.9		418.8	91.9	45.4		418.3	89.3	-39.0	
727.1	2.36	34.4		716.7	9.4	-52.6		715.1	13.0	87.1	
871.8	48.9	-5.2		858.5	70.3	-14.7		859.3	4.7	-11.2	
926.1	20.0	75.4		921.7	11.8	-107.0		905.6	229.9	-99.3	
1024.0	209.0	193.2		1048.9	324.4	503.2		1036.6	21.1	-34.7	
1100.6	26.6	-75.2		1064.4	95.6	409.2		1116.5	94.8	26.2	
1141.1	162.1	-741.0		1144.7	97.4	-520.0		1144.5	76.2	-460.9	
1248.7	187.3	814.4		1270.4	24.1	-227.1		1251.1	268.6	92.2	
1350.1	15.6	-100.2	-112	1331.2	9.9	213.4	13	1329.8	33.1	325.7	110
1353.2	10.2	-114.6	16	1367.2	24.5	3.5	-82	1368.1	19.1	103.3	27
1396.2	18.3	-20.1	9.2	1382.6	123.3	-196.4	-62	1375.3	29.3	35.8	11
1443.6	14.1	15.6		1447.9	15.1	-6.8		1447.8	14.7	-8.1	
1459.1	4.8	-6.0		1455.7	6.2	-7.8		1456.4	7.7	-12.3	
2125.6	76.4	-0.1	-5.8	2126.8	76.6	39.5	36	2176.2	39.9	39.9	-13
2889.2	78.6	-4.9	26	2910.2	26.8	1.4	18	2890.7	77.4	-51.2	-139
2923.7	24.8	-21.0	-28	2948.3	8.9	40.7	72	2911.1	27.0	-11.7	90
2983.5	47.8	-27.8	-45	2970.4	91.1	-70.4	24	2963.8	54.1	-48.9	-15
2992.1	47.5	30.8	53	2986.4	42.9	5.0	43	2987.0	52.8	53.7	65
3675.8	52.5	-0.7		3668.4	43.7	-17.2		3668.3	44.5	19.4	

[a] Ref 68; units cm^{-1}.
[b] Ref 68; units D $esu^2cm^2 \times 10^{-40}$, R $esu^2cm^2 \times 10^{-45}$; Vibrinic coupling theory, 6-31G~ basis set.
[c] Ref 66 The signs are reversed to coincide with the (*S*) enantiomer, *P* corresponds to "gauche-B", *M* corresponds to "gauche-A"; Stephens' theory, 6-31G* basis set.

Table 8. Frequencies, Rotational and Dipole Strengths of (*S*)–1,3 dideuterioallene *11*[a]

Frequency[b] cm^{-1}	D[c] TZ/2P[e]	D[c] 6–31G(ext)[e]	R[d] TZ/2P[e]	R[d] 6–31G(ext)[e]
3061	0.2	0.1	–3.3	–6.2
3053	0.2	0.3	3.3	6.3
2261	1.3	1.1	0.4	2.2
2252	0.2	0.2	–0.5	–2.4
1944	21.5	21.9	0.1	0.2
1335	0.0	0.0	1.1	1.7
1262	0.4	0.4	–4.1	–5.3
1052	0.3	0.4	9.7	11.6
867	1.7	2.3	–27.6	–33.1
847	16.3	17.1	91.0	93.6
838	1.1	0.8	–7.4	–6.7
758	24.6	24.5	–98.3	–96.7
636	7.9	7.9	35.1	34.3
328	9.3	9.7	30.0	27.1
324	9.6	10.0	–29.6	–26.6

[a] Data extracted from Table 1 of ref 74.

[b] Scaled force field of ref 75, at the experimental geometry.

[c] In 10^{-39} esu^2 cm^2

[d] In 10^{-44} esu^2 cm^2

[e] See reference 74 for description of the basis sets.

CHIRAL FEATURES OF PROTEINS

G.M. Maggiora, B. Mao, and K-C. Chou
Computational Chemistry
Upjohn Research Laboratories
Kalamazoo, MI 49001

1. Introduction

Scientists have been fascinated by the complexity and subtlety of the structure of proteins for more than half a century. And the relatively recent capability to determine their three-dimensional structures by x-ray crystallographic, and lately by 2D-NMR methods, has led to a flurry of activity in the development and application of new methods to analyze their structural features [1, 2]. A striking feature of proteins, and in fact of all biomacromolecules, is the many ways in which chirality (i.e. *handedness*) arises at all levels of the protein structure hierarchy (*vide infra*).

As will emerge from the discussion in this chapter, proteins possess chiral features due both to their geometric and topological characteristics. Examples of *geometrically* and *topologically chiral objects*, which correspond to many of the chiral features found in proteins, are given in Figure 1 and 2, respectively. Although many of these features are well known, there exist a number of other less well known ones, as well as those yet to be discovered.

The present chapter is meant to be an exploration rather than a definitive treatise. Thus, no claim is made that all chiral features will be treated -- some undoubtedly will be missed. Rather, the purpose of this chapter is to stimulate thinking about chirality in proteins, and in this way, perhaps, to aid in the discovery of new relationships that may help describe the structure, folding, and functional properties of proteins in useful new ways.

P. G. Mezey (ed.), New Developments in Molecular Chirality, 93–118.

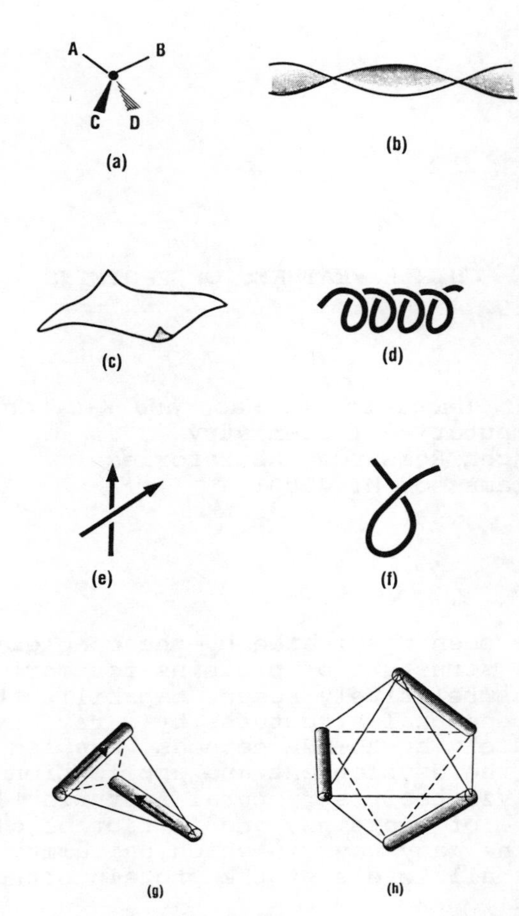

Figure 1:

Examples of geometrically chiral "objects" found in typical proteins: (a) asymmetric tetrahedral α-carbon atom of the L-amino acids found in proteins, (b) "twisted ribbon" representation of the right-handed twist observed in extended peptide chains, (c) "twisted sheet" representation of the right-handed twist observed in both parallel and anti-parallel ß-sheets, (d) right-handed helix, (e) right-handed crossover observed when directed polypeptide chains in protein cross, (f) right-handed, non-oriented loop, (g) "oriented cylinders" representation of the packing of two directed helices, (h) "non-oriented cylinders" representation of the packing of three undirected helices.

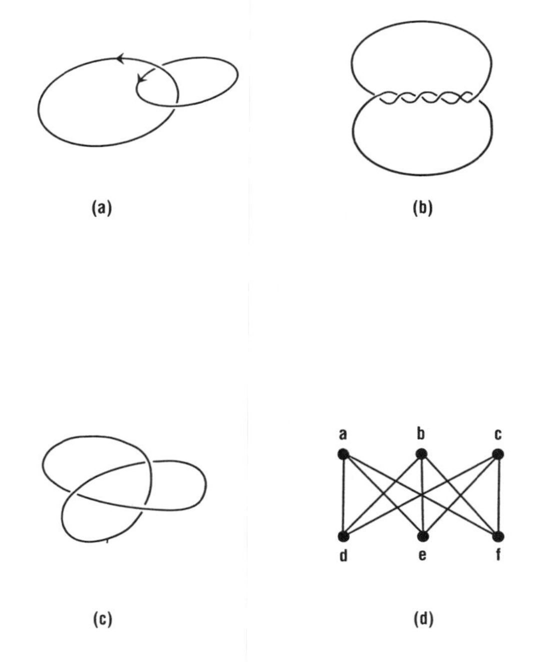

Figure 2:

Examples of topologically chiral "objects" that may be found in proteins: (a) "oriented link" representation of two directed, closed loops connected by a single link with right-handed crossings, (b) two multiply-linked undirected closed loops with left-handed crossings, (c) trefoil knot with right-handed crossings, and (d) non-planar Kuratowski K(3,3) graph. None of the first three objects have been observed in proteins (*Cf.*, however, the discussion presented in [35-37, 41]. Although quite rare, molecular graph representations of proteins with disulfide bonds exist that are non-planar as the discussion in Section 6 will show.

2. Geometric vs. Topological Chirality

Chirality or handedness as it is usually treated in chemical systems is related to the non-superimposeability of a given molecule with its mirror image. And while there are subtleties associated with non-rigid molecular systems, the concept is essentially geometric (Euclidean) in nature (*Cf.* [3]). Not unexpectedly, topological chirality differs markedly from that derived from the geometric characteristics of molecules [4].

The topological properties of an object relate to its invariant features with respect to particular types of very general transformations. Consider first a *continuous* deformation of an object. Under such transformations, an object cannot be torn and rejoined in any way. Objects related in this manner are said to be *homeotopic*.[1] A classical example is found in the topological equivalence between a teacup and a donut, which results from the fact that either object can be deformed into the other without tearing and rejoining.

When considering objects that are not homeotopically related, the dimension of the space into which they are embedded also becomes important. For example, it is impossible by any homeotopically allowed transformation to change the orientations of the two *oriented* closed curves depicted in Figure 3, which are embedded in the two-dimensional space (*i.e. "2-space"*) of the plane of the page. The two curves possess opposite orientations, as indicated by the arrows, and thus they exhibit a type of handedness which is independent of their individual shapes. Such non-homeotopic and chiral objects are said to be *topologically chiral*. In 3-space, however, the orientation of either curve can be changed quite simply by a rigid rotation. This argument can be extended to higher dimensions. For example, the two oriented, linked rings depicted in Figure 2a cannot be unlinked in 3-space without cutting and resealing at least one of the rings, although this can be accomplished in 4-space. *Thus, the topological chirality of an object depends critically on the dimension of the space into which it is embedded, a property that also applies in the case of geometric chirality* (*Cf. [5]*).

More general transformations, called *homeomorphisms*, which describe "point-to-point" mappings between objects but do not require continuity are also of great importance in topology. Under such transformations the oriented curves in Figure 3, and the objects depicted in Figures 2a-2c are all topologically equivalent to non-oriented, closed curves. In this sense, homeomorphisms are not very useful for characterizing the topological chirality of objects. The type of topological equivalence possessed by homeomorphic objects is usually termed *intrinsic* topological equivalence, while that which pertains to homeotopically related objects embedded in, for example, 3-space is usually termed *extrinsic* topological equivalence [4, 6]. *In this chapter, the focus will be on extrinsically topologically equivalent objects that are homeotopically related*; distinctions between homeotopic and homeomorphic as well as those between intrinsic and extrinsic will not be noted explicitly, except in cases where such distinctions become necessary.

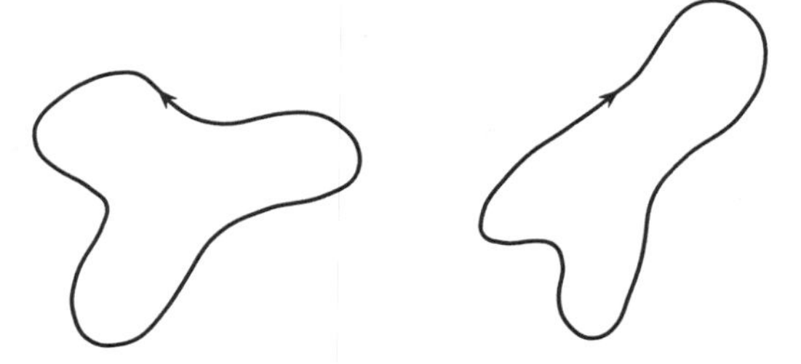

Figure 3:

An illustration of two oppositely oriented (indicated by the arrows), closed curves embedded in 2-space. As described in the text, neither curve can be continuously transformed into the other: thus, these objects are not homeotopically related and manifest topological chirality.

3. Elements of Protein Structure

An essential structural feature of all proteins is its hierarchical character. Figure 4, adapted from the work of Schulz and Schirmer [1], depicts the principal levels of protein structure schematically. Primary structure describes the "order" or sequence of the amino acids (see Figure 5a) in a polypeptide chain (see Figure 5b), beginning at the N-terminal amino acid, which contains the free amine, and proceeding towards the C-terminal amino acid, which contains the free carboxyl group. Secondary structure describes the quasi-regular features of a polypeptide chain. These features include α-helices (Figure 5c) and ß-sheets (Figure 5d). As indicated in the figures, hydrogen bonding (dashed lines) plays a significant role in stabilizing these structures. Super-secondary structure describes the packing of secondary structural elements with each other as illustrated in Figure 5e and 5f for helix-helix and helix-sheet packings, respectively. Domain structure represents specific regions of proteins which can be characterized as three-dimensional, geometrically separate entities.

Although, as seen in Figure 3, other, higher levels of protein structure are also important, only the first four will be considered in the present work. Robson and Garnier [2] present a very thorough description of the structural features associated with the different levels of protein structure (see in particular Tables 3.1-3.4), and their work and that of Schulz and Schirmer [1] should be consulted for additional details.

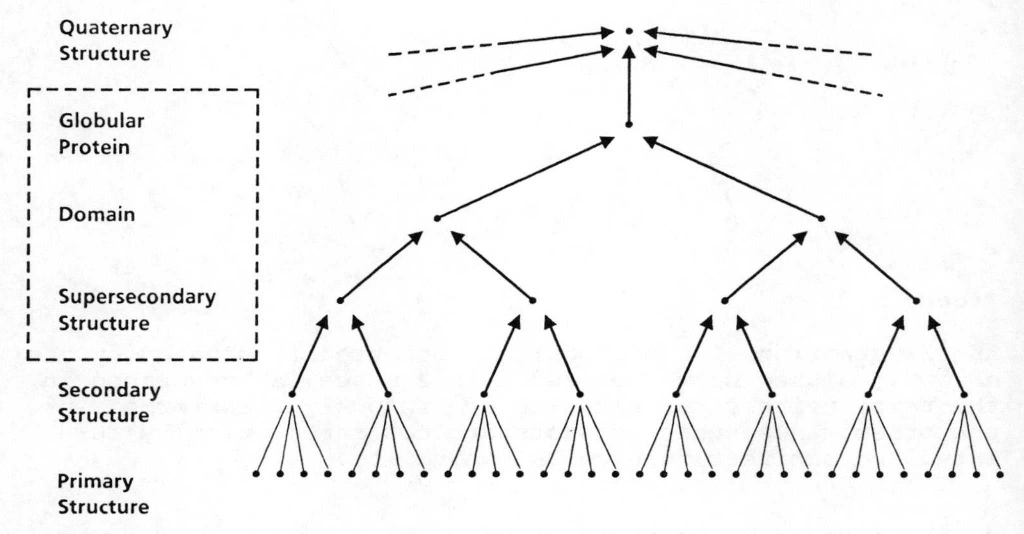

Figure 4:

A schematic depiction of the hierarchy of protein structure adapted from the work of Schulz and Schirmer [1]. The levels of structure designated in the dashed box on the left of the figure are generally subsumed under tertiary structure.

4. Geometrical Chirality in Proteins

As will become apparent in the forthcoming discussion, each level of protein structure possesses chiral features which are, at least to a limited extent, induced by chiral features of the next lower level of structure. For example, the L-stereochemistry of the α-carbon atom of each amino acid (see Figure 5a), except glycine, found in proteins tends to induce a right-handed twist in a fully-extended polypeptide chain, as illustrated in Figure 6 for poly-alanine.[2]

Chothia [7] has argued that the preference of an extended chain for a right-handed twist is at least partly responsible for the right-hand twist of ß-sheets as depicted in Figure 7. Such a right-handed twist is observed in essentially all ß-sheets [8, 9]. Chains of D-amino acids, on the other hand, tend to reverse this trend and produce extended chains and ß-sheets with left-handed twists.

(a)

N-Terminal C-Terminal

(b)

Figure 5:

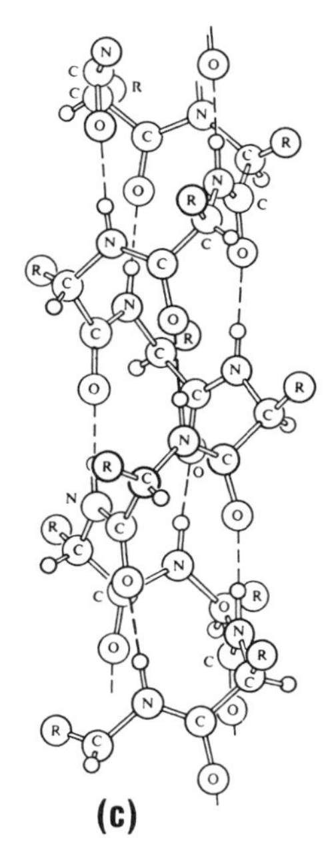

Examples of protein structural elements: (a) L-amino acid, (b) extended peptide chain with explicit designation of the N- (*i.e. amino*) and C- (*i.e. carboxy*) terminal regions -- by convention the chain is directed from the N-terminal to the C-terminal amino acid residue (*i.e. the amino acids are numbered from left to right*), (c) "ball-and-stick" representation of an idealized α-helix with hydrogen bonds indicated by dashed lines [*reprinted from* B.W. Low and J.T. Edsall, in D.E. Green (Ed.), *Currents in Biochemical Research*, Wiley (Interscience), New York (1956), p. 398. *with permission*], (d) "ball-and-stick" representation of an idealized anti-parallel ß-sheet with hydrogen bonds indicated by dashed lines [*reprinted from* L. Pauling and R.B. Corey (1951), *Proc. Natl. Acad. Sci.* (U.S.A.), **37**, 729 *with permission*], (e) relaxed-stereo view of the packing of two α-helices [*reprinted from* [17] *with permission*], and (f) relaxed-stereo view of the packing of an α-helix with a ß-sheet [*reprinted from* [19] *with permission*].

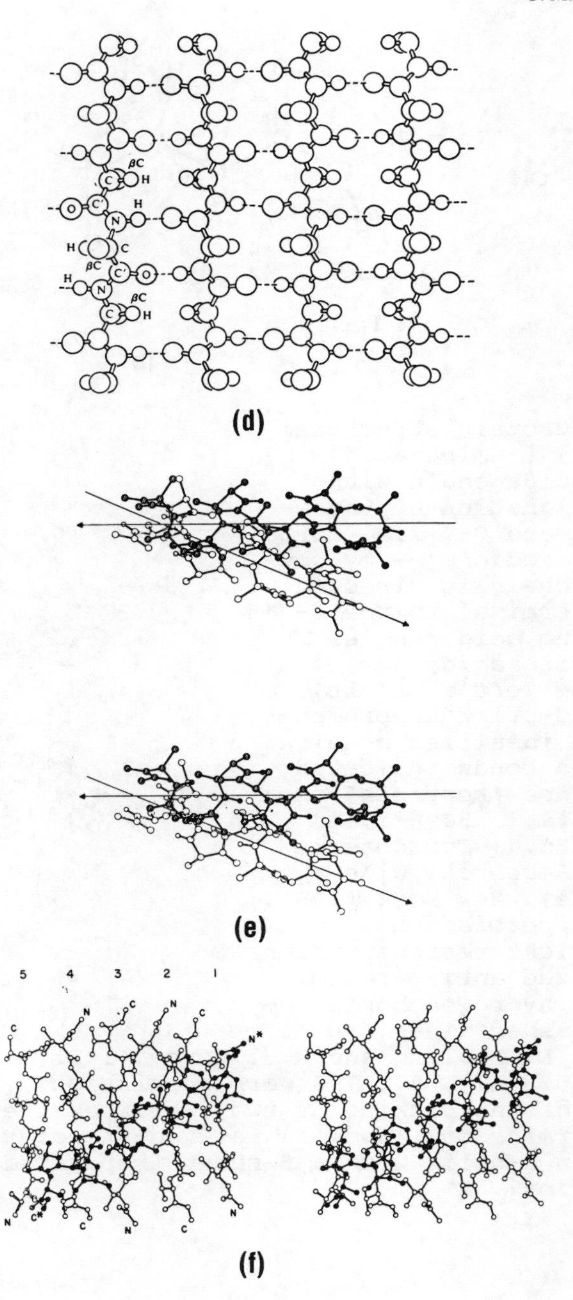

(d)

(e)

(f)

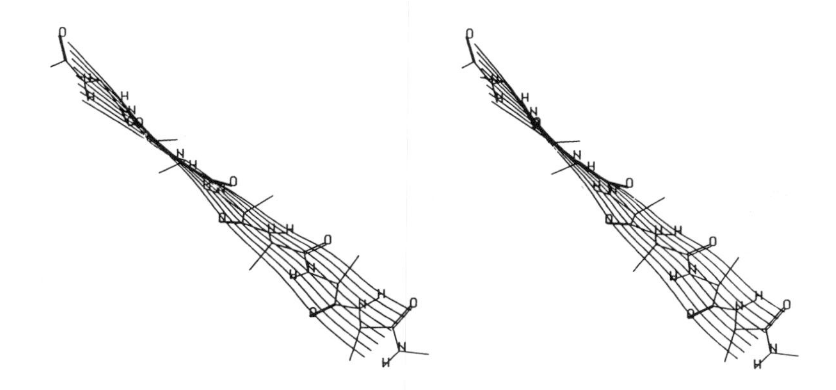

Figure 6:

Relaxed-stereo view of an extended poly-alanine chain illustrating the right-handed twist of the backbone induced by the L-stereochemistry of the alanine moieties.

L-amino acids also produce right-handed helices, while D-amino acids lead to the formation of left-handed helices.[3] Interestingly, a possible advantage of the right-handed twist of an extended polypeptide chain to the formation of a right-handed helix may be 'visualized' in the following manner. Twist a leather belt into a right-handed helix and then slowly pull the ends apart. If this is done carefully the belt will have a right-handed twist when it is fully extended: slowly reversing this process produces the desired result, as illustrated in Figure 8. Richardson [8, 10] has given a similar argument to explain the preference for the right-handed crossover of βXβ supersecondary structures (*Cf.* [11, 12].[4] Recent energy-based calculations on the handedness of the crossover in βαβ structures [13] also provide support for the observed preference.

The grooves of right-handed α-helices, which are made up of L-amino acids, have been implicated by numerous workers to explain the packing of α-helices [9, 14-18]. These unique chiral features in α-helices and β-sheets have also played an important role in the low-energy packing arrangements between α-helices [18], between α-helices and β-sheets [19], and between β-sheets [20].

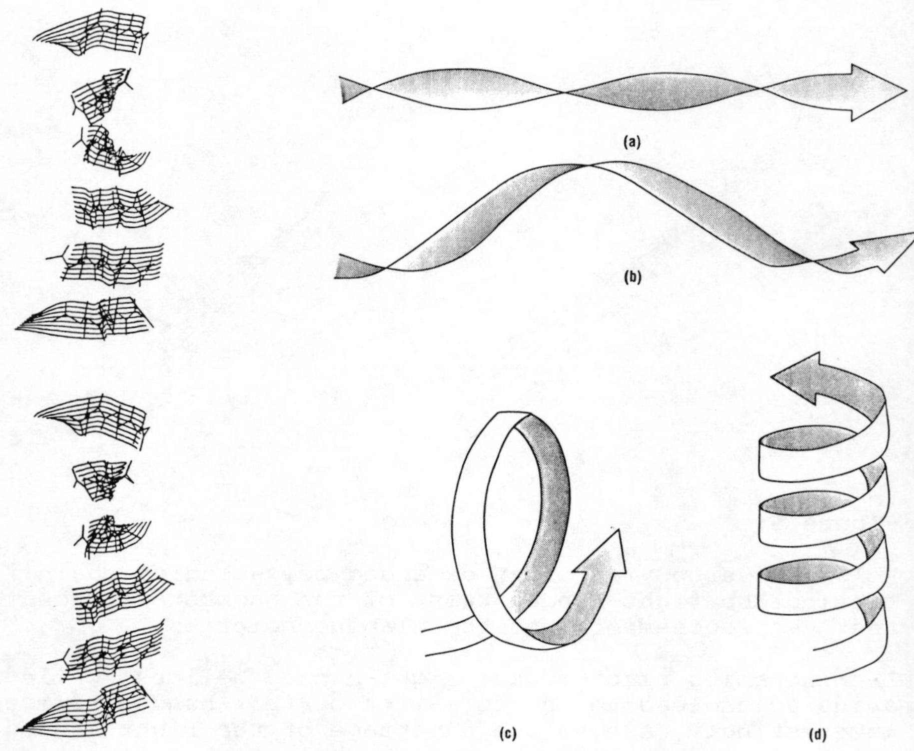

Figure 7:

Relaxed-stereo view of the right-handed twist of a ß-sheet in *Concanavalin* **A** [Hardman, K.D.; Ainsworth, C.F. (1972) *Biochemistry* **11**, 4910]. Note that the view is along the direction of the individual strands.

Figure 8:

"Twisted ribbon" representations of protein structural elements: (a) extended chain with right-handed twist, (b) "intermediate" twist structure which lies on the "folding pathway" from the extended chain depicted in (a) to either the right-handed ßXß-crossover found in parallel ß-sheets (c), or the right-handed twist sence of α-helices (d).

5. Chiral Properties of α-Helical Domains

5.1 Non-Oriented Helices

Determining the overall chiral properties of a set of secondary-structural elements in a given domain has not been attempted except in specific cases such as 4-α-helix bundle proteins [21]. One difficulty with this task is that it is not obvious how to decompose the problem into a set of simpler subproblems. Recently, Murzin and Finkelstein [22], presented an elegant solution to the helix packing problem for α-helical protein domains. In their work they were able to show that packing of α-helices in these domains can be described by the placement of *non-oriented* helices, represented as cylinders, along specific edges of regular, quasi-spherical polyhedra as depicted in Figure 9 for three, four, and five helices, respectively.

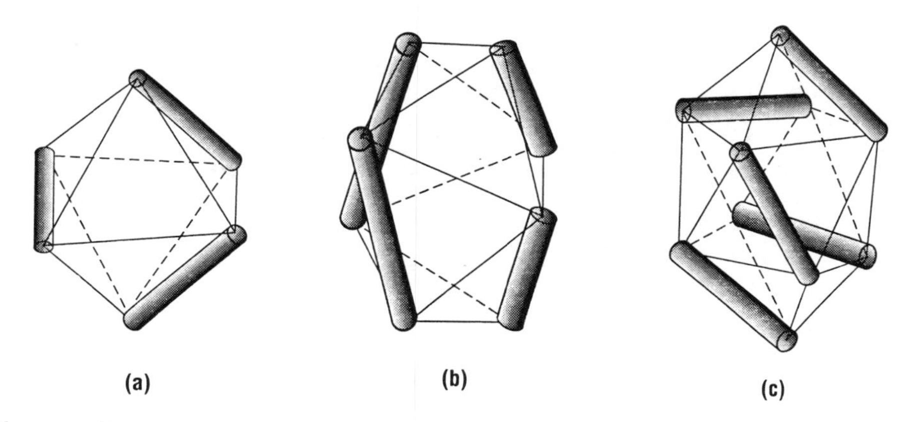

Figure 9:

Packing of non-oriented helices, portrayed as cylinders, on regular, quasi-spherical polyhedra after the method developed by Murzin and Finkelstein [22] for representing helix packing in α-helical domains of proteins: (a) three-helix packing on an octahedron, (b) four-helix packing on a dodecahedron, and (c) five-helix packing on a hexadecahedron. See descussion in Section 5 for further details.

Examination of the three-helix case shows that the two mirror-image packing arrangements depicted in Figure 10 are chiral, due to the fact that they are not superimposeable, *i.e.* the two packings form an enantiomeric pair. Also as indicated in the figure, a specific handedness can be assigned to each enantiomer as follows. First, consider the edges, excluding the one containing the cylinder, emanating from any vertex, say v or v′, of either octahedron in Figure 10. Second, label these edges "R", "M", and "L" for right, middle, and left, respectively, by considering an anti-clockwise order of edges as viewed from outside the octahedron. In the packing arrangement shown on the right side of Figure 10, edges "R" and "M" lead to the same helix, and hence v is referred to as an R-type vertex. All other vertices in this octahedron also are of the R-type, and consequently, the three-helix arrangement is designated as **R** to indicate its right-handed character. Applying an identical procedure to the packing arrangement depicted on the left-hand side of Figure 10 shows that at vertex v′ edges "L" and "M" lead to the same helix, and thus v′ is an L-type vertex. Again, examination of the other vertices shows that they are all of the L-type. Hence, this three-helix packing is designated as **L** to indicate its left-handed character. The octahedral characterization described here can be generalized for any polyhedral representation of three non-oriented helices. The details are presented in a recent work [24] where it is also shown that this procedure is entirely equivalent to that of tying a trefoil knot in the space into which the helix triple is embedded [24]. Such an approach was taken earlier by Mezey [25] as a means for characterizing the stereochemistry of chiral centers in organic molecules.

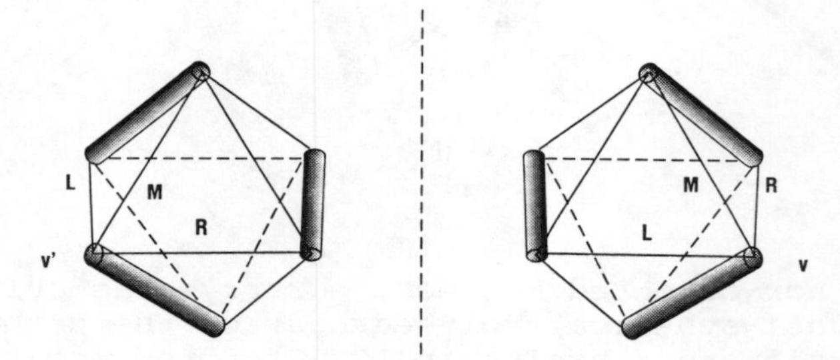

Figure 10:

Assignment of the "handedness" of mirror image three-helix packings. Note that the two packing arrangements are non-superimposable mirror images as required for the existence of chirality. See text in Section 5 for further details. The mirror plane is indicated by the vertical dashed line.

Assignment of the overall stereochemistry of four-, five-, and six-helix containing domains can be accomplished by considering all possible helix triples within a given α-helical domain. For example, a four-helix domain would possess 4!/3!1! = 4 helix triples. Such a situation is illustrated in Figure 11a which depicts, in the manner of Murzin and Finkelstein, and idealized four-helix packing arrangement common to the α-helical domains of a number of proteins, namely *Cytochrome c'* (dimer) [26], *Cytochrome* b526 [27], *Haemerythrin* [28], and *Melittin* (tetramer) [29]. The four helix triples derived from this arrangement are depicted in Figure 11b, and possess the L-stereochemistry of the prototypical left-handed helix triple located in the center of the figure.

An important feature of the above approach is that it does not require that the helices be embeddable on regular polyhedra in the fashion proposed by Murzin and Finkelstein, although their approach to helix packing did provide inspiration for the present work. Rather, the helix triples can be analyzed in their "native" protein environments as discussed in our earlier work [24]. A more complete discussion with additional examples will be presented in a forthcoming work [30].

5.2 Oriented Helices

Additional chiral features due to helix packing arrangements in α-helical domains are present if one provides each helix with an *orientation*. In such cases chiral information can be obtained from two oriented helices embedded along the edges of a tetrahedron as shown in Figure 12. The orientation of a given helix is directed from its N-terminal to C-terminal end, and thus is consistent with the standard convention used to define the orientation of the polypeptide chain [1]. The helices are then numbered in a manner consistent with their order in the polypeptide chain.

The handedness of the two possible helix arrangements can be obtained from an application of the "Right Hand Rule" used extensively in physics. In the present situation the thumb of the right hand is placed along the direction of one of the oriented helices, and if the orientation of the second helix is the same as that of the "curled" fingers, then the helix pair is right handed and is labelled **R**, otherwise the helix pair is left handed and is labelled **L**.

Figure 13 illustrates the chiral features for the three-helix domains of three different proteins. Specifically, Figures 13a, 13b, and 13c depict the oriented three-helix packings in, respectively, the N-terminal region of *Papain* [31], the interdomain contact region of *Aspartate Carbamoyl Transferase* [32], and the 'small' domain of *Adenylate Kinase* [33]. From the figures it is clear that both Papain and ACT

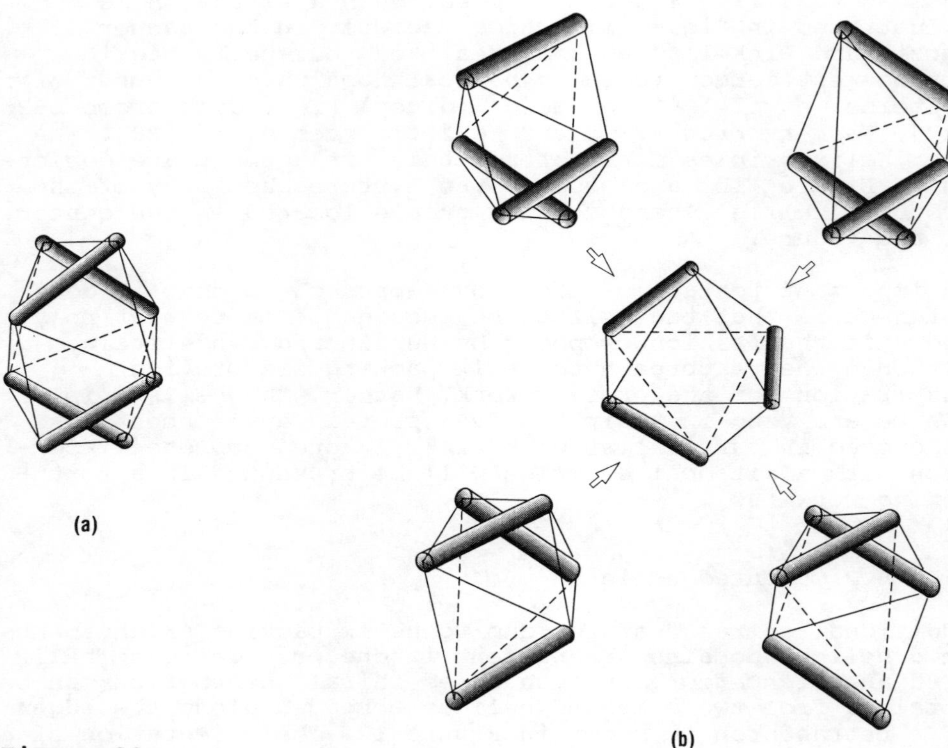

Figure 11:

(a) Idealized Murzin-Finkelstein packing of the four-helix systems of *Cytochrome c′* (dimer) [26], *Cytochrome b526* [27], *Haemerythrin* [28], and *Melittin* (tetramer) [29]. (b) Helix triples generated from the four-helix packing in (a). The idealized octahedral packing arrangement depicted in the center of (b) is *equivalent* to the helix triples located in each of the corners. See text for further details.

possess right-handed stereochemistry with respect to the packing of their non-oriented helices, while Adenylate Kinase possesses left-handed stereochemistry. The stereochemistry of the oriented helices in Papain [$helix_1$-$helix_2$: **L**, $helix_1$-$helix_3$: **R**, and $helix_2$-$helix_3$: **L**] is typical of all right-handed three-helix domains investigated to date, the only exception being ACT [$helix_1$-$helix_2$: **R**, $helix_1$-$helix_3$: **L**, and $helix_2$-$helix_3$: **L**]. Perhaps not surprisingly, the left handed three-helix domain in adenylate kinase has a stereochemistry opposite to that of papain [$helix_1$-$helix_2$: **R**, $helix_1$-$helix_3$: **L**, and $helix_2$-$helix_3$: **R**].

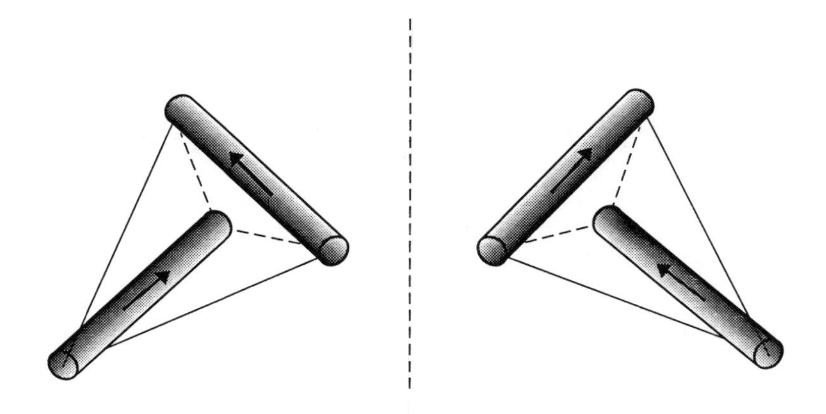

Figure 12:

An illustration of the handedness of mirror images of oriented pairs of cylinders (*i.e. helices*). As described in the text, the handedness can be determined using the familiar "Right-Hand Rule" of physics. Note that although the oriented cylinders are shown to lie along the non-adjacent edges of a regular tetrahedron, distorted tetrahedra are also allowed, the only restriction being that the cylinders not lie in the same plane. For purely practical reasons it is desireable that the cylinders be in close proximity, although they need not touch. The mirror plane is designated by the vertical dashed line.

While this examination of the chiral features of oriented helices is too limited to provide a basis for generalization, it does suggest that further work along this line is warranted. Currently, we are applying this approach in an analogous fashion to that for non-oriented helix packings in order to investigate the chirality of protein domains containing four, five, and six oriented helices.

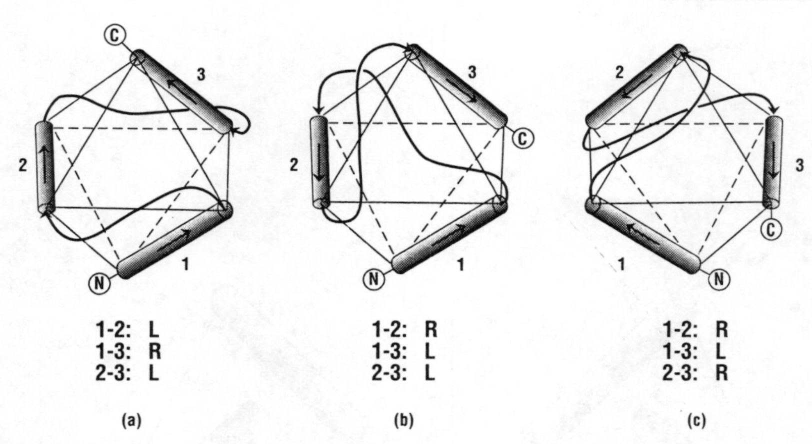

Figure 13:

An illustration of the additional chiral features due to *helix orientation (i.e. directionality)* possessed by "octahedrally packed" helix triples from (a) the N-terminal region of *Papain* [31], (b) the interdomain contact region of *Aspartyl Carbamoyl Transferase* [32], and (c) the "small" domain of *Adenylate Kinase* [33]. The numbering of the helices is consistent with the standard amino acid numbering convention in polypeptides and proteins (*Cf.* Figure 5). The designations **1-2: L, 1-3: R**, ... located below each figure indicate, respectively, that $helix_1$-$helix_2$ are related in a left-handed manner, $helix_1$-$helix_3$ are related in a right-handed manner, ... See text for additional details.

6. Topological Chirality in Proteins

Topological stereochemistry and chirality have been a subject of recent interest. As described by Walba [4], a number of the topological chiralities shown in Figure 2 have been realized in small synthetic organic molecules, and knots and linked-loops, in particular, have also been observed in DNA molecules [4, 34]. As depicted in Figures 2b and 2c, knots and linked-loops obviously require *closed* loops. Thus, in topologically chiral molecules covalent bonds naturally represent the topological connectivity as defined in the mathematical theory of topology. In proteins, closed loops require disulfide bonds in addition to the bonds making up the covalent backbone of a polypeptide chain (see Figure 5b). Complex loopings of covalent bonds that result in knots or linked loops have not been observed to date in proteins (see *e.g.* [35-37]). Patterns of disulfide bonds in a polypeptide chain may, however, require non-planar graphs[4] for their representation [38, 40], although topological stereochemistry in such proteins has been studied only recently [41].

The polypeptide chains of proteins (see Figure 5b) can be thought of as essentially linear structures except when disulfide bonds are present. In such cases, each disulfide bond links the side chains of two cysteine residues in the primary sequence. The covalent structure of a multi-disulfide polypeptide chain can be faithfully represented by a mathematical graph in which each vertex represents the α-carbon atom of a disulfide-linked cysteine residue, and each edge represents a covalent linkage between two such cysteinyl C_α atoms. Such a graph, called a molecular covalent graph, is planar[4] for protein molecules with three or fewer disulfide bonds, but may be non-planar when four or more disulfide bonds are present [38]. The four-disulfide chain shown in Figure 14 for instance is non-planar, and can be shown to be homeomorphic to the non-planar Kuratowski K(3,3) graph, which is depicted in Figure 2d. As the K(3,3) graph cannot be embedded in a plane [39], 3-space is required for its embedding. The mathematical concept of embedding a graph, which requires that edges not cross, is also chemically significant in that bonds representing interatomic interactions also do not cross in the three-dimensional structure of a molecule.

Planar graphs, which are embeddable in 2-space, are intrinsically topologically achiral in 3-space in that every embedding of the graph is homeomorphic to that in 2-space, which is necessarily achiral. The chirality of an embedding of a planar graph may be acquired extrinsically from certain complex loopings of edges which chemically correspond to the formation of topological stereoisomers of a molecule [41]. Embeddings of a labeled, non-planar graph, however, are intrinsically topologically chiral in that no homeomorphic achiral embedding can be found. Figure 15a shows the two different embeddings of the graph that are topological stereoisomers of the polypeptide chain. Walba [4] demonstrated that the K(3,3) graph is equivalent to a three-rung Möbius ladder and Simon [42] subsequently proved that Möbius ladders with three differentiated rungs are topologically chiral. Thus, the polypeptide chain of a stable globular protein can fold into one of the two enantiomorphic structures, which thus may be labeled ***D*** and ***L*** respectively [41]. In addition to the two structures shown above, the graph can be embedded in other ways which have additional, extrinsic topologically chiral features as shown, for example, in Figure 15b.

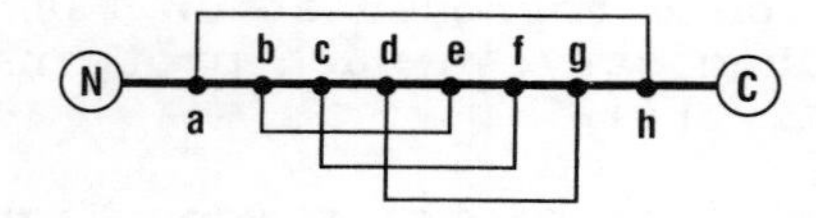

(a)

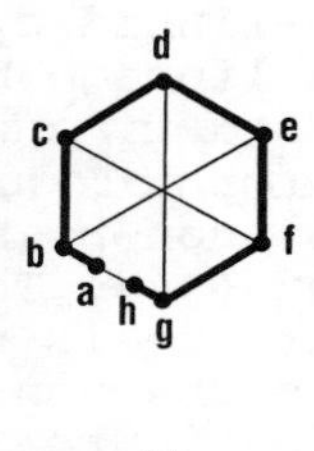

(b)

Figure 14:

A non-planar disulfide pairing scheme for a four-disulfide polypeptide chain [*reprinted from* [41] *with permission*]. (a) Normal representation of the polypeptide chain where C_α carbons of disulfide-linked cysteine residues are labeled alphabetically from the amino terminus. (b) An equivalent graph in the hexagonal form of Kuratowski K(3,3) graph, showing that the graph is homeomorphic to the complete bipartite graph of two edge sets (edges b, d, f and edges c,e,g); the graph is "homeomorphic" to the actual K(3,3) graph in that they are identical to within vertices of degree 2 (*i.e.*, vertices a and h).

Among the 105 possible pairing patterns for four-disulfide chains, four patterns are non-planar (*Cf.* [38, 41]), one of which was discussed explicitly above. The other three graphs are also homeomorphic to the K(3,3) graph and their topological stereochemistry is identical to the case discussed above [41].

For five-disulfide chains, the topological stereochemistry of one of 130 possible non-planar pairings is shown in Figure 15c. There again one finds two intrinsic topological stereoisomers that are chiral. Topological stereoisomers of five other five-disulfide pairings that are related to the one shown in Figure 15c have also been enumerated [41]. A more systematic enumeration of these topological stereoisomers has been developed [43].

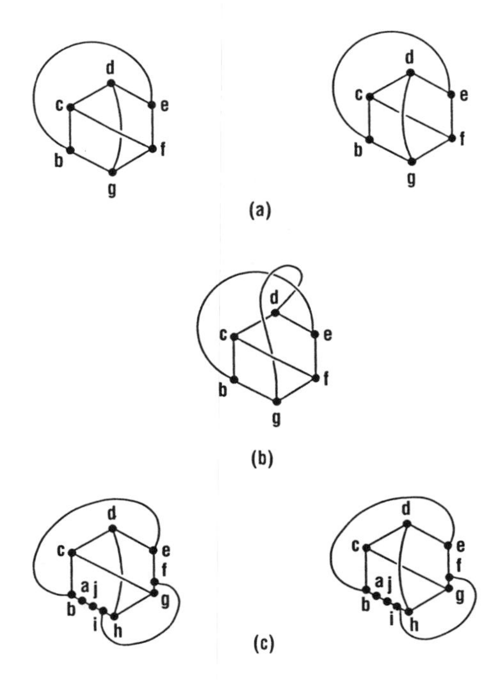

Figure 15:

Topological stereoisomers for some non-planar graphs.
(a) Two intrinsically topologically chiral embeddings of a graph for non-planar four-disulfide chains. The structure on the left is designated ***L***, while that on the right is designated ***D*** (*Cf.* [41]). These two structures are in fact topological enantiomorphs. (b) A topological stereoisomer for the same graph as in Figure 15a that has additional, extrinsic topological features; edge dg makes an additional looping around edge be. This structure is topologically diastereomorphic to those shown in Figure 15a. (c) Two intrinsic topological stereoisomers for a five-disulfide polypeptide chain.

The topological stereochemistry described above is observed in the four-disulfide chains of two mammalian-active scorpion neurotoxins (variant 3 toxin from *C. s.* Ewing and toxin II from *A. a.* Hector), the structures of which have been determined by x-ray crystallographic methods. These two molecules have overall sequence and structural homology [44, 45] and their structures belong to the ***D***-topology, as shown in Figure 16. The five-disulfide chain of mammalian pancreatic colipase can potentially be the second non-planar protein, although the exact disulfide pairing has yet to be established [46].

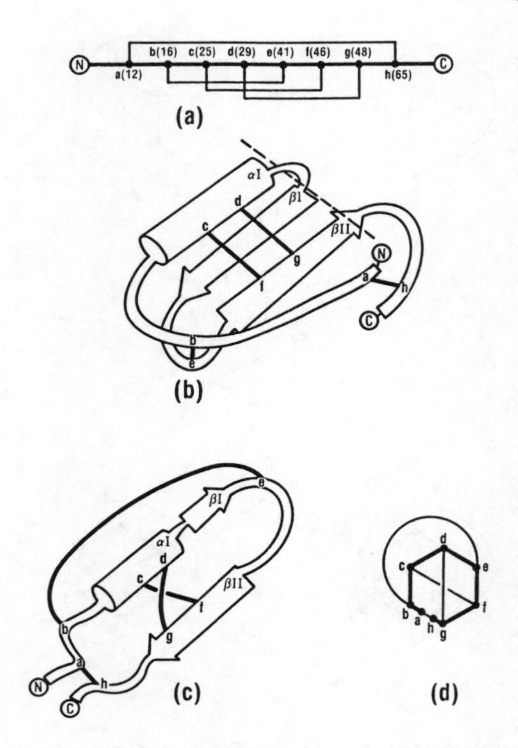

Figure 16:

Topological description of the polypeptide chain in variant 3 neurotoxin from *C. s.* Ewing [*reprinted from* [41] *with permission*]. (a) Disulfide pairings of variant 3 neurotoxin. (b) Schematic drawing of the three-dimensional structure of variant 3 neurotoxin. (c) A homeotopic representation of the neurotoxin structure for which the polypeptide chain is twisted and segments rearranged without breaking any covalent bond. (d) The graph representation of the four-disulfide chain to which the structure in Figure 16c corresponds.

Although covalent bonds are by far the strongest chemical interactions between atoms in a protein, a description of molecular structural topology need not be limited only to covalent interactions, and can be generalized to include other weaker but specific interactions in protein molecules. When hydrogen bonding is included in a more general "structural graph" representation (vs. the "covalent graph" representation described above for covalent linkages in a protein molecule), secondary structure elements, namely α-helices and ß-sheets, can also be described in topological terms [47]. In this work it is shown that α-helices and right-handed crossovers in parallel ß-sheets possess the *L* topology. This topological description of the hydrogen bonding in secondary structures may be extended to higher levels of protein structure and may provide a useful conceptual framework for studying the complexities of protein structure in general.

7. Concluding Remarks

As has been shown above, proteins exhibit both geometric and topologic chiral features. In the former case, it has been argued that lower order chiral features (*e.g.* the L-stereochemistry of individual amino acids) tend to induce higher order chiral features (*e.g.* the right-handed twist observed in extended polypeptide chains), which in turn induce even higher-order features (*e.g.* the right-handed twist observed in essentially all ß-sheets). Moreover, it can be argued that geometrically chiral features provide an additional means for characterizing and analyzing the underlying features of protein structure. The types of principles that can emerge from such investigations, while important in their own right, may also lead to practical procedures for analyzing mechanisms of protein folding or for predicting tertiary structure from primary structure. In this regard, the new approach to describing the chiral features of non-oriented and oriented packed helices presented in Section 5 may be exemplary. But considerable work needs to be done before it will become clear whether or not this is, in fact, the case.

Topological chirality is a more elusive property, and one whose application to protein structure and, for example, to folding problems is certainly less clear cut. Nevertheless, topological chirality may also provide useful descriptions of features difficult to characterize with more traditional approaches.

An interesting question which arises from the notion of protein chirality is "how can handedness be detected experimentally?" Certainly, the presence of the asymmetric α-carbon atom of each amino acid (excluding glycine) and the asymmetry due to interactions amongst peptide moieties can both be detected by circular dichroism (CD) spectroscopy [48]. Experimental confirmation of the existence and nature of "higher-level" chirality, whether geometric or topological, is not so straight forward, however. For example, it is yet to be established whether *Colipase* is a non-planar protein (see discussion in Section 6), and if so, which of the two possible topological stereoisomers it exists in. Currently, the crystallographic approach appears to be the only means for providing definitive answers to the questions.

As noted in the introduction, the material presented in this chapter was not meant to be complete. Rather, it was meant to provide a relatively broad overview of the chiral features of proteins, and thus stimulate creative thinking in this area that will ultimately lead to the development of new and novel ways for describing protein chirality specifically and protein structure in general.

Acknowledgements

The authors would like to thank Dr. Mark Johnson and Professor Paul Mezey for numerous helpful discussions during the course of this work, and Joe Moon with valuable assistance in the preparation of some of the figures.

Notes

1 Strictly speaking the correct mathematical usage would be *isotopic* rather than *homeotopic*. However, as pointed out by Walba [see for example Reference 3, p. 42 in "*Graph Theory and Topology in Chemistry*", R.B. King and D.H. Rouvray, Eds., Elsevier, Amsterdam, 1987] use of homeotopic appears to be acceptable to mathematicians, while use of isotopic is likely to cause confusion amongst chemists based on its widespread use in chemistry to denote the different isotopic forms of the elements.

2 Chothia [*J. Mol. Biol.* **75**, 295-302 (1973)] attributed the preference of ß-sheets for a right-handed twist to entropic factors. Weatherford and Salemme [*Proc. Nat. Acad. Sci.* (USA)**76**,19-23 (1979)], however, suggested that entropy had little to do with this preference; rather they attributed the right-handed twist to a partially tetrahedral deformation of bonds about the peptide nitrogen atoms. Subsequently, Salemme [*J. Mol. Biol.* **146**, 143-156 (1981)] pointed out that this deformation might be the result, rather than the cause, of the observed twisting. Chou et al. [*J. Mol. Biol.* **162**, 89-112 (1982); *J. Mol. Biol.* **168**, 389-407 (1983); *Biochemistry* **22**, 6213-6221 (1983); *Biochemistry* **24**, 7948-7953 (1985)] have investigated the formation and twisting of ß-sheets in an extensive series of molecular mechanics calculations. In their studies of ß-sheets they showed that the twist of

fully-extended peptide chains depends on the nature of the side chains: some side-chains possess minimum energy structures with a left-handed twist, although the difference in energy between right-handed or left-handed chain twists was quite small (on the order of 0.5 Kcal/mol). In any case, interchain interactions always were sufficient to insure the right-handed twist of the ß-sheet.

3 Ooi et al. [*J. Chem. Phys.* **46**, 4410-4426 (1967)] found that the observed preference of the right-handed screw sense of α-helices can be accounted for in terms of non-bonded interactions involving the C_β and H_β atoms.

4 Qualitatively, planar (mathematical) graphs are graphs which can be drawn on the plane of the page without the need for crossed edges. Conversely, non-planar graphs require that at least two edges cross.

Bibliograph

[1] Schulz, G.E.; Schirmer, R.H. (1979) "*Principles of Protein Structure*", Springer-Verlag, Berlin.

[2] Robson, B.; Garnier, J. (1986) "*Introduction to Proteins and Protein Engineering*", Elsevier, Amsterdam.

[3] Ugi, I.; Dugundji, J.; Kopp, R.; Marguarding, D. (1984) "*Perspectives in Theoretical Stereochemistry*", Springer-Verlag, Berlin.

[4] Walba, D.M. (1985) *Tetrahedron* **41**, 3161-3212.

[5] Mezey, P.G. (1990), *in press.*

[6] Gardner, M. (1968) *Sci. Am.* **219**, 112-115.

[7] Chothia, C. (1973) *J. Mol. Biol.* **75**, 295-302.

[8] Richardson, J.S. (1981) *Adv. Prot. Chem.* **34**, 167-339.

[9] Chothia, C.; Levitt, M.; Richardson, D. (1977) *Proc. Nat. Acad. Sci.* (USA) **73**, 2619-2623.

[10] Richardson, J.S. (1976) *Proc. Nat. Acad. Sci.* (USA) **73**, 2619-2623

[11] Sternberg, M.J.E.; Thornton, J.M. (1976) *J. Mol. Biol.* **105**, 367-382

[12] Sternberg, M.J.E.; Thornton, J.M. (1976) *J. Mol. Biol.* **110**, 2269-2283.

[13] Chou, K.C.; Némethy, G.; Pottle, M.; Scheraga, H.A. (1989) *J. Mol. Biol.* **205**, 241-249.

[14] Chothia, C.; Levitt, M.; Richardson, D. (1981) *J. Mol. Biol.* **145**, 215-250.

[15] Richmond, T.J.; Richards, F.M. (1978) *J. Mol. Biol.* **119**, 537-555.

[16] Efimo, A.V. (1977) *Dokl. Acad. Nauk SSSR, Biochem. Ser.* **235**, 699-702, Efimo, A.V. (1979) J. Mol. Biol. **134**, 23-40.

[17] Chou, K.C.; Némethy, G.; Scherage, H.A. (1983) *J. Phys. Chem.* **87**, 2869-2881.

[18] Chou, K.C,; Némethy, G.; Scherage, H.A. (1984), *J. Am. Chem. Soc.* **106**, 3161-3170.

[19] Chou, K.C.; Némethy, G.; Rumsey, S.; Tuttle, R.W.; Scherage, H.A. (1985) *J. Mol. Biol.* **186** 591-609.

[20] Chou, K.C.; Némethy, G.; Rumsey, S.; Tuttle, R.W.; Scherage, H.A. (1986) *J. Mol. Biol.* **188** 641-649.

[21] Weber, P.C.; Salemme, F.R. (1980) *Nature* **287**, 82-84.

[22] Murzin, A.G.; Finkelstein, A.V. (1988) *J. Mol. Biol.* **204**, 749-769

[23] Chothia, C. (1989) *Nature* **337**, 204-205.

[24] Maggiora, G.M.; Mezey, P.G.; Mao, B.; Chou, K.C. (1989) *Biopolymers*, submitted.

[25] Mezey, P.G. (1986) *J. Am. Chem. Soc.* **108**, 3976-3984.

[26] Finzel, B.C.; Poulos, T.L.; Kraut, J. (1984), *J. Mol. Biol.* **259**, 13027-13036.

[27] Lederer, F.; Glatigni, A.; Bethge, P.H.; Bellamy, H.D.; Matthews, F.S. (1981) *J. Mol. Biol.* **148**, 427-448.

[28] Stenkamp, R.E.; Sieker, L.C.; Jensen, L.H.; McQueen, J.E., Jr. (1978) *Biochemistry* **17**, 2499-2504.

[29] Terwilliger, T.C.; Eisenberg, D. (1982) *J. Biol. Chem.* **257**, 6010-6015.

[30] Maggiora, G.M.; Mezey, P.G.; Mao, B. (1989) *in preparation.*

[31] Kamphuis, I.G.; Kalk, K.H.; Swarte, M.B.A.; Drenth, J. (1984) *J. Mol. Biol.* **179**, 233-257.

[32] Honzatko, R.B.; Crawford, J.L.; Monaco, H.L.; Lardner, J.E.; Edwards, B.F.P.; Evans, D.R.; Warren, S.G.; Wiley, D.C.; Ladner, R.C.; Lipscomb, W.N. (1982) *J. Mol. Biol.* **160**, 219-263.

[33] Schulz, G.E.; Elzinga, M.; Marx, F.; Schirmer, R.H. (1974) *Nature* **250**, 120-123.

[34] Dröge, P.; Cozzarelli, N.R. (1989) *Proc. Natl. Acad. Sci.* (USA) **86**, 6062-6066.

[35] Crippen, G.M. (1975) *J. Theoret. Biol.* **51**, 495-500.

[36] Connolly, M.L.; Kuntz, I.D.; Crippen, G.M. (1980) *Biopolymers* **19**, 1167-1182.

[37] Kikuchi, T.; Némethy, G.; Scherage, H.A. (1986), *J. Comput. Chem.* **7**, 67-88.

[38] Klapper, M.H.; Klapper, I.Z. (1980) *Biochim. Biophys. Acta* **626**, 97-105.

[39] Chartrand, G.; Lesniak, L. (1986) "*Graphs and Digraphs*", Wadsworth and Brooks/Cole Advanced Books, Monterey, California.

[40] Kikuchi, T.; Némethy, G.; Scherage, H.A. (1989) *J. Comput. Chem.* **10**, 287-294.

[41] Mao, B. (1989) *J. Am. Chem. Soc.* **111**, 6132-6136.

[42] Simon, J. (1986) *Topology* **25**, 229-235.

[43] Mao, B.; Mezey, P.G. (1989) *in preparation.*

[44] Almassy, R.J.; Fontecilla-Camps, J.C.; Suddath, F.L.; Bugg, C.E. (1983) *J. Mol. Biol.* **170**, 497-527.

[45] Fontecilla-Camps, J.C.; Habersetzer-Rochat, C.; Rochat, H. (1988) *Proc. Natl. Acad. Sci.* (U.S.A.) **85**, 7443-7447.

[46] Charles, M.; Erlanson, C.; Bainchetta, J.; Joffre, J.; Guidoni, A.; Rovery, M. (1974) *Biochim. Biophys. Acta* **359**, 186-197.

[47] Mao, B.; Chou, K.C.; Maggiora, G.M. (1989) *Eur. J. Biochem.*, submitted.

[48] Woody, R.W. (1977), *Macromol. Rev.* **12**, 181-321.

A Topological Hierarchy of Molecular Chirality and other Tidbits in Topological Stereochemistry

David M. Walba

Department of Chemistry and Biochemistry
University of Colorado, Boulder, Colorado 80309-0215

Chemists have long been intrigued by the molecular basis of isomerism. Indeed, much of the powerful paradigm of structure based upon the molecular graph was first invented to explain isomerism, including the classifications used in modern stereochemistry. Thus, constitutional isomers describe pairs of isomeric molecular structures possessing non-homeomorphic molecular graphs, while the classical stereoisomers (enantiomers and diastereomers) possess molecular graphs which are homeomorphic and also homeotopic (interconvertable by continuous deformation in 3-space). This means that classical stereoisomerism is derived from the Euclidean properties of molecular graphs, being a manifestation of some kind of molecular rigidity.

Wasserman in his classic papers entitled "Chemical Topology" first focused the attention of chemists on the relevance of low dimensional topology in chemistry by proposing a possible approach to the synthesis of a molecular trefoil knot in addition to providing the foundation for the Biochemistry sub-discipline dealing with isomerism in circular DNAs.[1] The trefoil, recently realized in a small molecule structure for the first time by Dietrich-Buchecker and Sauvage,[2] serves as an example of a third type of isomerism which we first defined as topological stereoisomerism,[3] as shown in Figure 1.

Figure 1. Topological stereoisomers.

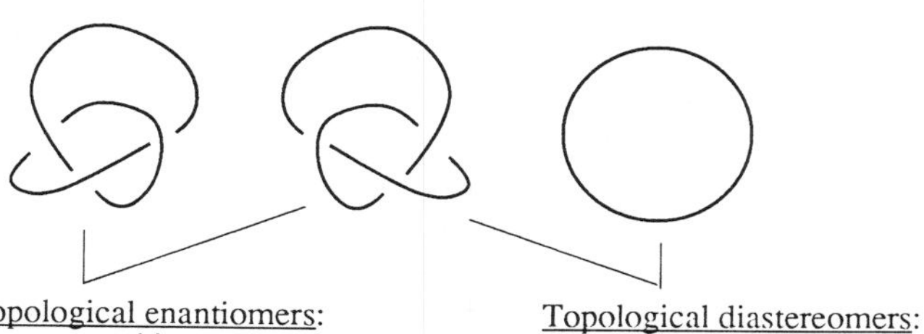

Topological enantiomers:
homeomorphic;
non-homeotopic;
may achieve mirror image presentations.

Topological diastereomers:
homeomorphic;
non-homeotopic;
cannot achieve mirror image presentations.

P. G. Mezey (ed.), New Developments in Molecular Chirality, 119–129.

Topological stereoisomers possess molecular graphs which are homeomorphic but non-homeotopic. The trefoil has long been known to be topologically chiral, thus the two trefoils serve as the prototypical topological enantiomers (non-homeotopic mirror images). In addition, either trefoil and the unknot are homeomorphic, non-homeotopic, and unable to achieve mirror image presentations, and thus serve as prototypical topological diastereomers.

It was not until our synthesis of the first molecular Möbius strip,[4] and the mathematical interest that followed, however, that the concepts of low dimensional topology were applied to molecular graphs more complex than simple circuits (linked and knotted rings). In this paper, a brief discussion of some chemical implications of this fascinating interface between mathematics and chemistry are presented.

THE UNCOLORED MÖBIUS LADDERS: AN APPLICATION OF LOW DIMENSIONAL TOPOLOGY IN CHEMISTRY

A fundamental question one may ask about the relationship between two molecular structures is whether the structures represent compounds which are, in principle, isolable under some stated conditions; in the following discussion room temperature and the "human time scale". This is equivalent to asking whether the molecular graphs are interconvertable, and when applied to knotted molecular graphs with the same molecular formula and constitution (homeomorphic) and a large number of atoms in the ring (flexible), is similar to the isotopy problem in knot theory. Of course, in small molecule chemistry, where only the simplest knot and simple links have been prepared to date, no new mathematics is required. It is well known, for instance, that even given infinite flexibility, the enantiomeric trefoils are indeed isolable (resolvable) under any conditions preserving constitution.

The novelty of modern topological stereochemistry is illustrated by the fact that a molecular graph may be topologically chiral even when possessing no knotted or linked subgraph. Thus, to our knowledge, the first graphs shown to be capable of supporting topological chirality independent of knots and links were the Möbius ladders.[5] In the context of topological stereochemistry, Jonathan Simon and Erica Flapan proved several theorems on the low dimensional topological properties of the Möbius ladders.[6,7] Here we describe some chemical implications of these theorems, illustrating an important application of low dimensional topology in chemistry.

First, consider the structures shown in Figure 2. While the arguments presented here hold for any compounds possessing the appropriate molecular graph, we chose to discuss the problem in terms of the homologous series of polycyclic amines shown in the Figure. We have discussed these particular compounds in the literature in terms of their intrinsic topology,[8] and here extend the discussion to include extrinsic properties of the embedding in 3-space.

Note first that these graphs are uncolored in the sense that each edge (-$CH_2CH_2OCH_2CH_2$-) and each vertex (N) is identical when taken independently of the graph. A chemically relevant question would be: which of these compounds is resolvable?

Figure 2. The uncolored Möbius ladders.

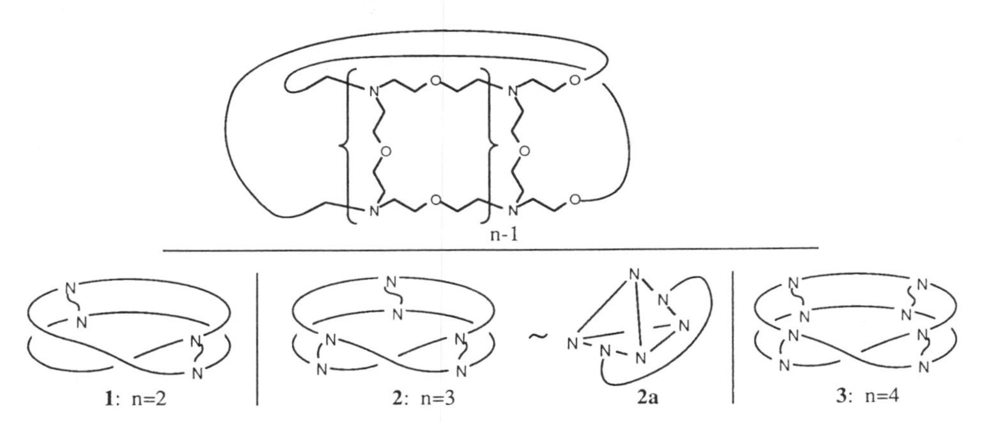

The lowest member of the series (**1**) has been prepared and characterized by Lehn and Graf.[9] The "two-rung Möbius ladder" is, in fact, not a Möbius ladder, but a tetrahedral graph. Since the graph is planar (may be embedded in a plane with no crossings),[10] it must be topologically achiral. Also, as may be seen in the rigidly achiral drawing of the structure indicated in the Figure (**1a**), the compound represented by the structure is clearly not resolvable at room temperature since chemistry tells us that even if the achiral conformation indicated is not an accessible minimum, at room temperature enantiomeric conformers would be rapidly interconverting. This illustrates the typical approach to determining whether a molecular structure is resolvable: attempting to find an achiral conformation.

Now consider the 3-rung uncolored Möbius ladder **2** (to our knowledge no such structure has yet been synthesized). While the presentation **2** is certainly rigidly chiral, it would be possible to prove using physical models that, in fact, the graph is topologically achiral, since it is possible to deform **2** into its mirror image! Indeed, the uncolored 3-rung ladder possesses rigidly achiral presentations (one is indicated in structure **2a**). Finally, chemical knowledge suggests that the compound is not resolvable at room temperature, since under these conditions the deformation taking **2** to its mirror image should be allowed.

So, using only the typical tools of the chemist (models and some chemical knowledge) we have managed to answer the question of resolvability for structures **1** and **2**. This, of course, assumes that the model manipulator is able to find the pathway for racemization of the 3-rung ladder or an achiral presentation. However, now consider graph **3**. One could make a model of this graph and try to find a pathway for racemization or an achiral presentation, but here no such pathway or presentation would be found. Never-the-less, this does not prove that the compound is resolvable, since a negative result with physical models is not rigorous.

One might argue that with enough work, a model manipulator would eventually find any racemization pathway existing in a graph as simple as **3**, but somehow this seems unsatisfying, since it is often surprisingly difficult to find specific continuous deformations (i.e. from **3** to its mirror image) in this type of system. Clearly, it would be useful to know whether **3** is resolvable. This is the role of low dimensional topology. By application of the mathematical machinery of knot theory to graphs, one may prove unequivocally whether a graph is topologically chiral. If the graph is topologically achiral, then application of chemical knowledge is necessary to determine whether the interconversion between mirror images is possible in the chemical system. If, however, the graph is topologically chiral, then any molecular realization of that graph must be chemically chiral under conditions preserving constitution.

Thus, Simon provides an elegant answer to the resolvability question in regard to compound **3** with his proof that the uncolored 4-rung Möbius ladder is, indeed, topologically chiral.[6] Therefore, compound **3** must, in principle, be resolvable based upon rigorous geometrical proof rather than a negative result using model manipulation. It is also interesting to note, as we have pointed out,[8] that conventional graph theory shows that all nine edges of the 3-rung ladder are in fact homotopic (the deformation of graph **2** into its mirror image takes a "rung" to an "edge"), while the rungs of the 4-rung ladder are heterotopic with the edges, i.e. the rungs essentially color themselves in the 4-rung case. Thus, as shown by Lehn experimentally, the 2-rung ladder has only 2 peaks in the carbon NMR spectrum, and the 3-rung ladder would also show only 2 peaks. However, the 4-rung ladder (racemic or enantiomerically pure) would exhibit in principle (and probably in practice with available spectrometers) 4 peaks in the carbon spectrum, two each for the edges and the rungs.

A TOPOLOGICAL HIERARCHY OF MOLECULAR CHIRALITY

Consideration of the low dimensional topological properties of graphs more complex than simple circuits has led to the discovery of new ways to classify chiral and achiral molecular graphs. This classification is completely rigorous, being based upon Euclidean and topological properties of the graphs. But, it is also possible to rank the classes of molecular graphs by "degree of chirality", from most chiral to least chiral. While this ranking is to some extent ad hoc (a compound is either resolvable or not, and one cannot really say that one molecule is "more chiral" than another), it is reasonable, and affords an interesting framework for discussion of the classification scheme, as described here.

Classes of Resolvable Graphs

As shown in Table 1, our hierarchy of chirality has six classes: three resolvable and three non-resolvable. At the top of the list, representing the most chiral molecules, are those possessing intrinsically chiral molecular graphs. An intrinsically chiral graph is defined as a graph possessing no achiral embeddings in 3-space.[11] That is, any embedding of an intrinsically chiral graph is homeomorphic only to topologically chiral graphs. Note that prior to the exploration of topological stereochemistry, no such object had every been described, since knot theory dealt with knots and links, and all knots and links are homeomorphic to achiral embeddings (unknotted or unlinked rings).

Table 1. Topological Hierarchy of Molecular Chirality.

Most Chiral	Intrinsically chiral graph : No topologically achiral embeddings
↑	Topologically chiral embedding of an intrinsically achiral graph
Resolvable	Nominal Euclidean chirality : Rigidly chiral presentation (or set of presentations) of a topologically achiral graph with no pathway for racemization
Non-resolvable	Topological rubber glove : Topologically and chemically achiral but rigidly chiral in every presentation
↓	Euclidean rubber glove : Topologically and chemically achiral but rigidly chiral in every accessible presentation (conformation)
Least Chiral	Nominally achiral : Rigidly achiral presentation accessible

Stimulated by topological stereochemistry, and in particular the low dimensional topology of the Möbius ladders, Erica Flapan first proved the existence of intrinsically chiral graphs in 1987.[7] Thus, Simon proved that the 3-rung Möbius ladder with colored rungs, realized chemically in the 3-rung molecular Möbius strip **4** shown in Figure 3, (a tetrahydroxymethylethylene-fused (THYME) poly crown ether), is topologically chiral.[6] Flapan then showed that indeed this graph, and any Möbius ladder with an odd number of colored rungs, possesses no achiral embedding. This makes compound **4** the most chiral organic molecule known, at the top of the hierarchy shown in Table 1.

Interestingly, Flapan showed that the 4-rung Möbius ladder **5**, also realized chemically in the THYME polyethers, and also proven by Simon to be topologically chiral, is in fact intrinsically achiral. This rather surprising result is easily proved using physical models. Thus, two alternative presentations of the 4-rung ladder embedded as in **5** are indicated by **6** and **7**. In the latter, the coloring of the rungs is indicated not by a double bond, but rather by different line weights.

It is easily seen that drawing **8** is in fact an alternative embedding of the 4-rung ladder. However, note that **8** is rigidly achiral, possessing an S_4 axis normal to the plane of the ring as indicated. Thus, we have a proof that the 4-rung ladder **5** is homeomorphic to an achiral embedding, and is therefore not intrinsically chiral. An alternative presentation of graph **8** is shown in **9**. It is interesting to note that cursory examination suggests that **9** is perhaps even more chiral than **5**, although a chemical realization of **9** with enough flexibility (probably necessary for the synthesis anyway), would not be resolvable.

Figure 3. Intrinsically chiral 3-rung Möbius ladder with colored rungs, and chiral and achiral embeddings of the 4-rung ladder.

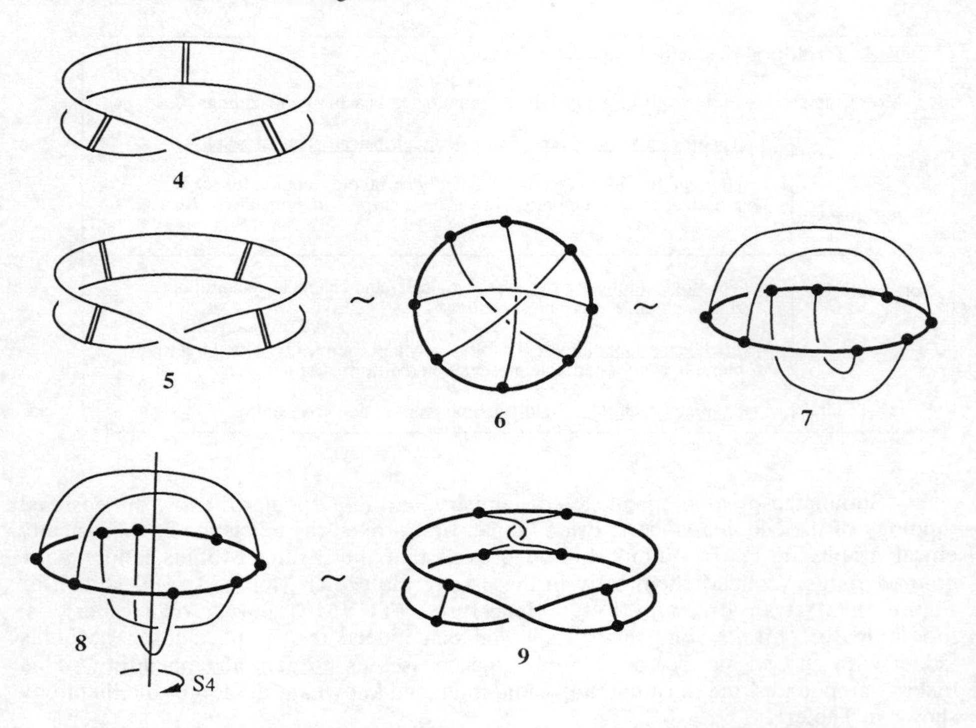

Thus, compound **5** joins the molecular trefoil knot, the topologically chiral link of Sauvage,[12] and the 4-rung THYME Möbius ladder synthesized in our laboratories[13] as examples of the second most chiral molecules: topologically chiral embeddings of intrinsically achiral graphs. Of course to date an achiral colored Möbius ladder such as **8/9** has not yet been realized chemically.

Lowest on the list of resolvable molecular graphs are those topologically achiral graphs which are resolvable due to Euclidean properties, the typical scenario in chemistry for resolvable compounds.

Classes of Non-Resolvable Graphs

At the bottom of the hierarchy of chirality are, of course, nominally achiral molecules with accessible rigidly achiral conformations. In some accessible minimum on the conformational hypersurface, the molecular graph of such molecules possesses an improper axis of symmetry, and is congruent with its mirror image. It has been known within the chemical community for quite some time, however, that non-resolvable (as opposed to achiral) molecules need not possess accessible rigidly achiral conformations. Mislow first pointed this out in 1954,[14] and actually synthesized the very elegant example shown in Figure 4.[15]

Figure 4. The Mislow Euclidean rubber glove.

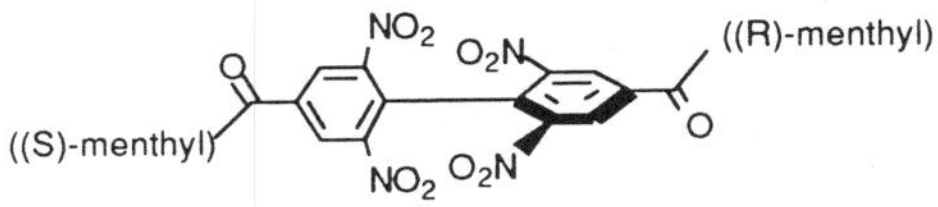

10

Thus, the biphenyl derivative **10** possesses four nitro substituents meta to the Ph-Ph bond, and chiral carbonyloxy groupings at the p-p' positions. The chiral alcohol units are enantiomorphic. If the biphenyl unit could become coplanar, then the molecule would represent a typical meso compound, with a mirror plane of symmetry bisecting the Ph-Ph bond. However, with the nitro substituents in place, the structure with coplanar phenyls is not accessible at room temperature, and in fact the molecular graph can not become rigidly achiral. Concurrent rotation about the p-p' bonds by 90°, however, interconverts mirror image conformations, thus providing an obligatory chiral pathway for racemization.

Given specified conditions, the lack of an accessible achiral conformation or intermediate in non-resolvable molecules is a well defined Euclidean property of the molecular graph. Such molecules share this property with Euclidean rubber gloves, which also "racemize" via a chiral pathway (by turning "inside-out") and are referred to as molecular Euclidean rubber gloves in the hierarchy. Note that Euclidean obligatory chiral pathways for racemization of non-resolvable molecules is not restricted to exotic structures such as **10**. Indeed, cis-1,2-dimethylcyclohexane also racemizes by a chiral pathway at room temperature.

The biphenyl **10**, cis-1,2-dimethylcyclohexane, and a rubber glove all racemize by chiral pathways due to the Euclidean properties of the objects. Topologically, the molecules, and even the rubber glove, have rigidly achiral presentations. For the glove, one such presentation is simply a flat disk. In the context of topological stereochemistry it is natural to ask whether any object exists which is topologically achiral but rigidly chiral in every possible presentation. We have dubbed such objects topological rubber gloves by analogy with their Euclidean counterparts.

It may be argued that a molecular topological rubber glove represents the most chiral non-resolvable class of compounds, and thus we put this class at the top of the non-resolvable classes in the hierarchy. But, do such objects exist? Indeed, the answer to this question was unknown when first posed by us in 1983. At that time, using a negative result with physical models, we had proposed a candidate topological rubber glove: the figure-of-eight knot **11** (Figure 5).[3a]

Figure 5. The figure-of-eight knot and a topological rubber glove.

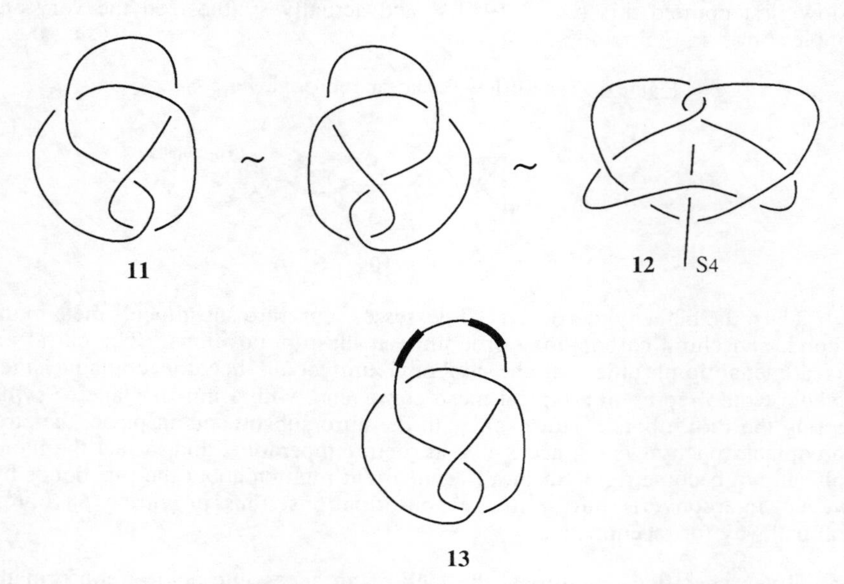

However, in this case the negative result was in fact false. The figure-of-eight does possess the rigidly achiral presentation **12** with an S_4 axis of symmetry. It was not until later that Erica Flapan proved that indeed topological rubber gloves do exist, and showed that one of the knots with a minimum of 10 crossings represents such an object.[16] Later still, Simon and Flapan showed that a figure-of-eight knot with a single colored point (or line segment), two colored points (**13**) or three colored points is also a topological rubber glove. To our knowledge this is structurally the simplest example known.[3c] It is interesting to point out that the topological rubber glove is the only class in the hierarchy which has not yet been realized in a molecular structure. Work aimed at remedying this situation is in progress in our group.

CUBANE DIYL AND A TOPOLOGICAL MESO COMPOUND

There are now several examples known of topological enantiomerism and topological diastereoisomerism. But, to date there is still no example of a topological meso compound. Indeed, there is no definition of such an object, though certain structures certainly fit the intuitive meaning of the phrase. Thus, the sum of two enantiomeric trefoils (the square knot **14** illustrated in Figure 6) possesses rigidly achiral presentations, and is obviously topologically achiral, although it possesses two topologically chiral parts. One possible definition of a topological meso compound would be as follows: A topologically achiral object which can be converted to a topologically chiral object by simplifying moves, i.e. removing crossings of a knot, or removing vertices and/or edges of a graph.

We have recently discovered a surprisingly simple graph representing a topological meso compound, and describe it as a fascinating illustration of how seemingly simple graphs can possess topologically interesting properties.

Figure 6. The square knot; a prototypical topological meso compound.

To introduce the topological meso compound, we will first consider the 1,4-dehydrocubane species recently discussed in the literature by the groups of Eaton, Michl and Borden.[17] This structure is composed of an array of carbon atoms at the vertices of a cube, with valences filled by hydrogen at each carbon except at two diagonal corners. If a bond forms between these two atoms, an interesting nonplanar graph results. 1,4-dehydrocubane itself almost surely exists as a biradical as shown in Figure 7 (**15**; cubadiyl). However, a molecule with the same topology as the hypothetical 1,4-bonded dehydrocubane can be easily envisioned by linking the 1,4 positions of the cube by a long chain of atoms as indicated by structure **16**.

Consider the topological properties of this relatively simple molecule. A more illuminating homeotopic presentation of the graph is shown in **17**. From this presentation it seems clear that the graph is non-planar, and the proof of this assertion is given in **18**, showing a $K_{3,3}$ subgraph.

In fact, the $K_{3,3}$ graph in **18** is a 3-rung Möbius ladder with two colored rungs. By Simon's proof of the topological chirality of the colored 3-rung ladder, the graph **18** is topologically chiral. Thus, the 1,4-linked cubane possesses a topologically chiral subgraph. However, the graph itself is obviously achiral, possessing rigidly achiral presentations! How can this be? Of course the homeotopic graphs **16** and **17** possess two 3-rung Möbius ladder subgraphs which are topologically enantiomeric (**18** and **19**), and graph **16** is a topological meso compound.

It may be stated with a high degree of confidence that structure **16** will be much more easily prepared than a molecular square knot. In addition, while none of the achiral Möbius ladders discussed above (the uncolored 3-rung ladder or the achiral embedding of the colored 4-rung ladder (**9**)) has yet been prepared, a synthesis of **16** would in fact represent realization in covalent bonds and atoms of an achiral Möbius ladder (actually two enantiomeric Möbius ladders in the same structure). To our knowledge, no such synthesis has yet been accomplished.

Figure 7. Cubadiyl and a topological meso compound.

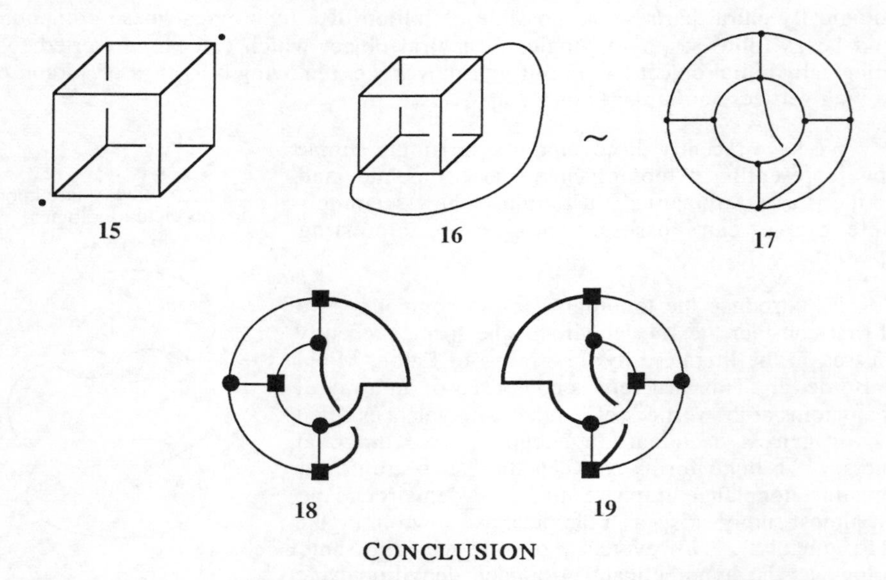

CONCLUSION

While we consider the topological results described above highly interesting, the real chemical implications seem restricted to the important task of determining the limits on resolvability (and all this implies) of compounds with topologically non-trivial molecular graphs. We have yet to discover a purely chemical means of differentiating intrinsically chiral molecules from Euclideanly chiral molecules, for instance. Topology does, however, serve as a rich source of new targets for total synthesis, and synthetic topological stereochemistry is quite active. Sauvage's beautiful synthesis of the first molecular knotted ring serves as the most recent landmark result in this area. However, examples such as the topological rubber glove and achiral Möbius ladders, in addition to more classic problems such as the Borromean rings, provide additional challenges in topologically stereocontrolled synthesis.

Acknowledgements

The author is greatly indebted to the Office of Naval Research for partial support of this work.

References

1 (a) Frisch, H. L.; Wasserman, E. *J. Am. Chem. Soc.* **1961**, *83*, 3789-3795. (b) Wasserman, E. *Sci. Amer.* **1962**, *207*, 94-100.

2 Dietrich-Buchecker, C. O.; Sauvage, J. P. *Angew. Chem.* **1989**, *101*, 192-194.

3 (a) Walba, D. M. "Stereochemical Topology," in *Chemical Applications of Topology and Graph Theory*, King, R. B. (Ed.); Elsevier, Amsterdam, 1983; pp 17-32. (b) Walba, D. M. *Tetrahedron* **1985**, *41*, 3161-3212. (c) Walba, D. M. "Topological Stereochemistry: Knot Theory of Molecular Graphs," in *Graph Theory and Topology in Chemistry*, King, R. B. (Ed.); Elsevier, Amsterdam, 1987; pp 23-42.

4 Walba, D. M.; Richards, R. M. *J. Am. Chem. Soc.* **1982**, *104*, 3219-3221.

5 The Möbius ladder graphs were first described in: Guy, R. K.; Harary, F. *Can. Math. Bull.* **1967**, *10*, 493-496.

6 (a) Simon, J. *Topology* **1986**, *25*, 229-235. (b) Simon, J. "A Topological Approach to the Stereochemistry of Nonrigid Molecules," in *Graph Theory and Topology in Chemistry*, King, R. B. (Ed.); Elsevier, Amsterdam, 1987; pp 43-75.

7 (a) Flapan, E. "Chirality of Non-Standardly Embedded Möbius Ladders," in *Graph Theory and Topology in Chemistry*, King, R. B. (Ed.); Elsevier, Amserdam, 1987; pp 76-81. (b) Flapan, E. *Mathematische Annalen* **1989**, *283*, 271-283.

8 Walba, D. M.; Simon, J.; Harary, F. *Tetrahedron Lett.* **1988**, *29*, 731-734.

9 Graf, E.; Lehn, J.-M. *J. Am. Chem. Soc.* **1975**, *97*, 5022-5024.

10 Harary, F. *Graph Theory*; Addison-Wesley: Reading, 1969.

11 Intrinsic chirality as used here was first described by Flapan in reference 7, and referred to as "inherent chirality".

12 Mitchell, D. K.; Sauvage, J. P. *Angew. Chem.* **1989**, *100*, 985-987.

13 Walba, D. M.; Armstrong, J. D., III; Perry, A. E.; Richards, R. M.; Homan, T. C.; Haltiwanger, R. C. *Tetrahedron* **1986**, *42*, 1883-1894.

14 Mislow, K. *Science* **1954**, *120*, 232.

15 Mislow, K.; Bolstad, R. *J. Am. Chem. Soc.* **1955**, *77*, 6712-6713.

16 Flapan, E. *Pacific Journal of Mathematics* **1987**, *129*, 57-66.

17 (a) Eaton, P. E.; Tsanaktsidis, J. *J. Am. Chem. Soc.* **1990**, *112*, 876-878. (b) Hassenrück, K.; Radziszewski, J. G.; Balaji, V.; Murthy, G. S.; McKinley, A. J.; David, D. E.; Lynch, V. M.; Martin, H.-D.; Michl, J. *J. Am. Chem. Soc.* **1990**, *112*, 873-874. (c) Hrovat, D. A.; Borden, W. T. *J. Am. Chem. Soc.* **1990**, *112*, 875-876.

CHIRALITY ALGEBRA

R.B. King
Department of Chemistry
University of Georgia
Athens, GA, USA

1. Introduction

The differentiation of an object from its mirror image leads to the property of chirality. The experimental observation of chirality in molecules arises from pseudoscalar measurements which have the following features:
(1) They depend upon the molecule but not upon its orientation in space;
(2) They have identical absolute values but opposite signs for mirror images, i.e., the two enantiomers of a chiral molecule.
Examples of pseudoscalar measurements of chemical significance include optical rotation, circular dichroism, and optical yields in asymetric syntheses.

This chapter summarizes an algebraic approach for the study of molecular chirality. Such chirality algebra provides qualitative concepts which lead to general insight into the chirality property as well as understanding of more detailed theoretical methods for the treatment of specific systems. The use of chirality algebra requires the dissection of the molecules of interest into a collection of ligands attached to an underlying skeleton. Group theoretical methods are then introduced which make use of both the symmetric group of all n! possible permutations of the n ligands as well as the point group of the underlying skeleton. Chirality algebra is most readily applied to molecules having the following structural features:
(1) The underlying skeleton is achiral, i.e., its point group contains an equal number of proper and improper rotation symmetry elements;
(2) The individual ligands are also achiral;

P. G. Mezey (ed.), New Developments in Molecular Chirality, 131–164.

(3) A sufficiently asymmetrical pattern of ligand substitution (ligand partition) onto the underlying skeleton can destroy all of the improper rotation symmetry elements of the skeleton thereby leading to a chiral molecule.
Within this general context chirality algebra has the following objectives:
(1) Determination of the ligand partitions for a given molecular skeleton which lead to chiral molecules, namely how asymmetrical must a ligand partition be in order to destroy all improper rotation symmetry elements (including reflection planes and inversion centers) of an achiral skeleton. Such ligand partitions are called *chiral ligand partitions*.
(2) Determination of mathematical functions which have suitable transformation properties under the symmetry operations of the skeletal point group for estimation of the magnitude and sign of a given pseudoscalar property (the dependent variable) for a given skeleton using parameters which depend only upon the ligands located at specific sited of the skeleton (the independent variables), the particular skeleton, and the particular pseudo-scalar property. Such mathematical functions are called *chirality functions*. If chirality functions are polynomials, they can be called more descriptively *chirality polynomials*. Such chirality polynomials of lowest possible degree may be regarded as initial terms in Taylor series for estimating the magnitudes of pseudoscalar properties.

Although the idea of chirality polynomials can be traced as far back as 1890 [1,2], the modern study of chirality algebra arises from the stereochemical analogy model ("Stereochemisches Strukturmodell") of Ruch and Ugi [3,4]. Increasingly sophisticated mathematical methods were then developed for the determination of chirality functions [5,6,7] eventually applying methods based on the induction of representations of finite groups, which had previously been shown to be useful for the very different problem of proving the Jahn-Teller theorem [8]. In 1972 Ruch [9] presented a general overview of the development of chirality algebra at this stage. Subsequently Mead published a review [10] providing the necessary mathematical background. Novel ideas of varying chemical and physical significance arising from this stage of the development of chirality algebra include homochirality [11,12], qualitative completeness [4,9,10], qualitative supercompleteness [13], and hyperchirality [14-19]. Subsequent work has related the ideas of chirality algebra to algebraic invariant theory [20,21] and to Pople's framework groups [22]. In addition, earlier suggestions [15,16,18,19] of the physical irrelevance of hyperchirality have been confirmed by a different method [23]. Polynomials found by the methods of chirality algebra have been tested experimentally for a number of important transitive skeletons including the polarized triangle (e.g., phosphines and phosphine oxides) [24], methane [25,26], disphenoid (e.g., allenes [27] and 2,2'-spirobiindanes [28,29]), polarized rectangle (e.g., [2,2]-metacyclophanes [30,31,32]), polarized pentagon (e.g., cyclopentadienylmanganese tricarbonyl derivatives [33]), cyclopentane [34], and ferrocene [35] skeletons. The success of these polynomial approximations to the experimental observations generally decreases with skeletons of increasing complexity in accord with the expected need for more terms in the Taylor series approximations of the more complicated systems. Thus the complexities of determining large numbers of necessary empirical ligand parameters for such higher terms would appear to be insurmountable. Nevertheless, the success of chirality algebra for the simpler skeletons such as the allene and methane skeletons justifies consideration of this approach.

This chapter surveys some of the important underlying ideas of chirality algebra. Many of the mathematical details are abbreviated or omitted for the sake of clarity. For further details of the underlying general mathematics including relevant areas of permutation group theory and group representation theory, the review of Mead [10] is useful although difficult for a typically chemically oriented reader with only an average mathematical background. In addition, I have presented many of the methods pertaining to chirality polynomials in a recent review elsewhere [23].

2. Skeletons and Their Symmetries

Consider a molecule of the type ML_n in which M is a metal or other central atom and the n ligands L may or may not be equivalent. For simplicity only molecules containing *achiral* ligands L will be considered although chirality algebra has also been applied to molecules containing chiral ligands [10,36]. Removal of the n ligands L from ML_n leaves what may be called its *skeleton* [14]. The following symmetry groups are required to describe the chirality of this ML_n molecule:

(1) The symmetric group P_n containing all n! different possible permutations of the n ligands;

(2) The point group G of the skeleton containing all of its $|G|$ symmetry elements;

(3) The rotation subgroup R of G containing only the $|R|$ proper rotations of the skeleton. For achiral skeletons R is a subgroup of G of index 2 (i.e., $|G|/|R|=2$). For a skeleton with a finite number of sites n, the groups G and R may be considered to be permutation groups of the n sites so that G is a subgroup of P_n. This idea is expressed in a more formal way by Ruch and Schönhofer [6].

The groups G and R can also be considered to be framework groups [37], which specify the symmetry of bodies containing a finite number of particles. Such framework groups may be classified into four types as follows [22]:
(1) *Linear*: Framework groups in which all sites are located in a straight line, i.e., in a one-dimensional subspace of three-dimensional space;
(2) *Planar*: Non-linear framework groups in which all sites are located in a flat plane, i.e., in a two-dimensional subspace of three-dimensional space;
(3) *Achiral*: Non-planar framework groups whose underlying point group contains at least one improper rotation $S_n(n \geq 1)$(including $S_1 \equiv \sigma$ and $S_2 \equiv i$) so that $|G|/|R| = 2$;
(4) *Chiral*: Non-planar framework groups whose underlying point group contains no improper rotations so that $G = R$ (i.e., $|G|/|R| = 1$).
Chiral framework groups are of no interest in chirality algebra since they lead to chiral systems for *any* ligand partition including the trivial ligand partition with all ligands identical. Linear framework groups are also of no interest in chirality algebra but for the opposite reason: they can never be chiral even if all ligands are different.

Chirality algebra thus involves the study of ML_n structures derived from skeletons having either achiral or planar framework groups. Achiral framework groups are clearly susceptible to the methods of chirality algebra since they can lead to either chiral or achiral molecules by varying the ligand partitions. In planar framework groups all sites are coplanar; the plane containing all of the sites can be called the *major plane* of the framework group. Any ligand partition of a planar framework group (even that having all ligands different) retains the major plane as a symmetry plane thereby never leading to a chiral structure. However, a chiral structure can be obtained from a planar framework group by destroying the symmetry of the major plane. Since this can be done by conceptually placing a positive charge on one side of the major plane and a negative charge on the other side, this process is conveniently called *polarization* [22]. From a chemical point of view polarization can involve polyhapto complexation of a planar skeleton (e.g., benzene) to a metal atom (e.g., chromium) on one side of the skeleton.

The symmetry of a skeleton also may be related to its transitivity. Thus a skeleton having all sites equivalent is called a *transitive* skeleton; otherwise the skeleton is called *intransitive*. A set of equivalent sites in a skeleton is called an *orbit*; the number of sites in an orbit is called the length of the orbit. A transitive skeleton thus has only one orbit consisting of all of its sites.

A skeleton may be characterized by its *site partition*, which may be represented by a symbol of the type

$$(a^{b_a}[a-1]^{b_{a-1}}[a-2]^{b_{a-2}}\ldots 1^{b_1})$$

in which b_k refers to the number of orbits of length k and all non-existent orbit lengths are omitted from this symbol. A transitive skeleton with n sites will have a site partition (n^1) (or (n) suppressing the exponent "1"). Transitive skeletons play a fundamental role in chirality algebra.

A general notation has been developed to describe skeletal framework groups [37]. In this notation the underlying skeletal point group is specified in Schoenflies notation [38] followed by an indication of the location of each of the sites in terms of subspaces relating to the symmetry elements of the underlying point group. The subspaces can be classified by their dimensionalities as follows:
0-dimensional: a central point (e.g., center of inversion (i) or intersection of a rotation axis with another rotation axis or a perpendicular plane of symmetry) designated as O;
1-dimensional: a rotation axis (C_n) designated by C_n where n is the order of the rotation;
2-dimensional: a reflection plane (σ) designated by σ_h, σ_v, or σ_d depending upon its location in the point group;
3-dimensional: the remaining part of three-dimensional space external to any of the symmetry elements of the point group, designated as X.
The location of any given site in the framework group is specified in terms of the subspace of the *lowest* possible dimensionality. This leads to the preference order $O>C_n>\sigma>X$.

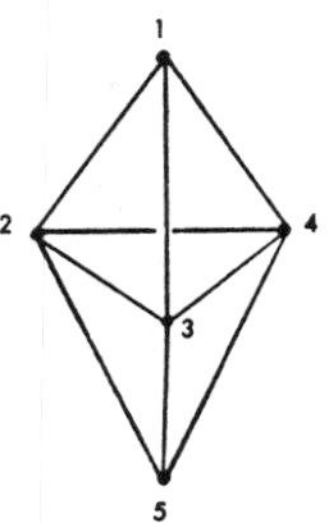

Figure 1
The trigonal bipyramid skeleton

The methods outlined above for describing various aspects of skeletal symmetry can be illustrated for the trigonal bipyramidal skeleton (Figure 1) found in many five-coordinate metal complexes ML_5. The symmetric ligand permutation group P_5 contains all possible 5!=120 permutations of the five ligands. The point group G of this skeleton is D_{3h} consisting of the 12 permutations {e, (234), (243), (15)(34), (15)(42), (15)(23), (15), (15)(234), (15)(243), (34), (42), (23)} using standard permutation notation [39] and e for the identity. The rotation subgroup R of G is D_3 consisting of the six proper rotations {e, (234), (243), (15)(34), (15)(42), (15)(23)}. The framework group of the trigonal bipyramid can be expressed as $D_{3h}[C_3(L_2), 3C_2(L)]$ in which the $C_3(L_2)$ means that the two axial sites are located on the C_3 axis and the $3C_2(L)$ means that each of the three equatorial sites are located on a different C_2 axis. The trigonal bipyramid is intransitive with a (32) site partition corresponding to the three equatorial sites and the two axial sites.

Chirality algebra also depends on the treatment of the point group G of the skeleton as a permutation group as noted above. Thus the action of any operation of G on the sites of the skeleton leads to a characteristic cycle structure, which may be represented as a cycle index term [40,41,42] or more simply as a *cycle partition*, which is a symbol of the type

$$(k^{c_k}[k-1]^{c_{k-1}}[k-2]^{c_{k-2}}\ldots 1^{c_1})$$

in which c_m refers to the number of cycles of length m and all non-existent cycle lengths are omitted from the symbol. For example, the symmetry operations E, C_4, C_3, and i of the six-site $O_h[3C_4(L_2)]$ regular octahedron framework group lead to the cycle partitions (1^6), (41^2), (3^2), and (2^3). All operations in the same conjugacy class of G lead to the same cycle partition.

The distribution of ligands on the skeleton is important in chirality algebra. The *ligand partition* can be depicted by a symbol of the type

$$(a^{b_a}[a-1]^{b_{a-1}}[a-2]^{b_{a-2}}\ldots 1^{b_1})$$

in which b_k refers to the number of sets of k identical ligands. In addition, the ligand partition can also be depicted by a collection of boxes called Young diagrams, which have the following features:
(1) The rows of boxes represent identical ligands;
(2) The top row is always the longest row and the lengths of the rows decrease monotonically from top to bottom;
(3) The left column is always the longest column and the lengths of the columns decrease monotonically from left to right.

Thus the ligand partition of the octahedral metal carbonyl complex $(C_6H_5)_3PMo(CO)_5$ is (51) and is represented by the Young diagram

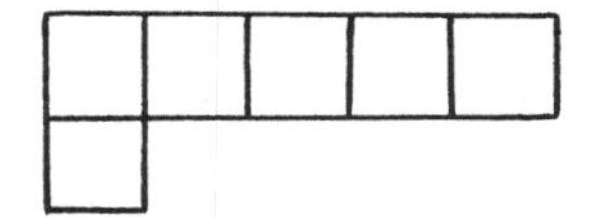

in which the "5" and the first row of five boxes represent the five CO ligands and the "1" and the bottom row of a single box represent the single $(C_6H_5)_3P$ ligand.

Young diagrams can be characterized by the following three parameters:
(1) Order (o): This represents the maximum number of identical ligands in the ligand partition and is simply the length of the top row. The order also corresponds to the number of columns in the Young diagram;
(2) Index (i): This represents the number of different ligands in the ligand partition and is simply the length of the left column. The index also corresponds to the number of rows in the Young diagram;
(3) Degree (g): This provides a basis for ordering Young diagrams and represents the minimum degree of the chirality polynomial for the corresponding ligand partition. The degree of a Young diagram can be calculated by the following equation in which c_k represents the length of column k:

$$g=\frac{1}{2}\sum_{k=1}^{k=order} c_k(c_k-1) \tag{1}$$

For Young diagrams having six of more boxes, there are cases where two or more different Young diagrams with the same number of boxes have the same degree as determined by equation 1. For example, in the case of Young diagrams having six boxes, both the (41^2) and (3^2) Young diagrams have degree 3 and both the (2^3) and (31^3) Young diagrams have degree 6. In general, Young diagrams having high degrees depict relatively asymmetrical ligand partitions and Young diagrams having low degrees depict relatively symmetrical ligand partitions. Thus the degree of a Young diagram may be viewed as a measure of the "asymmetry" of the corresponding ligand partition.

The permutation groups P_n, G, and R are used to characterize a skeleton as discussed above. The partition of ligands on the skeleton is characterized by the following two additional permutation groups V and H relating to the corresponding Young diagram:
(1) The *vertical permutation group* V consists of the direct product [43] of symmetric groups representing all permutations *within* each *column* of the Young diagram.

For a Young diagram of order o, this direct product is

$$V=P(c_1)xP(c_2)x...xP(c_0)$$

in which c_k is the length of column k (see equation 1) and $P(c_k)$ is the symmetric group consisting of all c_k! permutations within column k.
(2) The *horizontal permutation group* H consists of the direct product [43] of symmetric groups representing all permutations *within* each *row* of the Young diagram. For a Young diagram of index i, this direct product is

$$H=P(r_1)xP(r_2)x...xP(r_i)$$

in which r_m is the length of row m and $P(r_m)$ is the symmetric group consisting of all r_m! permutations within row m.

A number of the concepts described above can be illustrated for the octahedral metal complex $(C_6H_5)_3PMo(CO)_5$ with the Young diagram depicted above for its (51) ligand partition. The octahedral skeleton of this complex is characterized by the ligand permutation group P_6 of order 6! = 720, the point group $G=O_h$ of order 48, and the rotation subgroup R = O of order 24. The (51) ligand partition corresponds to a Young diagram with order 5, index 2, and degree 1. The vertical permutation group V of this Young diagram is P_2 and its horizontal permutation group H is P_5. Note that columns/rows of length 1 do not contribute to the degree or to the permutation groups V and H.

A specific feature of the symmetry of achiral skeletons of importance in chirality algebra is the relationship of the reflection planes of the skeleton to its sites. Such reflection planes can be classified into two types: separating planes and non-separating planes [22].
A *separating plane* in a skeleton having n sites contains exactly n-2 of these sites. Conversely, a reflection plane containing less than n-2 sites in a skeleton having n sites is a non-separating plane. Chiral molecules with underlying achiral skeletons in which *all* reflection planes are separating planes can be classified into left-handed and right-handed enantiomers with different enantiomers of the same "handedness" being called *homochiral* [11,12]. Such achiral skeletons have been called "category a" or *shoe-like* by Ruch [11,12], since left and right shoes can readily be distinguished regardless of their size, shape, or color. However, chiral molecules with underlying achiral skeletons having one or more non-separating planes have been called "category b" or *potato-like* by Ruch [11,12], since even though chiral, which member of an enantiomeric pair is "left-handed" and which is "right-handed" cannot be distinguished as is the case for potatoes. The T_d tetrahedron in which all six

reflection planes (σ_d) contain two (=n-2 for n=4) sites is a simple and important example of a category a or shoe-like achiral skeleton. Conversely, the C_{4v} polarized square in which two of the four reflection planes contain no sites is an example of a category b or potato-like achiral skeleton. The comparison of absolute configurations of different molecules is meaningful only for molecules derived from a category a or shoe-like skeleton. Fortunately, this is the case for the tetrahedral skeleton so important in organic chemistry. In addition, the lowest degree chirality polynomials appear to approximate experimental pseudoscalar measurements much more closely for category a skeletons than for category b skeletons.

3. Group Representations and Chiral Ligand Partitions

Consider a symmetric group P_n representing all n! possible permutations of n ligands on the skeleton of a ML_n derivative. This group contains exactly one conjugacy class [44] for each set of positive integers m_1, m_2 ..., m_k, whose sum

$$\sum_{i=1}^{i=k} m_i = n$$

Such a set of integers is called a *partition* of n and can be depicted by a Young diagram as outlined above. Furthermore, in any finite group there is a one-to-one correspondence between conjugacy classes and irreducible representations [38,44] as is readily evident from inspection of group character tables. The Young diagrams containing n boxes thus correspond not only to conjugacy classes of the symmetric group P_n but also to the irreducible representations of P_n. In addition, the partitions of n thus relating to the conjugacy classes and irreducible representations of P_n can correspond to partitions of ligands on a skeleton having n sites, as discussed in the previous section.

A chirality function is characterized by the so-called pseudoscalar property, which means that it is unaffected by ligand permutations corresponding to proper rotations in the skeletal point group G but undergoes a change in sign with no change in absolute value under ligand permutations corresponding to improper rotations in G. Consequently, the representation of the chirality function (i.e., the chiral representation) must contain the antisymmetric representation A^-, which is the one-dimensional irreducible representation of all *achiral* point groups having +1 characters for all proper rotations and -1 characters for all improper rotations. Thus, the chiral representations of P_n which correspond to chiral ligand partitions are those irreducible representations of P_n which occur in the representation A^{-*} of P_n induced [45,46] by the irreducible antisymmetric representation A^- of the skeletal point group G. Such a group induction process relates the representations of a group G_1 to the representations of a larger group G_2 of which G_1 is a subgroup. The dimensionality of the induced representation dim A^{-*} can be related to the maximum number of enantiomer pairs for the skeleton in question corresponding to the case in which each site of the skeleton has a different achiral ligand. This leads to the relationship

$$\dim A^{-*}=X_d=n!/|G|$$

in which X_d can be called the chiral dimensionality. The induced representation A^{-*} of P_n can therefore be reduced to a sum of irreducible representations of P_n of total dimensionality X_d. Each irreducible representation of P_n in this sum corresponds to a different chiral ligand partition. The chiral ligand partitions of the lowest degree as defined by equation 1 are of particular significance in representing the most "symmetrical" ligand partitions capable of destroying all improper rotations thereby leading to a chiral molecule containing the skeleton in question. In the case of certain skeletons derived from regular polyhedra, there are two or more different ligand partitions of the same degree, which are the lowest degree chiral ligand partitions [47]. Examples include the degree 6 (31^3) and (2^3) chiral ligand partitions for the regular octahedron, the degree 4 (4^2) and (521) chiral ligand partitions for the cube, and the degree 4 (84) and (921) chiral ligand partitions for the regular icosahedron [47]. Such skeletons may be called *chirally degenerate*. The polarized pentagon skeleton found in cyclopentadienylmanganese tricarbonyl derivatives may also be regarded as chirally degenerate, since its lowest degree (and only) chiral ligand partition (31^2) corresponds to an irreducible representation of P_5 which occurs *twice* in the representation A^{-*} of P_5 induced by the irreducible representation A^- of the C_{5v} skeletal point group.

Conversion of the group induction problem indicated above to the corresponding group subduction using the so-called Frobenius reciprocity theorem [10,45,46,48] facilitates the actual process of determining the chiral representations of P_n which correspond to the chiral ligand partitions for a skeleton having n sites and point group G. A subduction process relates the representations of a group G_2 to those of one of its subgroups G_1 and the Frobenius reciprocity theorem relates induction to subduction which is easier to perform in practice. Thus the determination of chiral ligand partitions using group representation theory is stated as a group induction problem but is solved as a group subduction problem. The chiral representations of P_n become the representations of P_n subduced by G, which contain the antisymmetric representation A^- of G. This leads to the following subduction algorithm [22,23] for determining the chiral ligand partitions for a skeleton having point group G:

(1) The characters for representations of G subduced by each irreducible representation of P_n are determined from character tables of P_n by copying down the characters of each irreducible representation of P_n for the operations of P_n which occur in G. The required character tables for P_n with $n \leq 13$ can be obtained from several sources [49,50,51] and computer methods are available for determining the required characters for still larger symmetric groups [52]. In order to recognize which operation in P_n corresponds to a given operation in G, the cycle partition of the operation in G is determined and the characters of the operations in the unique conjugacy class of P_n with that particular cycle partition are used.

(2) The characters for the antisymmetric (chiral) representation A^- of G are determined by simply using +1 for the proper rotations (E, C_n) and -1 for the improper rotations (σ, i, S_n).

(3) Standard group-theoretical methods based on orthogonality relationships [38,45] are used to determine which irreducible representations of P_n when restricted only to operations in G contain the representation A^- of G. These irreducible representations of P_n correspond to the chiral ligand partitions for a skeleton having point group G.

Note that representations which are irreducible in full P_n symmetry are no longer necessarily irreducible when the symmetry is reduced to G.

Table 1 illustrates the characters obtained from the first two steps of this algorithm for the allene skeleton numbered as in 1. This example, although nontrivial, is particularly simple since the relevant symmetric group P_4 is isomorphic to the tetrahedral group T_d found in conventional character tables [38]. The characters for the five irreducible representations of $P_4 = T_d$ designated both by the corresponding ligand partitions and by the conventional T_d labels are indicated in Table 1 for the symmetry operations of D_{2d} organized into its five conjugacy classes and identified by their cycle partitions as well as their conventional designations. Note that the only conjugacy classes of $P_4 = T_d$ missing in D_{2d} correspond to the three-fold symmetry operations having a (31) cycle partition (i.e., the C_3 operations of T_d). Thus destruction of all of the three-fold symmetry elements in $P_4 = T_d$ leads to its subgroup D_{2d}. The antisymmetric (chiral) representation A^- of $G = D_{2d}$ has characters for the four proper rotations E, C_2, and $2C_2'$ and -1 characters for the four improper rotations $2S_4$ and $2\sigma_d$. Application of the standard orthogonality relations [38,45] to the characters in Table 1 indicates that the subduced representations $(2^2) = E$ of dimension 2 and degree 2 and $(1^4) = A_2$ of dimension 1 and degree 6 each contain once the antisymmetric representation A^- but the other three subduced representations $(4) = A_1$, $(31) = T_2$, and $(21^2) = T_1$ do not contain the antisymmetric representation A^-. The chiral ligand partitions of the allene skeleton are therefore (2^2) and (1^4) of degrees 2 and 6, respectively, corresponding to structures II_2 and II_6, respectively. The chiral dimensionality of the allene skeleton is $\dim(2^2) + \dim(1^4) = 2 + 1 = 3 = 24/8 = 4!/|D_{2d}|$ consistent with equation 4.

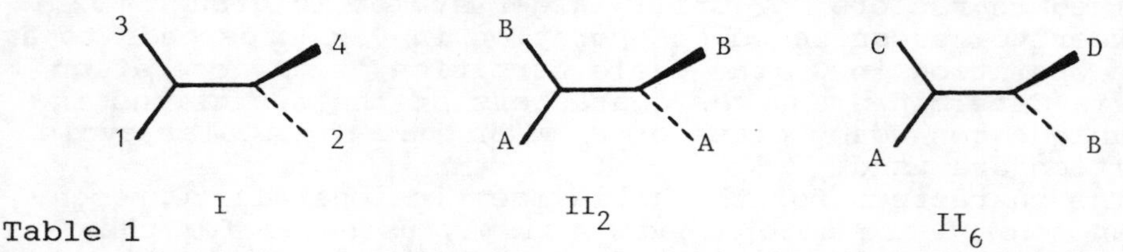

Table 1

CHARACTERS USED FOR THE DETERMINATION OF THE CHIRAL LIGAND PARTITIONS FOR THE D_{2d} ALLENE SKELETON

Symmetry operation of $G=D_{2d}$		E	$2S_4$	C_2	$2C_2'$	$2\sigma_d$
Cycle partition		(1^4)	(4)	(2^2)	(2^2)	(21^2)
Antisymmetric representation A^-		1	-1	1	1	-1
Characters of $G=D_{2d}$ subduced by irreducible representations of $P_4=T_d$						
$(4)=A_1$	(degree 0)	1	1	1	1	1
$(31)=T_2$	(degree 1)	3	-1	-1	-1	1
$(2^2)=E$	(degree 2)	2	0	2	2	0
$(21^2)=T_1$	(degree 3)	3	1	-1	-1	-1
$(1^4)=A_2$	(degree 6)	1	-1	1	1	-1

4. Chirality Functions and their Polynomial Approximations

As before consider molecules of the type ML_n in which M is a metal or other central atom and the n ligands L may or may not be equivalent. Now consider the problem of approximating a pseudoscalar measurement ψ with a function in n variables of the type

$$f(\lambda_1, \lambda_2, \ldots, \lambda_n)$$

i.e.

$$\psi \approx f(\lambda_1, \lambda_2, \ldots, \lambda_n) \quad (5)$$

In equation 5 the n independent variables $\lambda_1, \ldots, \lambda_n$ correspond to parameters associated with the ligands at sites 1,...,n, respectively, which depend only upon the ligand, the skeleton, and the pseudoscalar measurement. Thus for a given ligand k (e.g., methyl) the associated ligand parameter $\lambda(k)$ will be constant for a given skeleton and pseudoscalar measurement. Such ligand parameters are determined by fitting a series of psuedoscalar measurements of a given type (e.g., optical rotation of a given line) on compounds having the same skeleton and a restricted set of ligands using an appropriate chirality function determined as outlined in this section. The permutations of the symmetric group P_n interchange the variables $\lambda_1, \ldots, \lambda_n$ with the following effects if f is a valid chirality function:
(1) A proper rotation $r \in R \subset G$ leaves both the absolute value and sign of f unchanged.
(2) An improper rotation $s \in G$ changes the sign of f but leaves its absolute value unchanged. Thus $f \equiv 0$ for an achiral molecule having one or more improper rotations remaining in its point group after reduction of its skeletal symmetry by ligand substitution. Such a zero value for a chirality function of an achiral molecule is required by symmetry and is called an *achiral zero*.
(3) Any permutation in P_n not in the skeletal point group G may have any effect on the value of a corresponding chirality function f.

We now consider the problem of finding the simplest type of chirality function for a given skeleton and chiral partition having the required transformation properties. This turns out to be a polynomial $X(s_1, s_2, \ldots, s_n)$ of the same degree as that of the corresponding chiral ligand partition (see equation 1). If a skeleton has several chiral ligand partitions of degrees $g_1, \ldots, g_p$ where p is the number of chiral ligand partitions, a sum of chirality polynomials of degrees $g_1, \ldots, g_p$ with a separate set of ligand parameters for each polynomial may be required to describe all chirality phenomena for the skeleton in question. Such a sum of chirality polynomials is called a *qualitatively complete* chirality polynomial and the individual chirality polynomials making up the sum may be regarded as components of the qualitatively complete chirality polynomial. Since qualitatively complete chirality polynomials can become intractable very quickly for skeletons having several chiral ligand partitions, there is a natural reason to try abbreviated chirality functions using only the polynomials corresponding to the lowest degree chiral ligand partitions. Thus a qualitatively complete chirality polynomial for the allene skeleton has a degree 2 component for the (2^2) chiral ligand partition and a degree 6 component for the (1^4) chiral ligand partition. The degree 6 component vanishes identically for chiral allenes having (2^2) or (21^2) ligand partitions. Furthermore, comparison with experimental data on chiral allenes [27] as suggested below suggests that the degree 6 component for chiral allenes having a (1^4) ligand partition is so small that it can be neglected thereby simplifying considerably the application of this theory [27]. The further discussion on chirality polynomials in this chapter will focus on individual components of a qualitatively complete chirality polynomial.

The chirality function $f(\lambda_1,\lambda_2,...,\lambda_n)$ may also be regarded as a function of a vector $\vec{\lambda}$ in n-dimensional ligand parameter space Λ^n. The polynomial approximation $X(s_1,s_2,...,s_n)$ is then a function of a vector $\vec{s}$ in another n-dimensional ligand parameter space S^n with a smooth map $M: \Lambda^n \to S^n$ between the two ligand parameter spaces if the polynomial $X(s_1,s_2,.,s_n) = X(\vec{s})$ is a sufficiently close approximation to $f(\lambda_1,\lambda_2,...,\lambda_n)$. Furthermore, the lowest-degree polynomial $X(\vec{s})$ may be regarded as an initial term in a Taylor series approximation of a more accurate chirality function $F(\vec{s})$ of the same vector $\vec{s}$ in the same parameter space. The accuracy of the approximation of $F(\vec{s})$ by its first term $X(\vec{s})$ will naturally depend upon the skeleton and the pseudoscalar property and cannot be predicted *a priori* by theory alone.

An individual polynomial component $X(s_1,s_2,...,s_n)$ of a qualitatively complete chirality function is homogeneous and depends only on the s-differences s_i-s_k $(i,k = 1,2,...,n)$ in all important cases. It has the general form

$$X(s_1,s_2,...,s_n) = \sum_{k=1}^{k\leq|G|} a_k p_k(s_1,s_2,...,s_n) \qquad (6)$$

where $|G|$ is the number of operations in the skeletal point group G and the p_k's are homogeneous polynomials which are also functions of s-differences and are completely defined by the required transformation properties. For a shoe-like skeleton (category a) the chirality polynomial is a simple product of g s-differences, i.e.

$$X(s_1,s_2,...,s_n) = \prod (s_i-s_k) \qquad (7)$$

so that its only zeroes are achiral zeroes arising when the identity of ligands i and k lead to achiral molecules. The degree g of the lowest degree chirality polynomial for a shoe-like skeleton corresponds to the number of reflection planes, all of which are necessarily separating planes in the shoe-like skeleton. The indices i and k for each factor in the chirality polynomial of equation 7 correspond to a pair of sites outside a given reflection plane. The chirality polynomials for potato-like skeletons (category b) do not have the simple form of equation 7 since they contain sums as well as products of differences of ligand parameters. Also the degrees of the lowest degree chirality polynomials for potato-like skeletons are less than the total number of reflection planes. Polynomials for potato-like skeletons necessarily have chiral zeroes as well as achiral zeroes and appear to provide much poorer approximations of actual experimental data as discussed later in this chapter for the polarized rectangle skeleton of the [2,2]-metacyclophanes [30,31,32].

The general procedure for determining a chirality polynomial for a given chiral ligand partition of an achiral skeleton consists of the successive application of two projection operators to a monomial derived from the Young diagram corresponding to the chiral ligand partition of interest. The first of these projection operators is the Young operator, Y, which is derived from the permutation groups H and V of the Young diagram (equations 2 and 3) having equal numbers of positive and negative terms corresponding to the elements of V, the *vertical* permutation group, containing even and odd numbers, respectively, of two-element transpositions. The second projection operator is derived from the skeletal point group G having equal numbers of positive and negative terms corresponding to the proper and improper rotations, respectively, of G. Application of the Young operator Y can be omitted if only chirality polynomials corresponding to the lowest degree chiral ligand partitions are sought, since in such cases the projection operator derived from G has sufficient terms to generate the full chirality polynomial.

Even more drastically simplified procedures can be used to determine the lowest degree chirality polynomials of shoe-like skeletons (equation 7) or intransitive skeletons (see below). On the other hand, qualitatively complete chirality polynomials for a given skeleton consist of sums of chirality polynomial components corresponding to *each* chiral ligand partition using different sets of ligand parameters for each polynomial in the sum. Therefore, determination of qualitatively complete chirality polynomials in general requires determination of chirality polynomial components corresponding to chiral ligand partitions which are *not* those of the lowest degree for the skeleton of interest.

Such cases require the full procedure using both projection operators.

The full group-theoretical algorithm for determination of a chirality polynomial in ligand-specific parameters s_k corresponding to a given skeleton, ligand partition, and pseudoscalar measurement can be outlined as follows:

(1) Label the boxes in the Young diagram for the ligand partition of interest with the indices of the skeletal sites corresponding to the particular chiral species of interest with that ligand partition. These labels correspond to the indices k of the ligand specific parameters s_k.

(2) Determine from the columns of the labelled Young diagram the monomial arising from the following double product:

$$M(s_1, s_2, \ldots, s_n) = \prod_{e=1}^{e=order} \prod_{f=0}^{f=i-1} s_{k_{ef}}^{f} \qquad (8)$$

In equation 8 the following should be noted:

(a) The product over e contains one factor for each column of the Young diagram. The index e corresponds to the position of the column relative to the left column where e = 1.

(b) The product over f contains one factor for each box in the Young diagram in column e except for the top box where f = 0 and hence

$$s_{k_{ef}}^{f} \equiv 1$$

Thus if column e contains only one box, then the factor from column e is unity. Therefore columns of the Young diagram containing only a single box do not contribute to the monomial in equation 8. The variable f corresponds to the position of the box relative to the top box of the column e. In the top box f = 0.

(c) The subscript k_{ef} refers to the index entered in the Young diagram box corresponding to column e and row f.

(3) Determine from the Young diagram the corresponding horizontal (H) and vertical (V) permutation groups using equations 2 and 3.

(4) Apply the corresponding Young operator Y to the monomial M from equation 8 to give

$$Y[M(s_1, s_2, \ldots, s_n)] = \sum_{v \in V} \sum_{h \in H} e_v hv * M(s_1, s_2, \ldots, s_n) \qquad (9)$$

in which h and v represent elements in H and V, respectively; e_v = +1 or -1 if the vertical permutation element v consist of an even or odd number, respectively, of two-element transpositions; the summations are taken over all elements v of V and all elements h of H; and the star * means that the indices 1, 2,...,n of the variables s_k in $M(s_1,s_2,...,s_n)$ are permuted by the permutation hv arising from applying first the permutation v of V and then the permutation h of H. If the group H has |H| elements and the group V has |V| elements, then the Young operator Y expands the monomial $M(s_1,s_2,...,s_n)$ into a polynomial with |H||V| terms.

(5) Apply permutations corresponding to each of the elements g of the skeletal point group G to the indices of the variables s_k in the polynomial $Y[M(s_1,s_2,...,s_n)]$ from equation 9 to give the chirality polynomial $X(s_1,s_2,...,s_n)$ defined by

$$X(s_1,s_2,...,s_n) = \sum_{g\in G} e_g g*Y[M(s_1,s_2,...,s_n)]$$
$$= \sum_{g\in G} e_g g*[\sum_{v\in V}\sum_{h\in H} e_v hv*M(s_1,s_2,...,s_n)] \quad (10)$$

in which $e_g = +1$ if g is a proper rotation and $e_g = -1$ if g is an improper rotation. Equation 10 thus represents application of a projection operator corresponding to G. The polynomial $X(s_1,s_2,...,s_n)$ resulting from equation 10 is usually reduced to a product and/or sum of differences between pairs of ligand parameters

$$s_k^a - s_i^a$$

by standard algebraic methods, which can sometimes involve adding zero in the form

$$0 = K\sum_{k=1}^{k=n} (s_k^a - s_k^a) \quad (11)$$

followed by factoring into powers of the type $(s_k-s_i)^a$ in which a is a small integer.

Steps 3 and 4 of the above algorithm using the Young operator Y can be omitted for determining chirality polynomials corresponding to the *lowest* degree chiral ligand partitions for a given skeleton. In such cases the group theoretical algorithm, now consisting only of steps 1, 2, and 5 above, is simplified considerably [14,22]. However, for the determination of chirality polynomials for non-lowest degree chiral ligand partitions for a given skeleton, steps 3 and 4 may also be required. The net effect of these additional steps may frequently be regarded as the replacement of the projection operator derived from the skeletal point group G with that derived from a larger group L containing G as a subgroup.

Meinköhn [20,21] has used algebraic invariant theory [53] to show that only the chirality polynomials for the lowest degree ligand partitions are required to have the desirable property of depending only upon *differences* between the ligand parameters. The chirality polynomials for non-lowest degree chiral ligand partitions may also depend only upon differences between the ligand parameters if the ligand partition in question is the lowest degree chiral ligand partition for a larger permutation group L. In this connection the idea of signed permutation groups [48] may be used to extend the concept of chirality to permutation groups L which are not realizable as symmetry point groups in three-dimensional space. A simple example of a chirality polynomial for a non-lowest degree chiral ligand partition which is *not* a function solely of differences between ligand parameters is that for the (321) chiral ligand partition of the trigonal prism (cyclopropane) skeleton [23].

The algorithm for calculating chirality polynomials is illustrated with the allene skeleton of D_{2d} symmetry depicted in the previous section of this paper (structure I). The five steps of this algorithm for each of the two chiral ligand partitions (2^2) and (1^4) are summarized below:

(1) The following labelled Young diagrams correspond to the two chiral ligand partitions using the numbering of stucture I:

(2) These labelled Young diagrams lead to the following monomials M from equation 8:

$$M(2^2) = s_1 s_2 \quad (12a)$$

$$M(1^4) = t_1^3 t_2^2 t_3 \quad (12b)$$

(3) The vertical and horizontal permutation groups corresponding to the labelled Young diagrams are as follows:

$$V(2^2) = P_2(1,3) x P_2(2,4) = \{e, (13), (24), (13)(24)\} \quad (13aa)$$

$$H(2^2)=P_2(1,2)xP_2(3,4)=\{e,(12),(34),(12)(34)\} \quad (13ab)$$

$$V(1^4)=P_4(1,2,3,4)=T_d \quad (13ba)$$

(full tetrahedral group containing all possible permutations of the four ligands)

$$H(1^4)=P_1(1)xP_1(2)xP_1(3)xP_1(4)=\{e\} \quad (13bb)$$

(trivial group containing only the identity operation).
(4) Application of the Young operator (equation 9) from the vertical and horizontal permutation groups in equation 13 gives the following results:

$$Y(2^2)[M(2^2)]=Y(2^2)(s_1s_2)=4s_1s_2+4s_3s_4-2s_2s_3-2s_1s_4-2s_1s_3-2s_2s_4 \quad (14a)$$

$$Y(1^4)[M(1^4)]=Y(1^4)(t_1^3t_2^2t_3)=$$
$$(t_1-t_2)(t_1-t_3)(t_1-t_4)(t_2-t_3)(t_2-t_4)(t_3-t_4) \quad (14b)$$

Equation 14b has the same form as the chirality polynomial for a regular tetrahedron.
(5) The skeletal point group D_{2d} has the following eight permutations:

$$\{e,(13)(24),(14)(23),(12)(34),(13),(24),(1234),(1432)\} \quad (15)$$

with the first four permutations being positive and the second four permutations being negative. Application of the corresponding projection operator (equation 10) to the six-term degree 2 polynomial in equation 14a after suppression of a constant factor of 24 leads to

$$X(2^2) = s_1s_2 - s_2s_3 + s_3s_4 - s_4s_1 = (s_1-s_3)(s_2-s_4) \quad (16)$$

Note that the polynomial in equation 16 arises from the s_1s_2, s_3s_4, s_2s_3, and s_1s_4 terms of the polynomial in equation 14a and that application of the projection operator (equation 10) from the D_{2d} skeletal point group (equation 15) to the s_1s_3 and s_2s_4 terms in equation 14a leads identically to zero. Application of the projection operator from the D_{2d} skeletal point group to equation 14b results in no changes other than introduction of a constant multiplicative factor (which can be suppressed by redefining the ligand parameters). The two chirality polynomials for the allene skeleton thus are the degree 2 polynomial in equation 16 and the degree 6 polynomial in equation 14b.

The determination of chirality polynomials for intransitive skeletons can be simplified by considering each orbit individually and then multiplying together the chirality polynomials for each orbit to give the chirality polynomial for the entire skeleton. Thus the chirality polynomial for an intransitive skeleton having n sites and q orbits can be determined from the relationship

$$X(s_1, s_2, \ldots, s_n) = \prod_{k=0}^{k=q} X_k \tag{17}$$

in which X_k is the chirality polynomial for orbit k. Different sets of ligand parameters must be used for the chirality polynomials for each orbit. Failure to recognize this need for different sets of ligand parameters for different orbits of an intransitive skeleton was responsible for extensive confusion on hyperchirality [14-19] which thus appears to be a physically meaningless mathematical artifact. Also note that orbits of length 1 in intransitive skeletons (i.e., isolated sites invariant under all symmetry operations) can be ignored completely in the calculation of chirality polynomials.

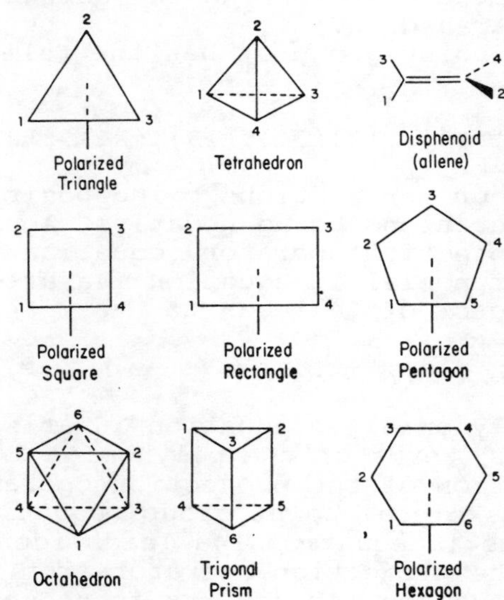

Figure 2
Transitive skeletons with six or fewer sites

Equation 17 for the chirality polynomials of intransitive skeletons suggests that for the purpose of calculating chirality polynomials the basic building blocks are the

Table 2

TRANSITIVE PERMUTATION GROUPS WITH SIX OR FEWER SITES AND THEIR CHIRALITY POLYNOMIALS

Skeleton	Point Group G[a]	Chiral Ligand Partition	Degree	Chirality Polynomial
Polarized triangle	C_{3v}	(1^3)	3	$(s_1-s_2)(s_1-s_3)(s_2-s_3)$
Tetrahedron	T_d	(1^4)	6	$(s_1-s_2)(s_1-s_3)(s_1-s_4)(s_2-s_3)(s_2-s_4)(s_3-s_4)$
Disphenoid (allene)	D_{2d}	(2^2)	2	$(s_1-s_3)(s_2-s_4)$
Disphenoid (allene)	D_{2d} (T_d)	(1^4)	6	$(t_1-t_2)(t_1-t_3)(t_1-t_4)(t_2-t_3)(t_2-t_4)(t_3-t_4)$
Polarized square	C_{4v}	(21^2)	3	$(s_1-s_3)(s_2-s_4)[(s_1-s_2)+(s_3-s_4)]$
Polarized rectangle	C_{2v}	(31)	1	$(s_1-s_2)+(s_3-s_4)$
Polarized rectangle	C_{2v} (C_{4v})	(21^2)	3	$(t_1-t_3)(t_2-t_4)[(t_1-t_2)+(t_3-t_4)]$
Polarized pentagon	C_{5v}	(31^2)	3	$(s_1-s_2)^3+(s_2-s_3)^3+(s_3-s_4)^3+(s_4-s_5)^3+(s_5-s_1)^3$
Polarized pentagon	C_{5v}	(31^2)	3	$(s_1'-s_2')^3+(s_2'-s_3')^3+(s_3'-s_4')^3+(s_4'-s_5')^3+(s_5'-s_1')^3$
Octahedron[b]	O_h	(2^3)	6	$f(s_n)[g(s_n)+h(s_n)]$
Octahedron[b]	O_h	(31^3)	6	$f(s_n')[g(s_n')+h(s_n')]$
Trigonal prism	D_{3h}	(42)	2	$(s_1-s_2)(s_4-s_6)-(s_1-s_3)(s_4-s_5)$
Trigonal prism	D_{3h} (T9')[c]	(41^2)	3	$(s_1'-s_2')(s_1'-s_3')(s_2'-s_3')-(s_4'-s_5')(s_4'-s_6')(s_5'-s_6')$
Trigonal prism	D_{3h} (none)[d]	(321)	4	$(u_1u_6-u_3u_4)(u_4-u_3)(u_1-u_6)+(u_1u_5-u_2u_4)(u_2-u_4)(u_1-u_5)+(u_2u_6-u_3u_5)(u_3-u_5)(u_2-u_6)$
Trigonal prism[b]	D_{3h} (O_h)	(2^3)	6	$f(t_n)[g(t_n)+h(t_n)]$
Trigonal prism[b]	D_{3h} (O_h)	(31^3)	6	$f(t_n')[g(t_n')-H(t_n')]$
Trigonal prism[b]	D_{3h} (T9')[c]	$(31^3)'$	6	$f(s_m'')[g(s_m'')-h(s_m'')]$[e]
Polarized hexagon[f]	C_{6v}			

(a) The group in parenthesis is a permutation group L larger than G so that $G \subset L$ and the chiral ligand partition is the lowest degree chiral partition for L.

(b) These chirality polynomials use the generation functions $f(x_n)=(x_1-x_6)(x_2-x_4)(x_3-x_5)$; $g(x_n)=(x_1-x_2)(x_2-x_3)(x_3-x_1)+(x_3-x_6)(x_6-x_4)(x_4-x_3)+(x_1-x_4)(x_4-x_5)(x_5-x_1)+(x_2-x_5)(x_5-x_6)(x_6-x_2)$ and $h(x_n)=(x_1-x_2+x_6-x_4)(x_1-x_5+x_6-x_3)(x_2-x_5+x_4-x_3)$ corresponding to the three perpendicular separating planes (σ_n), four distinct three-fold axes (C_3), and three perpendicular four-fold axes (C_4) of a regular octahedron having O_h point group symmetry.

(c) The T9' permutation group is of order 36 and is generated by adding a three-fold twist operation to D_{3h} (see reference 48 and G. Butler and J. McKay, *Commun. Algebra*, **11**, 863 (1983)).

(d) There is no permutation group L on six objects of which D_{3h} is a subgroup and (321) is the lowest degree chiral ligand partition. Therefore the corresponding chirality function is not a function solely of differences between the ligand paramaters u_n.

(e) The indices m are obtained from the indices n by interchanging the indices 3 and 6.

(f) The chiral ligand partition of the polarized hexagon are $(2^21^2)+(31^3)+(321)+2(41^2)+(3^2)$ of degrees 7, 6, 4, 3, and 3, respectively. For brevity their chirality polynomials are omitted from this table.

transitive skeletons. Thus an intransitive skeleton having q orbits can be decomposed into q transitive skeletons such that

$$\sum_{k=0}^{k=q} n_k = n \tag{18}$$

in which n_k is the number of sites in orbit k. The transitive skeletons with six or less sites are depicted in Figure 2. Their chiral ligand partitions and most of their chirality polynomials are listed in Table 2. Other transitive skeletons for which the lowest degree chirality polynomials have been determined include the C_{7v} polarized heptagon [23], the O_h cube [6,23,47], the D_{4h} planar cyclobutane skeleton [6], the D_{5h} pentagonal prism (eclipsed ferrocene) [6], the D_{5d} pentagonal antiprism (staggered ferrocene) [6], the I_h icosahedron (e.g., $B_{12}H_{12}^{2-}$) [47], and the D_{6h} hexagonal prism (eclipsed dibenzenechromium) [6].

Keller, Langer, Lehner, and Derflinger [54] have pointed out an important limitation of qualitatively complete chirality polynomials for the estimation of pseudoscalar measurements. Consider the allene skeleton discussed above. The qualitatively complete chirality polynomial $X(D_{2d})$ for the allene skeleton is the weighted sum of the degree 2 polynomial $X(2^2)(s_1,s_2,s_3,s_4)$ in equation 16 and the degree 6 polynomial $X(1^4)(t_1,t_2,t_3,t_4)$ in equation 14b, i.e.

$$X(D_{2d}) = aX(2^2) + bX(1^4) \tag{19}$$

However, application of the algorithm outlined in equations 12-16 to the (2^2) ligand partition using not the lowest degree monomial $M(2^2)=s_1s_2$ (equation 12a) leading to a non-vanishing chirality polynomial but instead a higher degree monomial such as the degree 3 monomials $s_1^2s_2$ or $s_1s_2^2$, the degree 4 monomials $s_1^3s_2$, $s_1^2s_2^2$, or $s_1s_2^3$, etc., can lead to chirality polynomials of degrees between 2 and 6 having the appropriate transformation properties. Thus the simple equation 19 having only terms of degrees 2 and 6 omits non-vanishing terms of intermediate degree in the relevant Taylor series for estimating the pseudoscalar property of interest for the skeleton in question. For this reason use of chirality polynomials for estimating pseudoscalar properties is questionable for skeletons having more than one chiral ligand partition except for ligand partitions for which all of the chirality polynomials except for those of the lowest degree chiral ligand partition must vanish identically, namely the ligand partitions of maximum symmetry to retain chirality such as the degree 2 (2^2) ligand partition in allenes [27], the degree 1 (31) ligand partition in the polarized rectangle skeleton (e.g., the [2,2]-metacyclophanes [30,31,32]) and the (42) ligand partition for cyclopropanes (i.e., the trans-disubstituted cyclopropanes). Fortunately, this limitation on the application of chirality polynomials does not exclude many of the chiral molecules of greatest chemical interest and accessibility.

The estimation of a chirality function by a polynomial of lowest degree is regarded as a "first approximation" ("erste Näherungsansatz") [6,7]. A chirality function may also be estimated by a function of unspecified type with the correct transformation properties using a linear sum of functions of the minimum number of ligand parameters. This procedure has been called the second approximation ("zweite Näherungsansatz"). The algorithm outlined in equations 8-10 can be used for the second approximation if the monomial $M(s_1,s_2,...,s_n)$ in equation 8 is replaced by a function of unspecified type $\emptyset(s_k)$ using the minimum number of different ligand parameters. The minimum required number of ligand parameters, h, for $\emptyset(s_k)$ for a given chiral ligand partition corresponds to the number of ligand parameters in the monomial $M(s_1,s_2,...,s_n)$ defined by equation 8, namely $h=n-o$, where n is the number of ligands and o is the order or length of the top row. The chiral ligand partitions leading to chirality polynomials of lowest degree and to those composed of sums of functions $\emptyset(s_k)$ of the minimum number of ligand parameters are not necessarily the same for all skeletons. Thus for the octahedron the two chiral ligand partitions (31^3) and (2^3) both lead to chirality polynomials of degree 6 by the first approximation. However, by the second approximation the chirality function from the (31^3) ligand partition can be constructed from functions $\emptyset(s_k)$ of three ligand parameters whereas the chirality function from the (2^3) ligand partition requires functions $\emptyset(s_k)$ of four ligand parameters.

Since the second approximation for chirality functions does not specify the form of the chirality function but only its ligand parameter dependence, it cannot be used for the direct calculation of pseudoscalar properties from empirically determined ligand parameters in contrast to the first approximation of chirality functions by chirality polynomials. However, the second approximation can be used to define sets of chiral molecules constructed from a given skeleton for which the sum of any of their pseudoscalar properties must vanish identically if the approximations implied by chirality algebra are valid. The second approximation minimizes the number of molecules that must be studied for experimental tests of chirality algebra for a given skeleton and has been applied successfully for the study of optical rotations of phosphines and phosphine oxides, which have the C_{3v} polarized triangle skeleton [24], as well as methane derivatives, which have the T_d tetrahedron skeleton [25]. In addition, the number of ligand parameters, namely h=n-o, in the functions $\phi(s_k)$ used for the second approximation for chirality functions has an important physical interpretation since interactions between at least h ligands must be considered in order to obtain a non-vanishing chirality function.

The procedure for applying the second approximation to chirality functions can be illustrated for the allene skeleton. Among its two chiral ligand partitions, namely (2^2) and (1^4), the (2^2) chiral ligand partition has the most equivalent ligands, namely 2 (the length of the first row of its Young diagram), so that o=2 and h=n-o=4-2=2. Using the same Young diagram as used above for determining the corresponding chirality polynomial (first approximation) leads to the function $\phi(s_1,s_2)$ corresponding to the monomial in equation 12a to which the operators in equations 9 and 10 must be applied. Application of the Young operator (equations 9, 13aa, 13ab) to $\phi(s_1,s_2)$ gives

$$Y[\phi(s_1,s_2)] \tag{20}$$
$$=4\phi(s_1,s_2)+4\phi(s_3,s_4)-2\phi(s_2,s_3)-2\phi(s_1,s_4)-2\phi(s_1,s_3)-2\phi(s_2,s_4)$$

which is clearly closely related to equation 14a. Subsequent application of the projection operator from the D_{2d} skeletal point group gives for the chirality function $f(2^2)$ after suppression of the constant factor of 12 the following sum of four $g(s_k)$ functions:

$$f(2^2)=g(s_1,s_2)+g(s_3,s_4)-g(s_2,s_3)-g(s_4,s_1) \tag{21}$$

in which $g(s_i,s_j)$ is a fully symmetric function so that

$$g(s_i,s_j)=g(s_j,s_i) \tag{22}$$

by making

$$g(s_i,s_j)=\phi(s_i,s_j)+\phi(s_j,s_i) \tag{23}$$

Note that the monomial $M(2^2)=s_1s_2$ (equation 12a) used to determine the chirality polynomial $X(2^2)$ in equation 16 is a special case of $g(s_1,s_2)$ since $s_1s_2=s_2s_1$. Thus the determination of chirality polynomials by the "first approximation" algorithm discussed above may be considered to be a special case of the "second approximation" algorithm.

Chirality functions obtained by the second approximation relevant to the discussion in the next section of this chapter on experimental tests of chirality functions are listed in Table 3.

Table 3

CHIRALITY FUNCTIONS AS LINEAR SUMS OF FUNCTIONS OF THE MINIMUM NUMBER OF LIGAND PARAMETERS FOR THE SIMPLEST TRANSITIVE SKELETONS AS DETERMINED BY THE "SECOND APPROXIMATION" PROCEDURE

Skeleton	Chiral Ligand Partition	Ligand Parameters per Term	Chirality Function: $f(s_1,\ldots,s_n)=$
Polarized triangle	(1^3)	2	$g(s_1,s_2)+g(s_2,s_3)+g(s_3,s_1)$; $g(s_i,s_j)=-g(s_j,s_i)$
Tetrahedron	(1^4)	3	$g(s_1,s_2,s_3)+g(s_1,s_3,s_4)-g(s_1,s_2,s_4)-g(s_2,s_3,s_4)$; $g(s_i,s_j,s_k)=g(s_j,s_k,s_i)=g(s_k,s_i,s_j)=-g(s_i,s_k,s_j)=-g(s_k,s_j,s_i)=-g(s_j,s_i,s_k)$
Disphenoid (allene)	(2^2)	2	$g(s_1,s_2)+g(s_3,s_4)-g(s_2,s_3)-g(s_4,s_1)$; $g(s_i,s_j)=g(s_j,s_i)$
Polarized square	(21^2)	2	$g(s_1,s_2)+g(s_2,s_3)+g(s_3,s_4)+g(s_4,s_1)$; $g(s_i,s_j)=-g(s_j,s_i)$
Polarized rectangle	(31)	1	$g(s_1)-g(s_2)+g(s_3)-g(s_4)$
Polarized pentagon	(31^2)	2	$g(s_1,s_2)+g(s_2,s_3)+g(s_3,s_4)+g(s_4,s_5)+g(s_5,s_1)$; $g(s_i,s_j)=-g(s_j,s_i)$
Octahedron	(31^3)	3	$g(s_1,s_2,s_3)+g(s_1,s_3,s_4)+g(s_1,s_4,s_5)+g(s_1,s_5,s_2)-g(s_6,s_2,s_3)-g(s_6,s_3,s_4)-g(s_6,s_4,s_5)-g(s_6,s_5,s_2)$; $g(s_i,s_j,s_k)=g(s_j,s_k,s_i)=g(s_k,s_i,s_j)=-g(s_j,s_i,s_k)=-g(s_k,s_j,s_i)=-g(s_i,s_k,s_j)$

5. Experimental Tests of Chirality Algebra

Two general methods are available for experimental tests of chirality algebra based on chirality functions determined by the two approximation procedures noted above:

(1) First Approximation (Chirality Polynomials): Experimental pseudoscalar measurements of a given type (e.g., optical rotation at a specific wavelength or circular dichroism) on molecules with a given skeleton and a limited set of different ligands are used to determine parameters for each ligand in the set to use in chirality polynomials of the types determined by the first approximation procedure. The self-consistency of the ligand parameters found by this method is a measure of the success of this approach.

(2) Second Approximation: Pseudoscalar measurements of a given type are made for a set of molecules with a given skeleton and with distributions of ligands chosen from a set not much larger than the number of sites so that the sum of the chirality functions determined by the second approximation is required mathematically to be identical to zero. The deviation from zero of the sum of the experimentally determined pseudoscalar properties for this set of molecules is then a measure of the errors arising from this approach.

The first approximation requires a much larger set of data on appropriate sets of molecules than the second approximation for a given skeleton in order to check the accuracy of the approach through determination of a sufficiently large set of self-consistent ligand parameters. However, the first approximation leads to chirality polynomials and sets of ligand parameters which can be used to calculate the same pseudoscalar property for unknown molecules with the same skeleton and with ligands chosen from the set for which parameters have been obtained. The second approximation, unlike the first approximation, does not assume a polynomial form for the chirality function and therefore may work in some cases where the first approximation fails.

Chirality functions determined by the methods of chirality algebra have been tested for a number of the simpler skeletons as discussed below. Since most of the sets of molecules studied by chirality algebra are hydrocarbon derivatives with many hydrogen atoms remaining as "ligands" at the skeletal sites, the ligand parameters for chirality polynomials as determined by the first approximation are frequently chosen so that the ligand parameter for hydrogen is zero.

(1) Polarized Triangle: Richter [24] has studied the molar rotations of optically active phosphines and phosphine oxides, which have the C_{3v} polarized triangle skeleton. Distribution of four different ligands (designated as a, b, c, and d) among the three skeletal sites leads to a quadruple of the following type (X=O or lone pair):

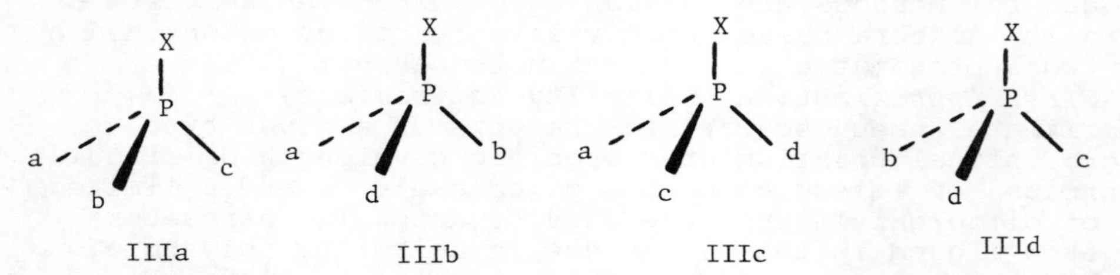

The sum of the molar rotations of the four members of such a quadruple is predicted to be identically zero for a chirality function either of polynomial type obtained by the first approximation (Table 2) or of more general type obtained by the second approximation (Table 3). Of the 7 quadruples of type III above obtained from the 7 optically active phosphines and the 16 optically active phosphine oxides, 5 had zero sums of molar rotations within a 90 % confidence limit whereas the sums of the molar rotations of the 2 remaining quadruples deviated significantly from zero. A satisfactory set of ligand parameters for the degree 3 chirality polynomial from the first approximation (table 2) could not be obtained from the available data.

(2) Tetrahedron: The T_d tetrahedron having a single chiral ligand partition (1^4) can be treated analogously to the C_{3v} polarized triangle with a single chiral ligand partition (1^3). In the case of methane derivatives based on the tetrahedron skeleton, the distribution of 5 different ligands, a, b, c, d, and e among the 4 skeletal sites leads to a quintuple of the following type:

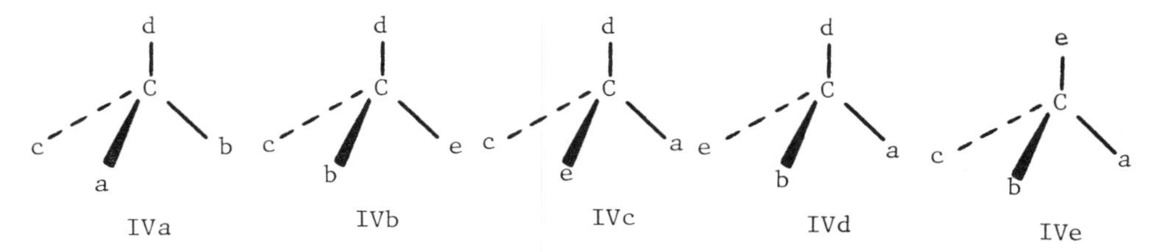

The sum of the values for a given pseudoscalar property of the five members of such a quintuple is predicted to be identically zero for a chirality function obtained by the second approximation (Table 3), i.e.:

$$f(T_d)(a,b,c,d)+f(T_d)(b,e,c,d)+f(T_d)(e,a,c,d)$$
$$+f(T_d)(b,a,e,d)+f(T_d)(b,a,c,e)\equiv 0$$

Studies on the optical rotations of 13 such quintuples obtained from 54 methane derivatives of known absolute configuration [25] indicate approximate agreement with the zero identity of equation 24 in most cases with significant deviations arising only in the cases of hydrogen bonding ligand pairs (e.g., NH_2/CH_2OH, $NHCHO/CH_2OH$, NH_2/CO_2H), bulky ligands (e.g. $CH_2OSi(CH_3)_3$), and derivatives containing the ligand $N{=}CHC_6H_5$ having significant absorption in the Na-D visible region where the optical activity measurements were performed. Meaningful ligand parameters for the degree 6 polynomial approximation (Table 2) were obtained for the ligands H, CH_3, C_2H_5, CH_2OH, $NH_3^+Cl^-$, $NH_2CH_3^+Cl^-$, $N(CH_3)_2$, $NH(CH_3)_2^+Cl^-$, and $NHCH_3$ thereby excluding consideration of strong hydrogen bonding ligands, bulky ligands, and light absorbing ligands.

(3) Disphenoid: The four-site D_{2d} disphenoid skeleton is found in allene derivatives and 2,2′-spirobiindane derivatives. This skeleton has two chiral ligand partitions, namely the degree 2 (2^2) partition as well as the degree 6 (1^4) partition corresponding to the single chiral ligand partition of the tetrahedron discussed above. A qualitatively complete chirality function for the disphenoid skeleton thus has two components. However, the component for the (1^4) chiral ligand partition vanishes identically for chiral disphenoids having (21^2) and (2^2) ligand partitions so that in these cases only the component derived from the (2^2) chiral ligand partition needs to be considered. The various chirality functions for the disphenoid (allene) skeleton have been presented in earlier sections of this chapter as illustrations of the methods of chirality algebra.

Ruch, Runge, and Kresze [27] have calculated a set of self-consistent ligand parameters for the H, C_6H_5, CO_2H, CH_3, and C_2H_5 ligands in allenes from optical rotation data on phenylallene carboxylic acid derivatives. Derivatives having two identical ligands, namely those with the (21^2) ligand partition, were used so that the degree 6 polynomial from the (1^4) chiral ligand partition (e.g., equation 14b) was identically zero and thus could be neglected. However, the same parameters could also be used to estimate the optical rotation in allenes having four different substituents (i.e., the (1^4) ligand partition) indicating that the degree 6 chirality polynomial from the (1^4) chiral ligand partition was negligible relative to the degree 2 chirality polynomial from the (2^2) chiral ligand partition. Thus in equation 19 for the qualitatively complete chirality polynomial of the chiral allenes, the coefficient b for $X(1^4)$ is found experimentally to be small relative to the coefficient a for $X(2^2)$. Difficulties in obtaining extensive series of chiral allene derivatives of high optical purity limited the scope of this study.

Neudeck, Richter, and Schlögl [28] have made an extensive study of the molar rotations of approximately 100 derivatives of the 2,2′-spriobiindane skeleton

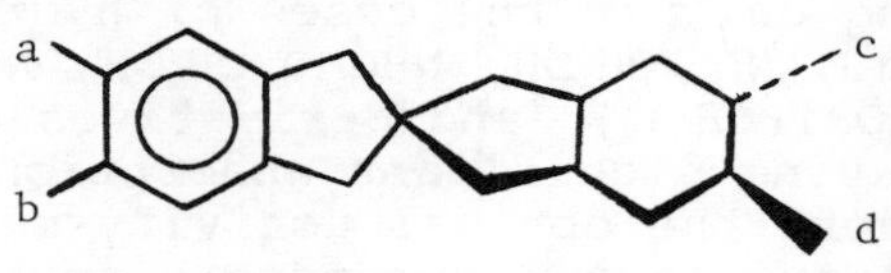

V

which like allene has the four sites and D_{2d} symmetry of the disphenoid so that the same chirality functions can be used. However, extensive series of chiral 2,2′-spirobiindanes of known optical purity are more readily accessible than such series of the chiral allenes discussed above. Molar rotation data on 5,5′-disubstituted 2,2′-spirobiindanes (V) having the (21^2) ligand partition were used to calculate self-consistent ligand parameters for the H, CH_3, C_2H_5, CH_2OH, CHO, CH_3CO, CO_2H, CO_2CH_3, CN, and OCH_3 ligands in the degree 2 $X(2^2)$ chirality polynomial (equation 16); for such derivatives with two remaining hydrogen ligands the degree 6 $X(1^4)$ chirality polynomial (equation 14b) vanishes identically and thus can be neglected. The magnitude of the degree 6 $X(1^4)$ chirality polynomial, which does not vanish for 5,5′,6-trisubstituted 2,2′- spirobiindanes having the (1^4) ligand partition (one ligand being hydrogen), can be estimated from the sums of the molar rotations of isomer triples schematically represented as

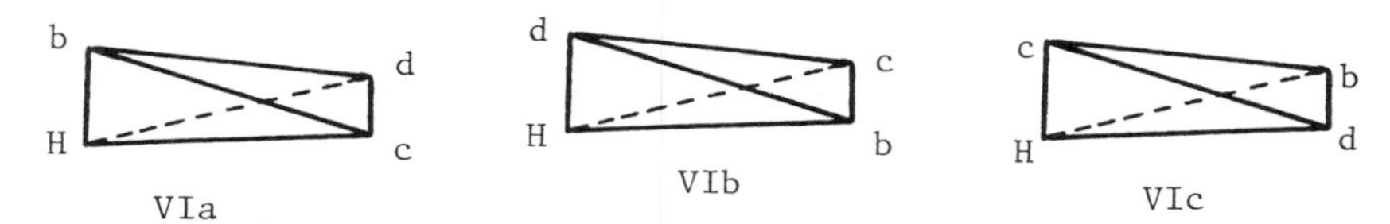

for which the degree 2 $X(2^2)$ chirality polynomial vanishes identically so that the deviation of this sum from zero represents exclusively the degree 6 $X(1^4)$ chirality polynomial. Available experimental data on such an isomer triple in which b = CH_3, c = C_2H_5, and d = CO_2CH_3 suggests that the degree 6 $X(1^4)$ chirality polynomial is responsible for about 25 % of the molar rotation, i.e., relatively small but far from negligible.

(4) Polarized Rectangle: The C_{2v} polarized rectangle skeleton is theoretically significant since it is the simplest potato-like (category b) skeleton having more than one chiral ligand partition. The C_{2h} [2,2]-metacyclophane skeleton (VII) is

VII

permutationally equivalent to the C_{2v} polarized rectangle so that the same chirality functions can be used. Thus a qualitatively complete chirality polynomial for the C_{2v} polarized rectangle skeleton and applicable to the C_{2h} [2,2]-metacyclophane skeleton has the form

$$X(C_{2v}) = aX(31) + bX(21^2)$$

in which

$$X(31) = s_1 - s_2 - s_3 + s_4 \tag{26a}$$

and

$$X(21^2) = (t_1 - t_4)(t_2 - t_3)(t_1 - t_2 - t_3 + t_4) \tag{26b}$$

where sites 1, 2, 3, 4 are labelled a, b, c, d, respectively, in structure VII. For 4-monosubstituted derivatives (a≠H) and for 4,14-homodisubstituted derivatives (a=b≠H) the degree 3 component (equation 26b) of the qualitatively complete chirality polynomial (equation 25) vanishes identically so that the resulting qualitatively complete chirality polynomial is simply the linear "quadrant rule" represented by equation 26a. The linear quadrant rule predicts that the magnitude of any pseudoscalar measurement of a 4,14-homodisubstituted [2,2]-metacyclophane derivative will be twice that of the corresponding 4-monosubstituted derivative with the same substituent.

Keller, Krieger, Langer, Lehner, and Derflinger have made detailed studies of the molar rotations [31] and circular dichroism [32] of 30 monosubstituted and disubstituted [2,2]-metacyclophane derivatives. The double values for the molar rotations and circular dichroism values of the 4,14-homodisubstituted derivatives relative to the corresponding monosubstituted derivatives predicted by equation 26a are found for only a few of the cases investigated indicating the insufficiency of this simple linear approximation. However, a more detailed analysis of possible chirality polynomials for the (2^2) chiral ligand partition of the polarized rectangle skeleton indicates the possibility of degree 2 and degree 3 chirality polynomials as higher terms in a Taylor type approximation arising from application of the relevant projection operators to monomials of the types s_1s_2, s_1^2, $s_1s_2^2$, etc. This could account for some of the observed discrepancies. An attempt was also made to evaluate the significance of the degree 3 polynomial for $X(21^2)$ in equation 26b by study of the optical rotations of the 4,12-heterodisubstituted derivatives having the (21^2) ligand partition in which a=c=H, d=Br, and B=CO_2H, CO_2^-, CO_2CH_3, $CONH_2$, CN in structure VII. Major discrepancies in the calculated ligand parameter for bromine for the molar rotations of these five compounds using only equation 26a for X(31) indicate the significance of the degree 3 equation 26b for $X(2^2)$ for determining the molar rotations of these derivatives for which $X(2^2)$ is not required by theory to vanish identically. However, the available range of heterodisubstituted derivatives having the (21^2) chiral ligand partition was not sufficient for the calculation of a meaningful set of t_k parameters for $X(21^2)$ in equation 26b.

In any case there appear to be some major difficulties in the application of the methods of chirality algebra for the [2,2]-metacyclophane derivatives having the polarized rectangle skeleton.

(5) Polarized Pentagon: The polarized pentagon skeleton can be used to study chiral heterodisubstituted ferrocene derivatives of the following types:

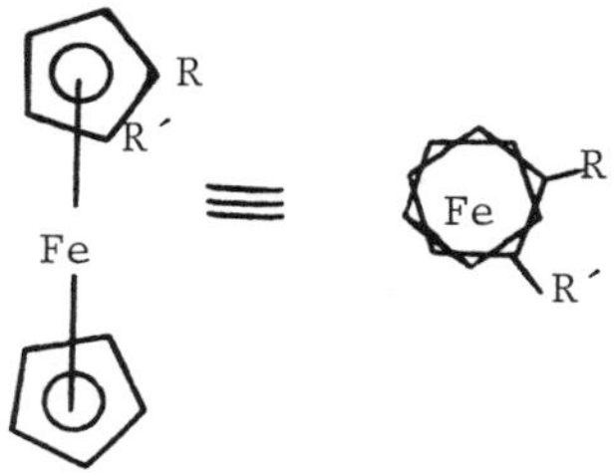

VIII$_{12}$ 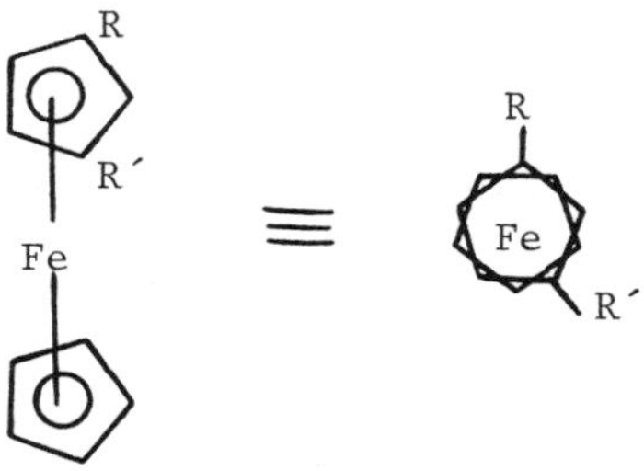 VIII$_{13}$

In this case, where three of the five skeletal sites are hydrogen atoms assigned the ligand parameter of zero as a reference point, the degree 3 chirality polynomial for the polarized pentagon X(31^2) in Table 2 reduces to

$$X(31^2) = s_a s_b^2 - s_a^2 s_b$$

in which a=1, b=2 for the 1,2-heterodisubstituted derivatives VIII$_{12}$ and a=1, b=3 for the 1,3-heterodisubstituted derivatives VIII$_{13}$. However, attempted application of equation 27 to 17 heterodisubstituted ferrocene derivatives by Rapić, Schlögl, and Steinitz [35] failed to give sets of ligand parameters that could reproduce the observed molar rotation in equation 27 without major discrepancies. However, in this case the 589 nm wavelength of the light used for the optical rotation measurements may be too close to the positions of the longest wavelength maxima in the 540-440 nm range in the electronic spectra so that the optical rotation measurements no longer fall in the transparent region required for the semiempirical equations of chirality algebra to be valid.

6. Summary

Chirality algebra applies methods of permutation group theory and group representation theory to the derivation of functions having suitable transformation properties for describing pseudoscalar measurements such as optical rotation and circular dichroism. Application of chirality algebra requires that the molecules in question be dissected into a collection of ligands attached to an underlying skeleton.

The independent variables of the so-called chirality functions are empirically determined parameters depending upon the ligand, skeleton, and pseudoscalar measurement but not on the skeletal site occupied by the ligand in question. The most useful chirality functions are the chirality polynomials of lowest degree.

Chirality functions determined by chirality algebra have been tested experimentally for the simpler skeletons for which a sufficiently wide range of chiral molecules of known optical purity and limited ligand variety are available. The success of such approximations depends greatly upon the complexity of the skeleton, particularly with regard to the numbers of sites and chiral ligand partitions. Thus chirality functions provide fair to good approximations of optical rotation data for chiral derivatives of simple shoe-like skeletons such as the polarized triangle skeletons of phosphines and phosphine oxides, the tetrahedral skeleton of methane derivatives, and the disphenoid skeletons of allene and 2,2'-spirobiindane derivatives. However, the approximations provided by chirality algebra deteriorate rapidly when attempts are made to apply them to chiral derivatives of more complicated skeletons having several chiral ligand partitions or to potato-like skeletons such as the polarized rectangle skeleton of [2,2]-metacyclophanes. Fortunately, the simple skeletons for which chirality functions provide the best approximations are those of greatest chemical significance.

Literature References

(1) Crum Brown, *Proc. Roy. Soc. Edin.*, **17**, 181 (1890).
(2) P.-A. Guye, *Compt. Rend.*, **110**, 714 (1890).
(3) E. Ruch and I. Ugi, *Theor. Chim. Acta*, **4**, 287 (1966).
(4) E. Ruch and I. Ugi, *Topics Stereochem.*, **4**, 99 (1969).
(5) E. Ruch, A. Schönhofer, and I. Ugi, *Theor. Chim. Acta.* **7**, 420 (1967).
(6) E. Ruch and A. Schönhofer, *Theor. Chim. Acta*, **10**, 91 (1968).
(7) E. Ruch and A. Schönhofer, *Theor. Chim. Acta*, **19**, 225 (1970).
(8) E. Ruch and A. Schönhofer, *Theor. Chim. Acta*, **3**, 291 (1965).
(9) E. Ruch, *Accts. Chem. Res.*, **5**, 49 (1972).
(10) C.A. Mead, *Top. Curr. Chem.*, **49**, 1, (1974).
(11) E. Ruch, *Angew. Chem. Int. Ed.*, **16**, 65 (1977).
(12) E. Ruch, *Theor. Chim. Acta*, **11**, 183 (1968).
(13) G. Derflinger and H. Keller, *Theor. Chim. Acta.*, **56**, 1 (1980).
(14) J. Dugundji, D. Marquarding, and I. Ugi, *Chem. Scripta*, **9**, 74 (1976).
(15) W. Hässelbarth, *Chem. Scripta*, **10**, 97 (1976).
(16) C.A. Mead, *Chem. Scripta*, **10**, 101 (1976).
(17) J. Dugundji, D. Marquarding, and I. Ugi, *Chem. Scripta*, **11**, 17 (1977).
(18) C.A. Mead, *Chem. Scripta*, **11**, 145 (1977).
(19) W. Hässelbarth, *Chem. Scripta*, **11**, 148 (1977).
(20) D. Meinköhn, *Theor. Chim. Acta*, **47**, 67(1978).
(21) D. Meinköhn, *J. Chem. Phys.*, **72**, 1968 (1980).
(22) R.B. King, *Theor. Chim. Acta*, **63**, 103 (1983).
(23) R.B. King, *J. Math. Chem.*, **2**, 89 (1988).
(24) W.J. Richter, *Theor. Chim. Acta*, **58**, 9 (1980).
(25) W.J. Richter, B. Richter, and E. Ruch, *Angew. Chem. Int. Ed.*, **12**, 30 (1973).
(26) W.J. Richter, H. Heggemeier, H.J. Krabbe, E.H. Korte, and B. Schrader, *Ber. Bunsenges. Phys. Chem.*, **84**, 200 (1980).
(27) E. Ruch, W. Runge, and G. Kresze, *Angew. Chem. Int. Ed.*, **12**, 20 (1973).
(28) H. Neudeck, B. Richter, and K. Schlögl, *Monatshefte für Chemie*, **110**, 931 (1979).
(29) E.H. Korte, P. Chingduang, and W.J. Richter, *Ber. Bunsenges. Phys. Chem.*, **84**, 45 (1980).
(30) H. Keller, C. Krieger, E. Langer, H. Lehner, and G. Derflinger, *J. Mol. Struct.*, **40**, 279 (1977).
(31) H. Keller, C. Krieger, E. Langer, H. Lehner, and G. Derflinger, *Liebigs Ann. Chem.*, **1296** (1977).
(32) H. Keller, C. Krieger, E. Langer, H. Lehner, and G. Derflinger, *Tetrahedron*, **34**, 871 (1978).

(33) H. Gowal and K. Schlögl, *Monatshefte für Chemie*, **99**, 972 (1968).
(34) W.J. Richter and B. Richter, *Isr. J. Chem.*, **15**, 57 (1976).
(35) V. Rapić, K. Schlögl, and B. Steinitz, *Monatshefte für Chemie*, **108**, 767 (1977).
(36) C.A. Mead, E. Ruch, and A. Schönhofer, *Theor. Chim. Acta*, **29**, 269 (1973).
(37) J.A. Pople, *J. Am. Chem. Soc.*, **102**, 4615 (1980).
(38) F.A. Cotton, *Chemical Applications of Group Theory*, Wiley-Interscience, New York, 1971.
(39) F. Budden, *The Fascination of Groups*, Cambridge University Press, London, 1972, Chapters 10 and 18.
(40) G. Pólya, *Acta Math.*, **68**, 145 (1937).
(41) N.G. de Bruin in *Applied Combinatorial Mathematics*, ed. E.F. Beckenbach, Wiley, New York, 1964, chapter 5.
(42) G. Pólya and R.C. Read, *Combinatorial Enumeration of Groups, Graphs, and Chemical Compounds*, Springer-Verlag, New York, 1987.
(43) M. Hamermesh, *Group Theory and its Application to Physical Problems*, Addison-Wesley, Reading, Massachusetts, 1962, Chapter 2.
(44) C.D.H. Chisholm, *Group Theoretical Techniques in Quantum Chemistry*, Academic Press, New York, 1976, Chapter 6.
(45) D. Gorenstein, *Finite Groups*, Harper and Row, New York, 1968, Chapter 4.
(46) S.L. Altmann, *Induced Representations in Crystals and Molecules*, Academic Press, London, 1977.
(47) R.B. King, *J. Math. Chem.*, **1**, 45 (1987).
(48) R.B. King, *J. Math. Chem.*, **1**, 15 (1987).
(49) F.D. Murnaghan, *The Theory of Group Representations*, Johns Hopkins, Baltimore, Maryland, 1938, Chapter 5.
(50) D.E. Littlewood and A.R. Richardson, *Phil. Trans. Roy. Soc, (London), Ser. A*, **A233**, 99 (1934).
(51) M. Zia-ud-Din, *Proc. London Math. Soc.*, **42**, 340 (1936).
(52) X. Liu and K. Balasubramanian, *J. Comput. Chem.*, **10**, 417 (1989).
(53) O.E. Glenn, *A Treatise on the Theory of Invariants*, Ginn and Company, Boston, 1915.
(54) H. Keller, E. Langer, H. Lehner, and G. Derflinger, *Theor. Chim. Acta*, **49**, 93 (1978).

Algebraic Topological Indices of Molecular Chirality

Kenneth C. Millett
Department of Mathematics
University of California, Santa Barbara
Santa Barbara, CA 93106, USA

1. INTRODUCTION

In this chapter we shall focus attention upon those aspects of the mathematical analysis of molecular graphs which relate to the geometric and topological properties of their spatial symmetries and of their knotting and linking. Interconnections with the theory of knots and links (which can also serve as models for small molecules) will also be presented. In this first section we shall briefly review the fundamental concepts which delineate the interrelationships between the molecular structures, the geometric structures, and the topological structures. In the second section we shall present several of the significant test families of molecular graphs which will be employed in the comparison of the stereotopological indices used to detect chirality. This discussion continues in the third section where some of the elements of the knot theory of molecular graphs are introduced. Chimerical graphs, i.e. graphs with a fixed structure near vertices, their elementary properties, and their structural properties are presented in the fourth section. In the fifth section we present a survey of the algebraic topological methods that have been employed to detect and quantify the chirality of classical knots and links as well as molecular graphs. This discussion concludes with the discussion of the development, by E. Witten, of stereotopological indices arising from interactions with statistical mechanics and gauge theories from mathematical physics.

The starting point is the consideration of graphs as models for molecular structures in which the nodes, or vertices, represent the nuclei of atoms which are connected by the edges of the graph representing the bonds of the molecule. One may differentiate between various nodes as they represent distinct atomic nuclei such as hydrogen, carbon, oxygen, etc. and differentiate between various edges according to the nature of the bond to which it is associated. Commonly, the hydrogen atoms are omitted and only covalent bonds are represented in these molecular graphs. One should also note that such models, being discrete structures

P. G. Mezey (ed.), New Developments in Molecular Chirality, 165–207.

representing these complex objects, do not typically indicate the relative significance of the bonds. Nevertheless such models are found everywhere in chemistry and, for certain uses, have become a part of the methodology and language of the field. The geometric or metric properties, as distinguished by the purely combinatorial or topological properties, associated to these models greatly enhance their utility. All of these various levels of structure have their place in chemistry and give rise to some of the fundamental questions concerning the chemical properties of various molecules.

These concerns have among their origins the 1847 demonstration, by Louis Pasteur, that certain crystals arise in association with two twin molecules which are mirror reflections of each other. Pasteur observes: "le paratratrate et le tartrate (doubles) de soude et d'ammoniaque ont la meme composition chimique, la meme forme cristalline, avec les memes angles, le meme poids specifique .... Mais le tratrate dissous tourne le plan de la lumiere polarisee et le paratratrate est indifferent.... Ici, la nature et le nombre des atomes, leur arangement et leurs distances, sont les memes dans les deux corps composes". How could chemically identical molecules enjoy such dissimilar optical properties? A proposed mirror reflection relationship gave rise to the concept of stereochemistry ...... the study of the 3-dimensional Euclidian symmetries of molecules. These theories were developed in the later researches of Kekule, Couper, and le Bel and van t'Hoff in the study of the carbon atom. From a historical perspective it is interesting to recall that Pasteur developed a method of separation of the left handed form from a racemic composition (one in which both right and left handed forms are present in equal quantities) utilizing a physiological method: "faire vivre des petites de penicillium glaucum, de cette moisissure qu'on trouve partout, a la surface de cendres et d'acide paratartrique, et a voir l'acide de tartrique gauche apparaitre", because, "pour sa nutrition, le petit etre s'accommode mieux de groupe tartrique droit que de groupe tartique gauche". Thus these micro-organisms became the tools of stereochemical analysis thereby pointing the way toward current techniques of molecular biological synthesis and analysis!

The Pasteur example reflects a (Euclidian) geometric distinction as contrasted to the topological differentiations that we wish to discuss. Thus we shall devote the remainder of this introduction to a review of the basic concepts and vocabulary that is employed. Recall that **isomers** are molecules which can be distinguished by virtue of being able to be separated and isolated, but which consist of the same number and type of atoms. Among these are found the **constitutional isomers**, those having distinct molecular graphs;

stereoisomers; isomers having homeomorphic molecular graphs; and **topological stereoisomers**; stereoisomers having molecular graphs whose placements in space can not be continuously deformed, one to the other. Among the possible molecular graphs it is helpful to give special attention to those having the property that each node meets precisely two edges. The spatial realizations of such graphs are called **knots**, if there is only one strand, or **links**, if there are several strands. Figure 2.1 (a) shows a representative of the trivial knot while (b) shows the simplest nontrivial knot, the trefoil knot. In this figure the broken line is used to indicate that one portion of the strand passes behind the other in space, from the perspective of the observer looking at the page.

It is necessary to be aware that the Euclidian geometry of the molecular graph imposed by virtue of the chemical bonds between atoms and associated energy considerations explains much isomerism (stereochemistry). This isomerism does not arise from the topological structure but, rather, from the symmetry properties of portions of the molecular graphs with the Euclidian geometry imposed, c.f. Figure 1.1. E. J. Wasserman, [W4], was the first to propose that there might be a connection between certain incidents of isomerism (the existence of isomers) and the extrinsic topological properties of the associated molecular graphs. **Intrinsic properties** of molecular graphs are those associated to the abstract graph irrespective of its specific spatial realization. **Extrinsic properties** of molecular graphs are those which are associated to some specific spatial realization.

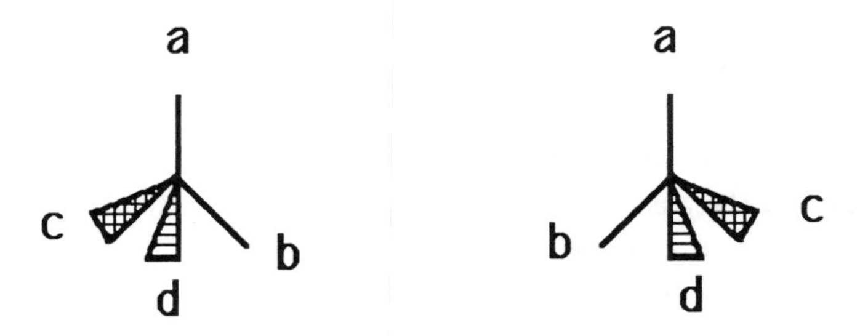

Figure 1.1

While the E and Z-isomers of alkenes shown in the Figure 1.1 have homeomorphic molecular graphs, indeed they are related by reflection in a mirror held perpendicular to the page and containing the "a bond", they are chemically distinct. They are stereoisomers by virtue of the Euclidian geometric structure: one can show that any spatial equivalence taking one to the other must reverse the orientation of space, i.e. they are **geometric diastereomers**. In the event that two isomers have homeomorphic molecular graphs by virtue of a mirror reflection one says that they are **geometrically chiral.** In some cases a molecular graph will be **intrinsically geometrically chiral** inasmuch as it contains a subgraph which always gives rise to isomerization. If, however, a molecular graph and its mirror reflection are not isomers we say that graph is **achiral**. For example, Figure 2.1 (b) and (c) show a trefoil knot and its mirror reflection (one uses the plane of the page as the mirror and simply changes all crossings in the picture) which are known to be topologically distinct and, therefore, gives an example of a chiral pair. Clearly any graph realized in the plane of the page is achiral but there are also non trivial knots which are achiral, c.f. Figure 3.2. The indices of chirality which we shall discuss in the last section of this chapter provide some measure of the phenomena.

The topological stereoisomers are different from these geometric stereoisomers in that they have homeomorphic molecular graphs which are embedded in space in topologically distinct ways, i.e. the spatial realizations are extrinsically distinct. The concept of extrinsic topological distinctness or sameness requires some effort to define precisely. On an informal level one can describe the difference between the geometric situation and the topological situation as follows: In the geometric case the molecular graph is realized in space as a configuration having straight edges of specific length and two realizations of a molecular graph are geometrically the same if one moves one configuration to the other without bending or stretching the edges nor changing the dihedral angles between the various edges at the nodes. If, however, one is allowed to continuously bend and stretch edges and change angles during the course of a deformation of the graph, the resulting relationship is extrinsic topological sameness and mathematicians say that the two placements are **isotopic** or **topologically equivalent**. We shall give a more formal definition of this concept in section 3.

If the spatial realization of a molecular graph is topologically distinct from its mirror reflection one says that it is a **topological diastereomer** or that it is **extrinsically topologically chiral**, i.e. topologically distinct from its mirror reflection. The stronger notion of **intrinsic topological chirality** requires that, in addition to the above, certain specified substructures of the graph be preserved during the deformation. If a spatial realization is topologically equivalent to its mirror reflection one says that it is a **topological enantiomer** and that it is **topologically achiral.**

D. Walba has written a couple of excellent review articles in which some of the basic material discussed in this chapter can be found developed to a greater extent, [W1,W3]. There are, in addition, several other recent review articles containing discussions of chirality and which are primarily devoted to chemistry, [F5,S2,S6,W4].

Although many instances of geometric stereoisomerism arise and their study is important, we shall focus our attention upon the case of topological stereoisomerism and specifically the recognition of occurrences of topological chirality. The principle goal of this chapter is the discussion of methods to determine the chirality or achirality of a given spatial realization of a specific molecular graph. We shall show that this is an extension of the search for criteria for determining the chirality of the classical knots and links, a goal which already has chemical significance.

2. THE MÖBIUS LADDERS, MOLECULAR KNOTS AND OTHER EXAMPLES

One of the goals of a synthetic chemist concerned with topological stereochemistry is the chemical synthesis of topological stereoisomers. This occurs for macromolecules, e.g. large polymers, DNA's etc, but the synthetic chemist is more concerned with the synthesis of small molecules, the smaller the better. One approach to this is by way of the synthesis of molecular knotted rings of atoms having homeomorphic molecular graphs which are embedded in topologically distinct ways. Some examples of this are discussed in the review article of Walba, [W1,W3]. Walba has attempted to carry out a proposal of Wasserman and, independently, van Gulick has attempted to use the synthesizes of three twisted Möbius ladders, shown in Figure 2.1 (3), as a pathway to the synthesis of a trefoil knot, shown in Figure 2.1 (b). In general, one syntheses molecular ladders and attempts to join their ends so as to create molecules with a variety of twists in them, [W2]. If this number of twists is even, one has a molecular cylinder and, if this number is odd, one has a twisted Möbius ladder. If these can be separated and the rungs removed, by some chemical process, various molecular knots and links are then synthesized.

The Walba approach, c.f. [W1], has been by way of the THYME polyethers in which the backbone of the ladder would be composed of polyethyleneoxy chains and the runs would be tetraethers of tetrahydroxymethylethylene (THYME). In addition

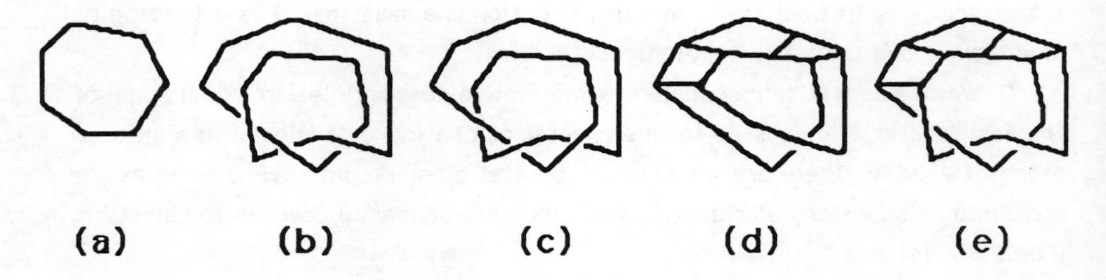

Figure 2.1

to the formidable problems of synthesis, separation, and identification, he was faced with the intermediate problem of determining whether or not the molecular Möbius band was chiral in order to achieve a small topologically chiral molecule. If successful, he would then be in a position to synthesize Möbius ladders with more twists and, ultimately, the molecular trefoil knot. His group has not yet been successful in achieving the synthesis of a molecular trefoil by this specific method but Christiane Dietrich-Buchecker and Jean-Pierre Sauvage, of the University of Strasbourg, have now achieved this first synthesis of a small knotted molecule by another method in which two cuprous ions are used as templates to control the synthesis, [D].

Historically, however, one of the first topological problems in this area of mathematical chemistry was the need to determine the chiral structure of the class of graphs known as Möbius ladders, shown in Figure 2.2. Each Möbius ladder,

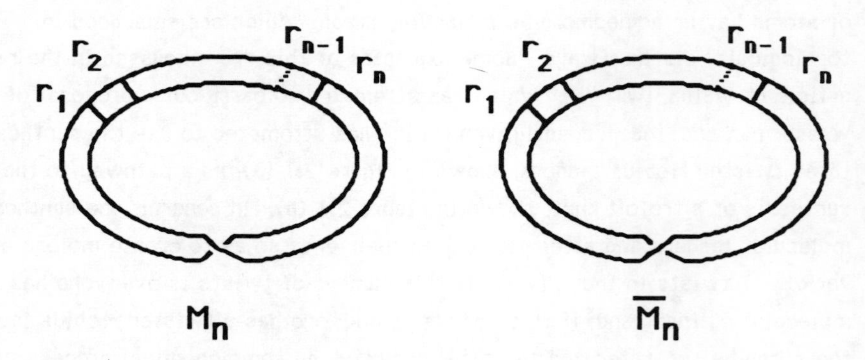

Figure 2.2. The Möbius ladders

M_n, is a regular graph having vertices of valence 3 and consists of 2n vertices with interconnecting edges as shown in the figure. The edges identified as r_i are called rungs of the graph. The complement of the rungs of the Möbius ladders shown in the figure are unknotted simple closed curves.

Examples of both intrinsic and extrinsic topological chirality are illustrated in the recent paper of Simon [S3] where he shows that the Möbius ladders are extrinsically chiral, if $n \geq 4$, and are intrinsically chiral if $n \geq 3$ and the set of "rungs" of the Möbius ladder is required to be preserved. In the Walba synthesis context, the edges of the Möbius ladder are of two distinct types: the rungs represent carbon-carbon double bonding while the others are polyethylenoxy chains. As a consequence, it is not unreasonable from the perspective of a mathematical chemist to require that the set of rungs be preserved under topological equivalences. For the case $n = 3$, Simon has provided an excellent visual way, shown in Figure 2.3, to demonstrate the utility of the assumption of this additional intrinsic structure and the requirement that it be preserved if one is to have a chiral configuration. Without this requirement one simply twists the

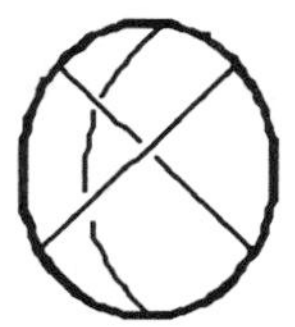

Figure 2.3 "twisting" equivalence ignoring rung identification

configuration to move the dark strand on the left into the configuration of the dark strand on the right side of the figure. The two rung Möbius ladder is achiral even with this requirement. This can be observed by suppressing the third rung in the figure and performing a π rotation of the resulting figure to take the configuration to its mirror image. Simon uses the topology of 2-fold cyclic branched covers of S^3, branched along the boundary curve associated to the Möbius ladder (consisting of edges which are not identified as rungs), and associated linking numbers in order to deduce the chirality of the configurations. These branched covers are again S^3 and the graphs lift to new graphs in a manner such that the vertices and edges correspond spatially, if there is an equivalence of the two original Möbius ladders. One finds in these graphs pairs of oriented circles which must correspond. Calculation of the linking numbers shows that this is not the case and, therefore, the embeddings are not equivalent. Observing that each homeomorphism of the graph M_n leaves the cycle determined by the "boundary" edges invariant, i.e. the notion of boundary edge is well defined, for $n \geq 4$ and induction on the number of rungs completes the argument.

These examples relate specifically to 3-valent graphs, i.e. graphs in which each vertex meets precisely either two or three edges, a subject to which we shall return in our discussion of stereotopological indices of chirality in section 5. The Simmons-Paquette K_5 molecule, shown in Figure 2.4, illustrates the same type of chirality phenomena in more general setting. The molecular graph shown in Figure 2.4, represents a molecule which was synthesized in 1981 by H. Simmons et al and L. A. Paquette et al, independently, [W1]. In this figure the dark circles represent carbon nuclei, the light circles represent oxygen, and, following the standard convention, the hydrogen nuclei are not represented in the graph. Simon, [S3], has

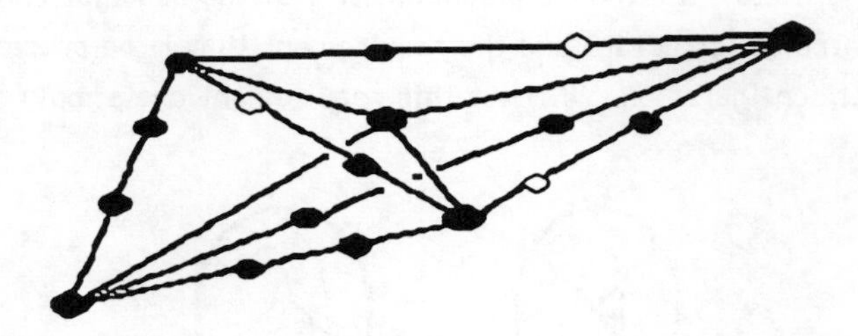

Figure 2.4 Simmons-Paquette K_5 molecule

used a 2-fold cyclic branched cover argument similar to the previous one to show that there does not exist an intrinsic equivalence if one requires that the circular subgraph containing the oxygen nuclei be preserved. This fact is due to the existence of a natural orientation on this subgraph which must be reversed in the mirror image.

3. TOPOLOGICAL CHIRALITY AND KNOTTING

In this section we shall consider the question of knotting and the relationship of knotting to chirality. Suppose that the molecular graph is **abstractly planar**, i.e. it can be realized as a subset of the plane. Examples of such graphs include collections of circles and 'θ' graphs, c.f. Figure 5.19, and are characterized by Kuratowski's Theorem to the effect that a finite graph is abstractly planar if and only if it contains no subgraph homeomorphic to K_5 or $K_{3,3}$. Since any planar realization of a graph is achiral, being fixed under the mirror reflection in the plane containing it, any realization which is topologically equivalent to a planar one is topologically achiral. For knots and links, the embeddings of collections of circles, one says that the knot or link is trivial if it is equivalent to a planar collection. Knot theory and its extension to include graphs is concerned with the development of methods to determine necessary and sufficient conditions for a knot or link to be trivial. One such theorem is due to Papakyriakopolous, [P]: a knot is trivial if and only if its complement has free fundamental group. The theorem asserts that one may associate to the complement, the exterior, of a knot an algebraic object, called the fundamental group, which has the special property of "freeness" exactly when the knot is topologically equivalent to a planar circle. Scharlemann and Thompson, [S1], have demonstrated an unknotting theorem in this spirit for abstractly planar graphs: A finite graph is topologically equivalent to a planar graph if and only if it is abstractly planar, every proper subgraph is topologically equivalent to a planar graph and, the fundamental group of its complement is free. In Figure 3.1 (a) is shown the planar, or trivial, theta-graph

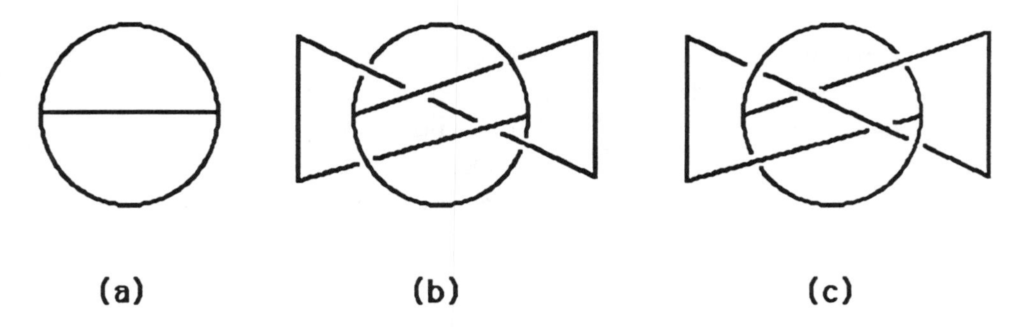

Figure 3.1 Kinoshita examples

and, in (b), Kinoshita's example of an embedding of the theta-graph having the property that every proper subgraph is topologically equivalent to a planar circle but whose complement does not have a free fundamental group and is, therefore, not topologically equivalent to a planar graph, [K5-7]. Wolcott has shown that the Kinoshita example is topologically chiral, i.e. (b) is not topologically equivalent to (c).

Shown in Figure 3.2 are two versions of a nontrivial knot, the "figure 8" knot, which is achiral. In 3.2 (a) one has a simplest representation of the knot in the sense that there does not exist a presentation having fewer crossings and, in (b), a topologically equivalent presentation is provided which demonstrates the achirality: reverse all crossings and rotate by 90 degrees regain the given presentation. If there is a such a presentation of a graph which can be interconverted with its mirror reflection by a rigid motion we say that the embedding is **rigidly achiral**. Flapan, [F1], has show that there are knots which are topologically achiral but which are not rigidly achiral.

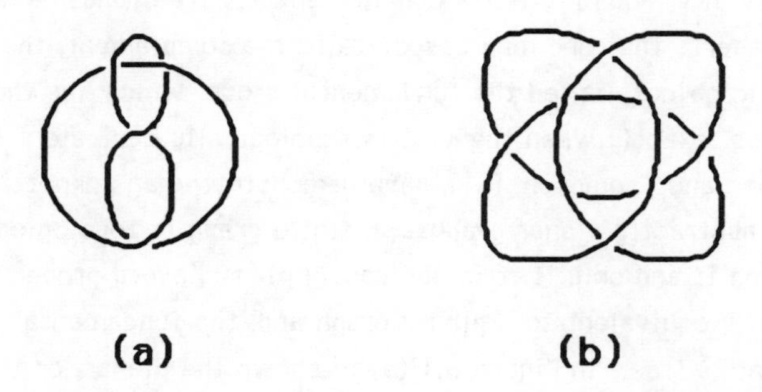

Figure 3.2 The figure 8 knot: two representations

How does one show that a knot or graph in space is achiral in the absence of a presentation which admits a visible rigid motion such as in the previous example or, how does one show that the assertion that the two knots shown in Figure 3.2 actually are topologically equivalent? Topologists have shown that one may always find a perspective with respect to which one has a presentation of the spatial configuration as a shadow of the projection of the graph on the plane in which there are no degenerate shadows. For many purposes one uses the graph model but for others, such as drawing pictures, the rigidity of the graph structure is inconvenient and one, informally, draws representations in which curved lines are employed as if Figure 3.2. Thus, we shall restrict our discussion to the properties of the graph situation although we might, where more appropriate, employ curved lines in the figures. What this graph restriction means is that one can define the concept of a non degenerate projection as one in which the vertices of the graph have images which miss the images of the edges and no three edges have images which have a point in common.

In addition, topologists have shown that if two spatial configurations are topologically equivalent then there must exist a sequence of elementary alterations of associated presentations, the Reidemeister moves and their generalizations as shown in Figures 3.4, 3.5, and 3.6, taking any presentation of the one configuration to the other. Furthermore, if there does exist such a sequence then the two spatial configurations are topologically equivalent.

The search for spatial invariants of position is the search for algebraic quantities which are unchanged by these movements, which are easily calculated from some description of the placement and which are, nevertheless, sufficiently sensitive so as to be able to distinguish between topologically inequivalent placements. We shall return to this question in section 5.

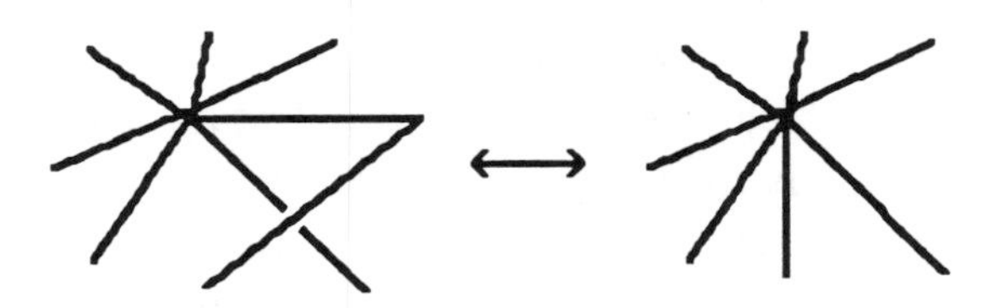

Figure 3.4 Generalized type I Reidemeister move

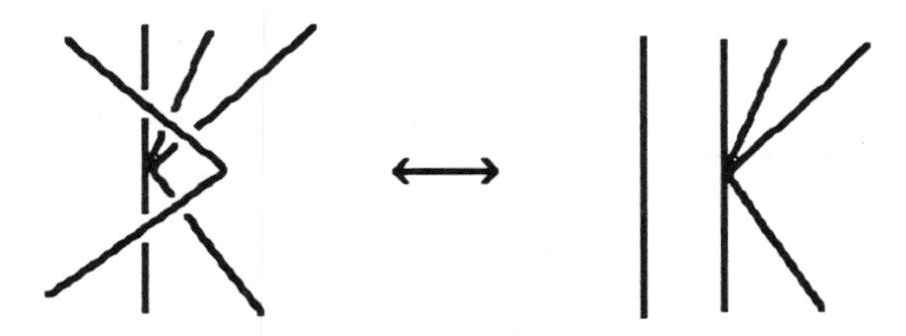

Figure 3.5 Generalized type II Reidemeister move

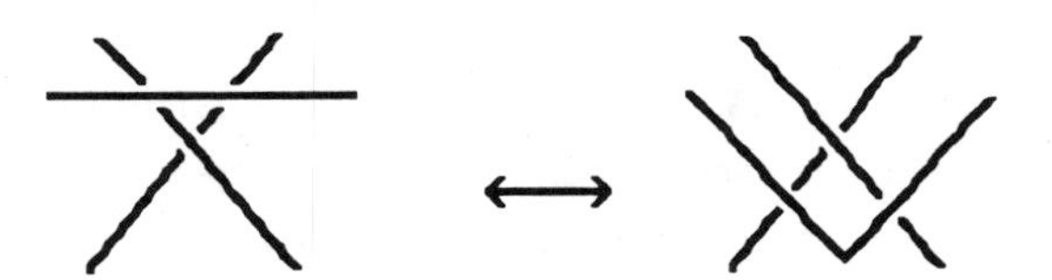

Figure 3.6 Type III Reidemeister move

4. CHIMERICAL GRAPHS

In order to develop the topological methods applicable to certain aspects of molecular graphs we identify several fundamental properties which facilitate the associated mathematical analysis. First, we shall restrict ourselves to 4-valent graphs, i.e. graphs in which each vertex meets exactly two or four edges as in

Figure 2.4. Second, a basic requirement is the provision of some structure to reflect an appropriate rigidity of the local molecular bonding structure in the vicinity of each atom. One method to accomplish this is by means of the notion of a **chimerical graph**, [J4,J5,M2]. The key point of this definition is the choice of a fixed model, or template, for the position of the edges in the vicinity of a specific vertex. Although the specific choice of template is <u>not</u> the key, we shall select those models that exhibit the maximal degree of symmetry possible when the local structure is projected to a plane. The mathematical requirement that the template structure be preserved will then have the effect preserving the distinction between their roles in the the resulting mathematical theory. A **chimerical graph**, Γ, is a 4-valent graph which, near each vertex, is endowed with a standard rigid planar template determining the local relative position of the adjacent edges. Thus, although other choices are possible, in the present research we shall study those 4-valent graphs having a planar template with equal angles between edges at each vertex exactly as shown in Figure 4.1.

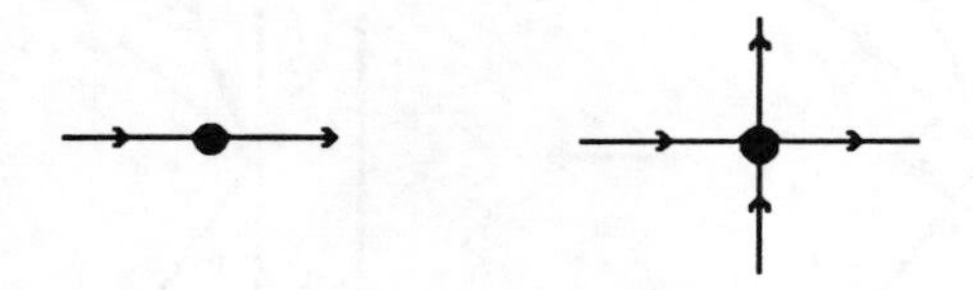

Figure 4.1 Oriented Graph Templates

For the example of the Simmons-Paquette molecule, Figure 4.2, it seems natural that the circuit (a concept which will be defined precisely later in this section) containing the three oxygen atoms corresponds to a circuit which can be traversed by "going straight across" at the relevant 4-valent vertices as shown in the representation of the vertex template above.

We define an **orientation** of a chimerical graph as a choice of direction on each edge of the graph subject to the requirement that the directions at 4-valent vertices are consistent for opposite edges at a vertex, as shown is Figure 4.1. In this example we see that a portion of the molecular graph has a natural choice of orientation determined by the sequence of the carbon and oxygen molecules and their bonding while, on the remainder, one can make an arbitrary choice of orientation consistent with the structure. For example, in this case, one can choose to orient the edge connecting the 2-valent carbon atom and the 2-valent oxygen in the carbon to oxygen direction. Next one observes that this choice extends to an orientation of the entire graph. It is important to remark that sometimes the choice of orientation will be inspired by some such aspect of the molecular structure while on other occasions it may be more useful to make a choice that reflects the spatial configuration and some other structure that one is attempting to detect. Both aspects are present in the oriented Simmons-Paquette molecule, shown in Figure 4.3, where on one of the circuits we have made a chemically natural choice of orientation inspired by the placement of the atoms and their bonding while for the other circuits there does not appear to be a natural choice so one simply makes one and tests whether the resulting mathematical structure can be used to reflect the spatial structure under study. As we shall see later in the paper, the mathematical theory which we shall describe requires and makes important use of the oriented structure on the graph. It is a mathematical fact the every chimerical 4-valent graph admits a least two distinct choices of orientation so that the specific choice is often a function of which aspect of the structure is being studied.

Figure 4.2 Simmons-Paquette K_5 molecule

It is traditional to amalgamate the edges connected by the valence two vertices of a graph in order to obtain a simpler graph which reflects the same essential topological structure. Thus one obtains the graph shown in Figure 4.3 from the Simmons-Paquette molecular graph.

In this context the local rigidity of the graph structure at the vertex of a chimerical graph seems physically appropriate. Indeed, one hopes that this will increase the chemical applicability of the proposed mathematical analysis. In order to be able to create an accessible mathematical theory, however, we add the further assumption that the edges are completely elastic away from the vertices. Although this later assumption may not be physically justified, recall that it is true that any collection of configurations which can be distinguished by means of the present elastic-edge/rigid-vertex theory will also be distinct under even more restrictive assumptions on the nature of the allowable spatial movements.

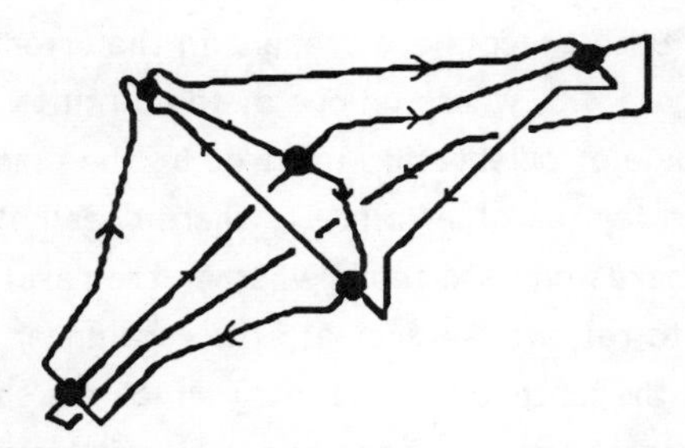

Figure 4.3 An oriented chimerical 4-valent graph

We shall also need again to discuss briefly the nature of the spatial movements of chimerical graphs because we desire mathematical indices of their placement which will be unchanged under allowable movements. Although these spatial movements can, in general, be quite complicated it is a theorem that any such movement can be described in terms of a finite sequence of rather simple movements, which we shall again call the "generalized Reidemeister moves", [J5], and of movements which preserve the qualitative structure of the generic projection of the object. The first three of these moves are exactly those which occur for the classical knots and links and which do not involve the structure of the chimerical graph. The typical examples are recalled in Figures 4.4, 4.5, and 4.6. Because we are concerned with the issue of orientation in a fundamental way it is necessary to indicate several versions of essentially the same spatial movement due to the various possible orientations. In Figures 4.7 and 4.8 we have shown the typical examples of the additional moves that are required to represent the spatial movements of chimerical graphs. A circle has been added to indicate the rigid

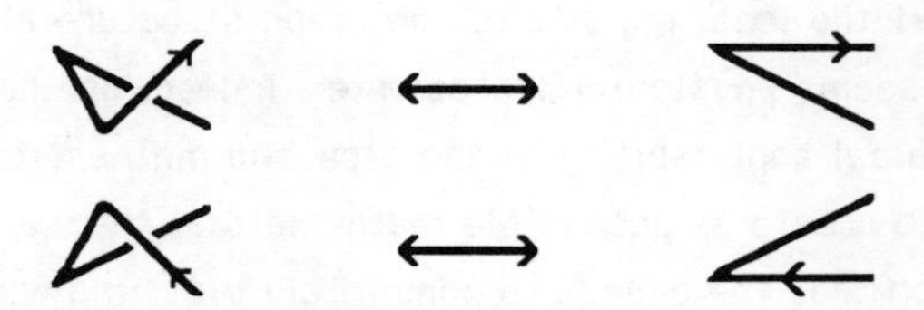

Figure 4.4 Type I Reidemeister moves

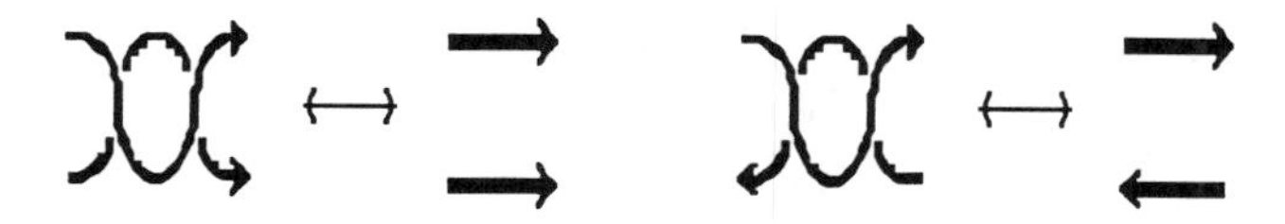

Figure 4.5 Type II Reidemeister moves

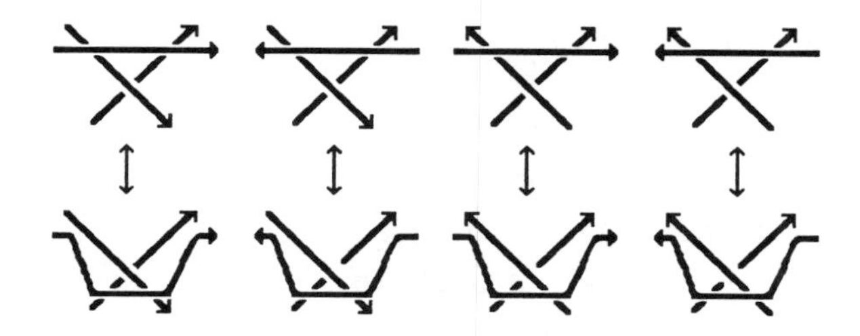

Figure 4.6 Type III Reidemeister moves

template structure at the vertex of the graph. Note that this rigid template structure has been preserved in each of these cases although it has been "turned over" in the case of the generalized type II move shown in Figure 4.8.

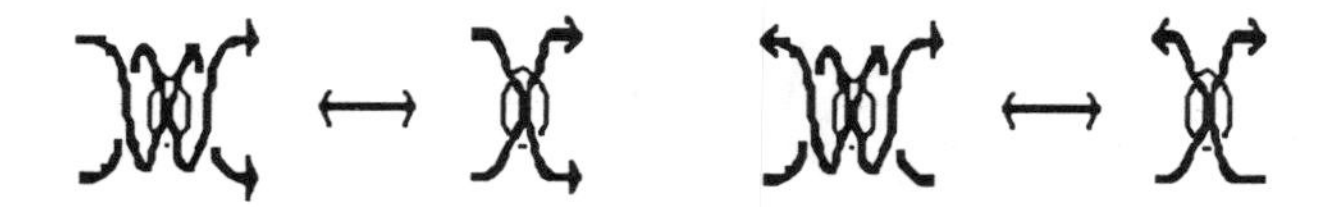

Figure 4.7 Chimerical type II Reidemeister moves

One of the fundamental concepts in topology is the notion of two objects linking in space. In the chapter on aspects of classical knot theory, the linking number associated to an oriented link of two components was defined. This concept plays such a crucial role in much of differential geometry and topology that it is not surprising that a modest generalization to the case of chimerical graphs in 3-space is also useful in the creation of new invariants of the extrinsic placement of chimerical graphs. Recall that a **circuit** in a graph is a sequence of edges, e_1, e_2, ..., e_m , joining distinct vertices, $\{v_0,v_1\}$, $\{v_1,v_2\}$, ..., $\{v_{m-1},v_0\}$. A **circuit** in a

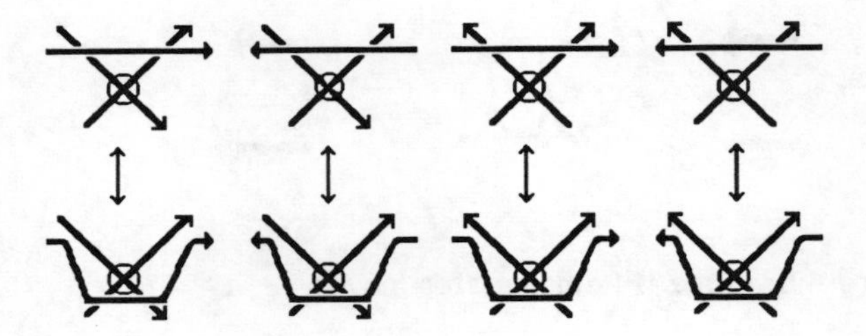

Figure 4.8 Chimerical type III Reidemeister moves

chimerical graph is a circuit which "goes straight ahead" at every vertex. Two examples, $\{e_i\}$ and $\{d_j\}$, are shown in Figure 4.10.

This concept is well defined and invariant because of the template structure at the vertices. Two circuits in a chimerical graph are said to be **transverse circuits** if their intersection consists only of vertices. The two circuits, $\{e_i\}$ and $\{d_j\}$, in Figure 4.10 are transverse circuits. Associated to two transverse circuits in an oriented chimerical graph is the concept of the **linking number** of the circuits. This linking number is defined to be one-half the sum of the crossing

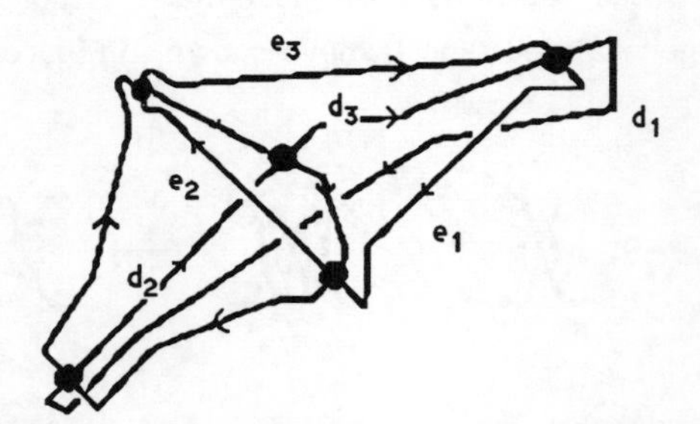

Figure 4.10 Two transverse circuits $\{e_i\}$ and $\{d_j\}$

numbers, where the **crossing number** of a crossing is ±1, according to the crossing convention recalled from classical knot theory in Figure 4.11 We define the "vertex

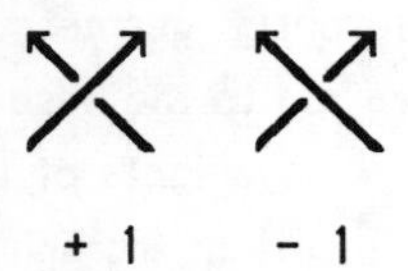

Figure 4.11 The algebraic crossing convention

crossing," being the average of the ±1 crossings, contribution to the linking number to be equal to 0. This definition of the linking number generalizes the classical concept of linking number for oriented links and introduces "fractional linking" for chimerical graphs, [J5] such as those shown in Figure 4.12.

Figure 4.12 Fractional linking in chimerical graphs

A fundamental fact is that this linking number for transverse circuits in a chimerical graph is an invariant of the spatial placement of the chimerical graph, i.e. any spatial deformation which leaves invariant the template structures at the vertices must leave the linking number unchanged. This is demonstrated by observing that each of the generalized Reidemeister moves described in section one leaves the linking number of the transverse circuits unchanged. An elementary application of this fact shows that the Simmons-Paquette K_5 molecule and its mirror image are spatially inequivalent, i.e. it is a chiral molecule. One may calculate the linking number between the two circuits, indicated in Figure 4.10, to find that they are -1/2 and 1/2 for the Simmons-Paquette K_5 molecule and its mirror reflection, respectively. Therefore the placement shown in Figure 4.10 and its mirror image are spatially distinct.

There is another situation in which chirality in chemistry has been studied for some time and in which there are also chirality polynomials which have been employed. We shall consider only two simple examples to illustrate the connection between these situations. The cyclopropane, or trigonal prism, is shown in its standard configuration in space in Figure 4.13. By employing group theoretic methods and the group of symmetries of the graph it has been shown that there are six chiral ligand partitions. This **intrinsic chirality problem** is the determination of the ligand (each of which is always assumed to be individually achiral) partitions for a specific achiral molecular graph which lead to an achiral system due to the lack of symmetry of the ligand partitions under equivalences of the molecular skeleton independent of any extrinsic considerations.

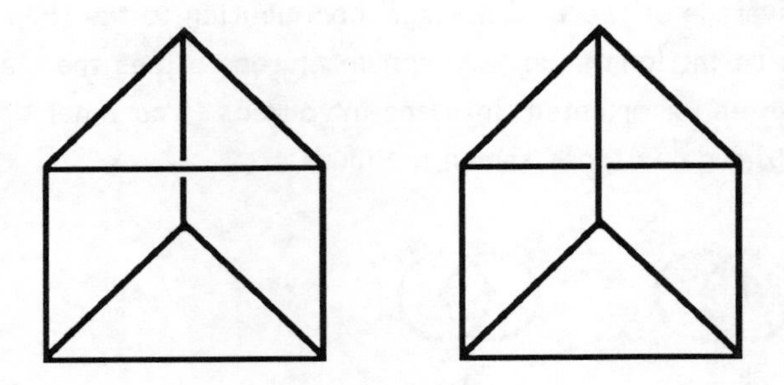

Figure 4.13 Cyclopropane

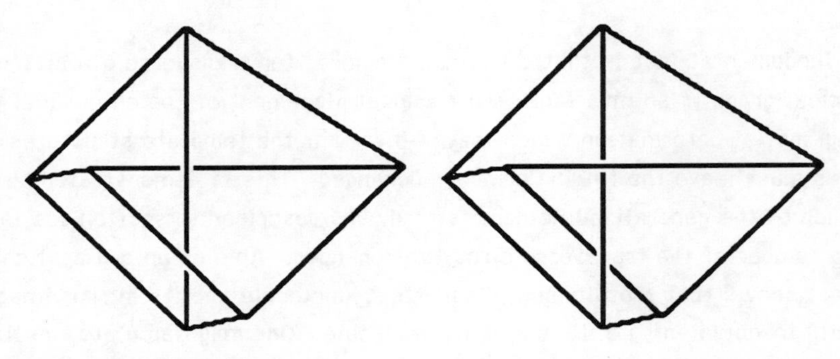

Figure 4.14

From the perspective of the extrinsic structure of the specific placement one sees that it and its mirror image are spatially equivalent. There are, however, spatially extrinsically distinct placements of this same graph in space such as those shown in Figure 4.14. To see that these are distinct and are, therefore, chiral one can use a linking number method, related to the one for chimerical graphs, applied to some specific choice of orientation on the pair of triangular subgraphs. One observes that this pair of subgraphs is a uniquely defined pair by virtue of the structure of this specific graph. It is not necessary to orient the remaining four edges as they play no role in the argument. Neither is it necessary to restrict attention, in this case, to a rigid vertex structure as the linking number of the two circuits will be invariant even with complete topological flexibility. In the case of this example we calculate that one of the associated linking numbers is 1 while the other is -1.

In the context of 4-valent graphs the regular octahedron, shown on the left in

Figure 4.15, is sufficiently complex to again illustrate the distinction between the two notions of chirality. Although we have not, for the sake of simplicity in the figure, preserved the rigid vertex structure at the central vertices we have shown

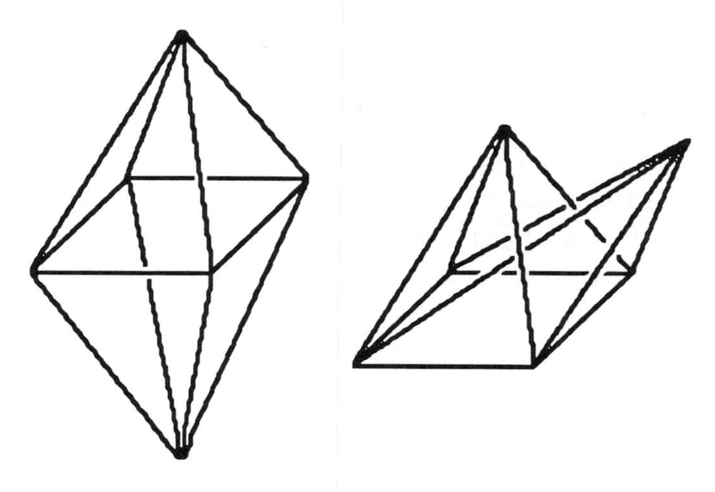

Figure 4.15. Octahedron placements

an extrinsically spatially distinct placement of the octahedron to its right. The same linking method can be employed to show that the right hand placement is chiral while, by a simple spatial rotation, one can show that the left hand placement is, in fact, achiral. It is often useful to remark that any planar placement of a graph is automatically extrinsically achiral because the plane of reflection of the mirror can always be taken to be precisely the plane of the graph and, therefore, the graph is unchanged by the reflection. This is the case for the first placement of the octahedron.

5. ALGEBRAIC INVARIANTS

The essential nature of the approach to the detection of chirality by means of algebraic invariants is the calculation of algebraic quantities such as numbers, equations, or more complicated structures such as vector spaces, rings, modules, etc. which enjoy the property that they will be the same so long as the knot, link, or embedding of a graph in space are equivalent with respect to spatial movements. In the previous section we have described these elementary spatial movements which describe all others and thus, if we wish to demonstrate this invariance property, we need only verify that the algebraic quantity is unchanged by these movements. Thus, for oriented classical knots and links or for oriented chimerical graphs, the linking number is an example of such an invariant. In certain cases we have seen that this number is sufficient to detect the chirality of the configuration. This method is not always successful and therefore researchers are lead to develop other methods of providing algebraic invariants which might be more sensitive to the chirality of the placement. In this section we shall present and discuss a number of such methods and associated examples.

The Alexander module and Alexander polynomial, [A], associated to a classical knot or link were first developed in the 1920's. In 1967, J. H. Conway, [C] presented a vision of the polynomial which emphasized a recursive method of calculation. Because the polynomial is determined by the fundamental group of the complement and because the mirror reflection takes the complement of one

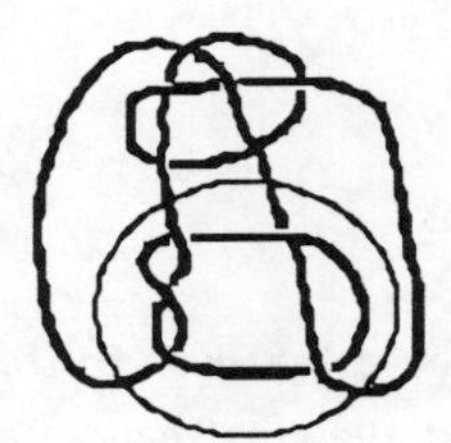

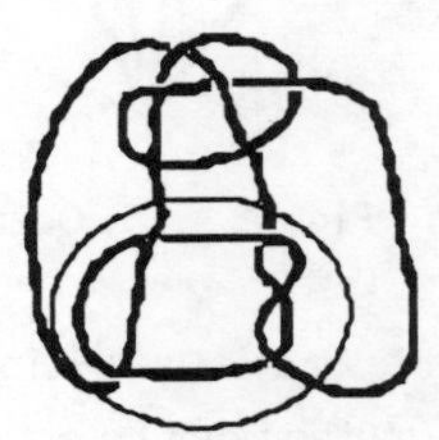

Figure 5.1 Two distinct non trivial knots with trivial Alexander polynomial (related rotation of circled region)

configuration to the complement of the mirror image configuration, the Alexander polynomial is insufficient to detect chirality. In addition, there are many knots whose "knottiness" it fails to detect and pairs of distinct knots having the same Alexander polynomial, c.f. Figure 5.1. Nevertheless, since the Alexander polynomial provides an excellent introduction to similar invariants which are quite successful in detecting chirality, we shall give a brief description of them. It is important to note that the Alexander ideals and the Alexander polynomial associated to knots and links were later extended to the case of spatial graphs by Kinoshita and others, [K1-7,S8]. They too suffer from the same inadequacy with respect to detection of "knottiness", c.f. Figure 5.2, and of chirality, Figure 3.1 (b and c) as does the classical Alexander invariants

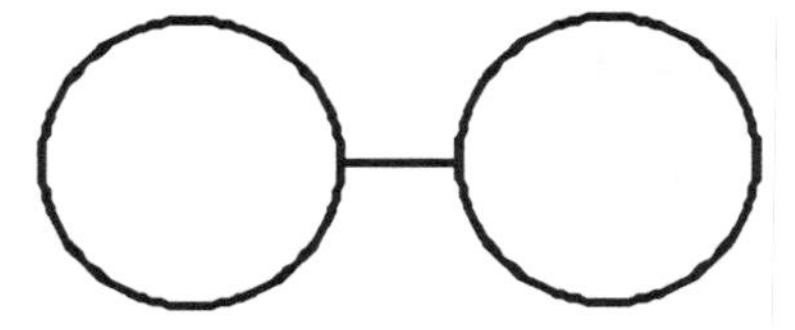
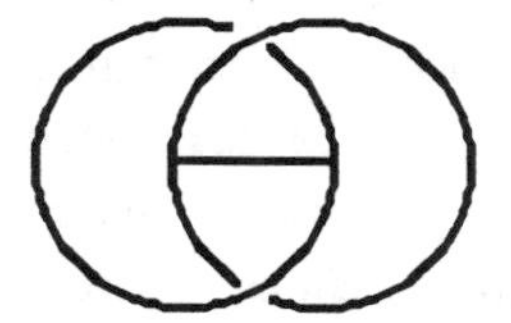

Figure 5.2

The **Alexander polynomial**, $\Delta(L)(t)$, L a classical link, can be calculated recursively by employing the following properties:

(i) $\Delta(U)(t) = 1$, where "U" denotes the trivial knot, and

(ii) if L_+, L_-, and L_0 are planar pictures of oriented links in each of which is identified a small circular region of the picture containing either a single crossing or, in the last case, no crossing at all, as shown in Figure 5.3, and such that outside these small circular regions shown in the figure below, the planar pictures are exactly the same, then the Alexander polynomial satisfies the formula:

$$\Delta(L_+)(t) - \Delta(L_-)(t) - (t^{-1/2} - t^{1/2})\Delta(L_0)(t) = 0.$$

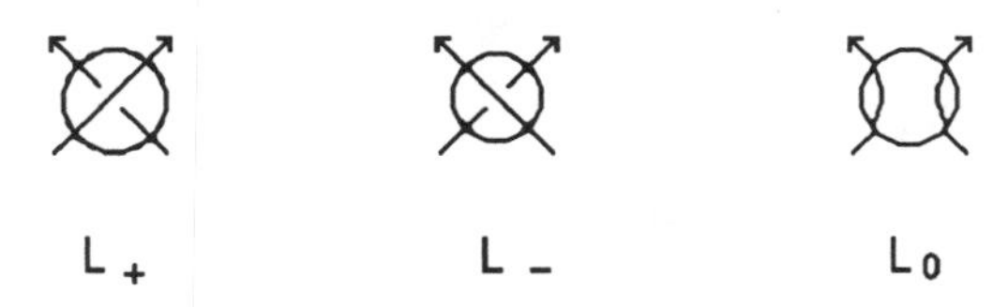

Figure 5.3

If one replaces $t^{-1/2} - t^{1/2}$ with "z" in the above formula, one has the definition of the Conway potential function which is equivalent but has fewer terms. From these formulae one can recursively calculate the Alexander polynomial since any knot can be changed to an unknotted presentation by changing crossings, in many different ways. One standard method is to order the components, 1, 2, ..., k and to select a starting point on each of these components. One can begin at the starting point, indicated by the dot in the figure, and proceed in the prescribed direction, indicated by the arrow, changing crossings as necessary so that the result agrees with that which be given by covering the given diagram with a cord laid upon the plane of projection proceeding in the same direction, as illustrated in Figure 5.4. One arrives, just before the dot, at the highest point of the circle and descends (in 3 space) to the beginning point to complete the circle. If there are several components, simply apply this process with each component in turn. The result is called the **standard ascending link associated to the oriented base-pointed ordered link presentation.** The computation proceeds by changing crossings to achieve the presentation of a trivial link with summands in terms of knots and links having presentations requiring fewer crossings and which, therefore, can be calculated independently. The major difficulty with this approach is the exponentially increasing number of terms that must be calculated as a function of the number of crossings in the presentation. This recursion method of calculation

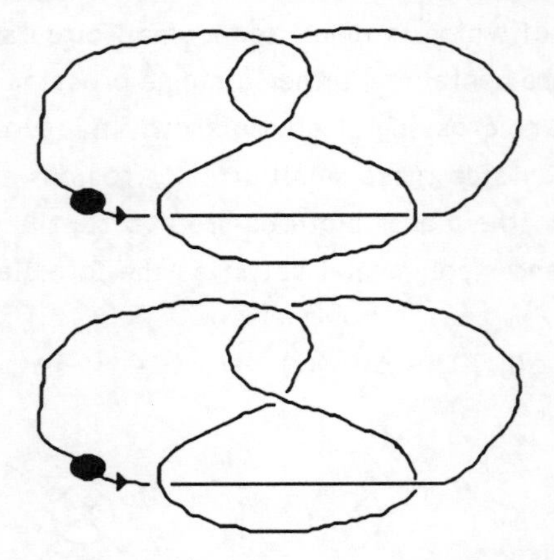

Figure 5.4 Trivial knot presentation associated to figure 8 knot

can be made sufficiently canonical and invariant under the Reidemeister moves so that it can be employed to demonstrate the existence and uniqueness of the polynomials satisfying these relations. The effect of taking the mirror reflection of a link upon the calculation is determined by interchanging positive and negative crossings. From this one observes that the Alexander polynomial is unchanged if there is an odd number of components and is taken to its negative if there is an even number of components and, therefore, fails to detect the possible chirality of knots. The Alexander polynomial for the trefoil knot, shown in Figure 5.5, is $t^{-1} - 1 + t$.

In the spring of 1984 the Jones polynomial, $V(L)(t)$, a completely new integral Laurent polynomial invariant of an oriented link, L, in a single variable, also called "t", was defined [J1-3]. There is a relationship between $V(L_+)$, $V(L_-)$ and, $V(L_0)$ which resembles the earlier relationship enjoyed by the Alexander polynomial and which, as in that case, could be employed to achieve its calculation. This is expressed by the following properties:

(i) if U denotes the standard unknotted circle in the plane, then $V(U)(t) = 1$, and

(ii) if L_+ , L_-, and L_0 are planar pictures of oriented links in each of which we have identified a small circular region of the picture containing either a single crossing or, in the last case, no crossing at all, and such that outside these small circular regions shown in Figure 5.3, the planar pictures are exactly the same, then the **Jones polynomial** satisfies the formula:

$$t^{-1} V(L_+)(t) - t V(L_-)(t) + (t^{-1/2} - t^{1/2}) V(L_0)(t) = 0.$$

The Jones polynomial provided an elementary means to distinguish between the positive and negative trefoil, shown in Figure 5.5, and thereby provided the first elementary method to detect chirality. In the Jones polynomial case, the mirror

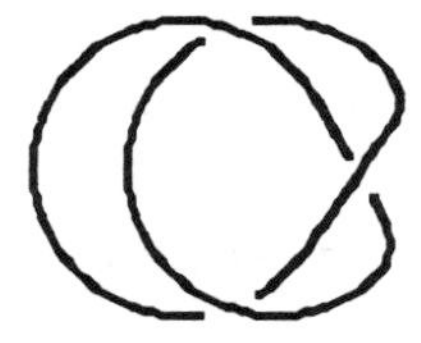

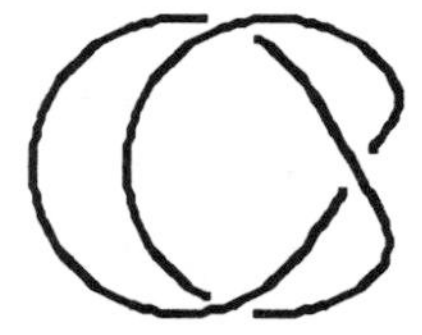

positive trefoil **negative trefoil**

Figure 5.5 The two trefoil knots

reflection acts on the Jones polynomial by the change of variables taking t to 1/t. Thus any knot or link whose Jones polynomial was changed under this change of variables must be chiral. In the case of the positive trefoil knot one has the Jones polynomial $t + t^3 - t^4$ while its mirror reflection has Jones polynomial $- t^{-4} + t^{-3} - t^{-1}$, showing that it is chiral. The nine crossing knot shown in Figure 5.6 shows that the Jones polynomial is not a perfect detector of chirality as the knot is known to be chiral but the Jones polynomial is unchanged under taking t to 1/t.

Figure 5.6 The knot 9_{42}

The similarity between the recursion formulae for the Alexander and Jones polynomials was too great to be accidental: a number of people believed that the two known cases must be both instances of a more general polynomial invariant for isotopy classes of oriented links. Indeed this is the case as was announced independently and simultaneously, in August 1984, by Lickorish and myself, Freyd and Yetter, Ocneanu, and by Hoste,[F4]; and also, in January 1985, by Przytycki and Traczyk. The **oriented polynomial**, $P(L)(\ell,m)$, of an oriented knot or link, L, is the unique Laurent polynomial in the variables, "ℓ" and "m", which satisfies the following fundamental formulae:

(i) if U denotes the standard unknotted circle in the plane, then $P(U)(\ell,m) = 1$, and

(ii) if L_+, L_-, and L_0 are planar pictures of oriented links in each of which we have identified a small circular region of the picture containing either a single crossing or, in the last case, no crossing at all, and such that outside these small circular regions shown in Figure 5.3, the planar pictures are exactly the same, then the **oriented polynomial** satisfies the formula:

$$\ell\, P(L_+)(\ell,m) + \ell^{-1}\, P(L_-)(\ell,m) + m\, P(L_0)(\ell,m) = 0.$$

It is easy to see that the oriented polynomial is a simultaneous generalization of the Alexander and Jones polynomials, i.e.;

$\Delta(L)(t) = P(L)(i, -i(t^{-1/2} - t^{1/2}))$, and

$V(L)(t) = P(L)(it^{-1}, i(t^{-1/2}-t^{1/2}))$.

Unlike the Alexander polynomial, the Jones polynomial and the oriented polynomial are each successively stronger in detecting the chirality of placements. Specifically, let $\overline{L}$ denote the mirror reflection of a link, L. Then, $P(\overline{L})(\ell,m) = P(L)(\ell^{-1},m)$. They are, nevertheless, not sufficiently sensitive to detect the chirality of the example shown in Figure 5.6.

Subsequent to the definition, another polynomial was discovered by Brandt, Lickorish, and myself, [B2], and Ho which required no orientations of any sort (hence the name 'absolute polynomial') but which involved the consideration of the other possible way of removing the crossing, denoted by L_∞ in the Figure 5.7. The **absolute polynomial**, Q(L)(z), is the unique Laurent polynomial in the variable "z" associated to the isotopy classes of unoriented knots and links, which satisfies the following fundamental formulae:

(i) if U denotes the standard unknotted circle in the plane, then Q(U)(z) = 1, and

(ii) if L_+ , L_-, L_0, and L_∞ are planar pictures of oriented links in each of which we have identified a small circular region of the picture in the figure above containing either a single crossing or, in the last cases, no crossing at all, according to the convention shown in Figure 5.7 (with orientations ignored), and such that outside these small circular regions the planar pictures are exactly the same, then the **absolute polynomial** satisfies the formula:

$$Q(L_+)(z) + Q(L_-)(z) = z\,[Q(L_0)(z) + Q(L_\infty)(z)].$$

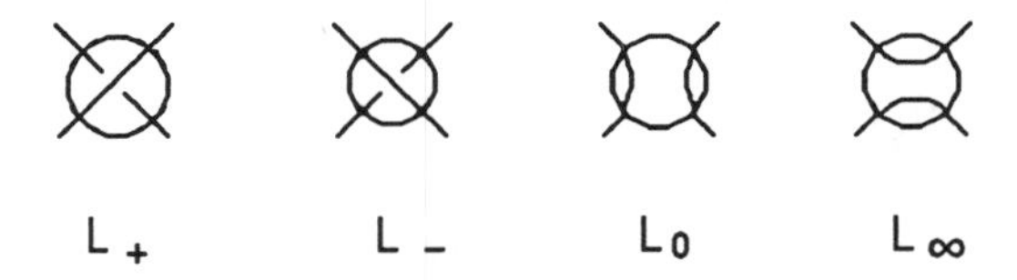

Figure 5.7

The form of the recursion relation indicates that the absolute polynomial suffers from the same weakness with respect to the detection of chirality as does the Alexander polynomial. Fortunately, the absolute polynomial was quickly extended by Kauffman, [K2], who explained how to introduce a second variable, "a", into the absolute polynomial. As a result, there is another polynomial, distinct from the previous ones and which satisfies a somewhat more complicated recursive formulation involving two versions of the crossing relation depending upon the number of distinct strands involved in the crossings and which is rather more, although not completely, successful in the detection of chirality. The number of strands, or components, in a link L will be denoted by **c(L)** and let **<X,Y>**, X and Y mutually disjoint oriented links, denote the algebraic linking number of X and Y. This number can be calculated from a presentation of the links as one-half the sum of +1's and -1's for each of the crossings between X and Y, according as whether the crossing is positive or negative as indicated in Figure 5.8. The orientation conventions that are to be employed in the recursive calculation change according to whether the crossing in L_+ involves the same or distinct strands of the link. These are shown in Figure 5.8.

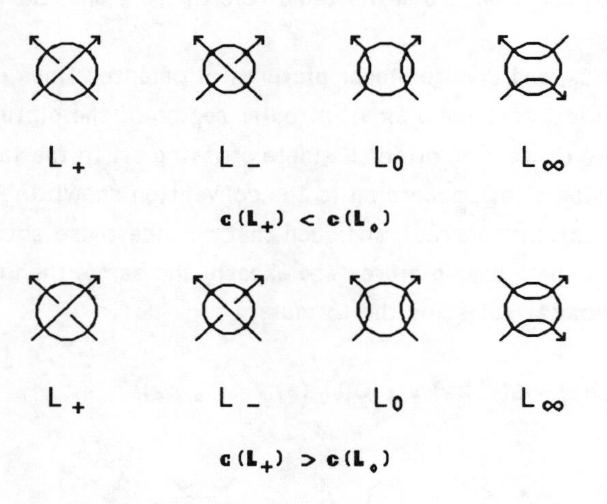

Figure 5.8

The **semioriented polynomial**, F(L)(a,x), of an oriented knot or link, L, is the unique Laurent polynomial in the variables, "a" and "x", which satisfies the following fundamental formulae:

(i) if U denotes the standard unknotted circle in the plane, then F(U)(a,x) = 1, and

(ii) if L_+, L_-, and L_0 are planar presentations of oriented links in each of which we have identified a small circular region of the picture containing either a single crossing or, in the last cases, no crossing at all, and such that outside these small circular regions shown in Figure 5.8, the planar pictures are exactly the same, then the **semioriented**

polynomial satisfies one of the formulae, according to whether the crossing in L_+ involves the same or distinct components:

(I) If $c(L_+) < c(L_0)$, let $\lambda = <x, L_0 - x>$ and

$$aF(L_+)(a,x) + a^{-1}F(L_-)(a,x) = x[F(L_0)(a,x) + a^{-4\lambda} F(L_\infty)(a,x)]$$

(II) If $c(L_+) > c(L_0)$, let $\mu = <x, L_+ - x>$ and

$$aF(L_+)(a,x) + a^{-1}F(L_-)(a,x) = x[F(L_0)(a,x) + a^{-4\mu + 2} F(L_\infty)(a,x)].$$

Kauffman's semi-oriented polynomial distinguishes different pairs of knots and links than does the oriented polynomial and therefore they are independent invariants. The discovery of the absolute polynomial provoked another very important advance, the creation of a state model for the Jones polynomial by Kauffman, [K3]. This provides an extraordinarily simple proof of the existence of the Jones polynomial and underlies many of the recent efforts to extend the polynomial invariants to the category of graphs as well as hinting at profound connections with theoretical physics. Thus, stimulated largely by questions of chirality in the context of spatial graphs arising in chemistry, a number of mathematicians have sought algebraic invariants generalizing those above which would apply to spatial graphs. By way of a conceptual introduction we should first consider another approach to the development of both topological and chimerical invariants of graphs which was proposed by Kauffman. Kauffman's idea was to associate to the spatial realization of a graph a collection of knots and links which would remain unchanged under either the topological or the chimerical equivalence, as appropriate. For example, if the spatial graph contains a nontrivial knot as a subgraph then the graph cannot be equivalent to a planar graph since all circuits in a planar graph are unknotted. Furthermore, if the graph contains a circuit which is a chiral knot but does not contain its mirror reflection one can conclude that the embedding is topologically chiral. This method is not always sufficient and one is lead to seek other methods to detect chirality that exploit the richer aspects of the graph structure. One can, if desired, associate the polynomial invariants discussed above to this collection and, thereby, extract an algebraic invariant for the graph.

Thus the extension of the theorems presented earlier in the chapter provide the existence of finite Laurent polynomials for the category of, for example, oriented chimerical graphs. In order to develop the subsequent invariants we first recall the Kauffman state model for the Jones polynomial. There is a function which associates to each unoriented link presentation, L, a finite Laurent polynomial, denoted $\langle L \rangle$, in one variable, A, satisfying the following three rules:

(i) $\langle \bigcirc \rangle = 1$

(ii) $\langle L \cup \bigcirc \rangle = -(A^{-2}+A^{2}) \langle L \rangle$

(iii) $\langle \times \rangle = A \langle)(\rangle + A^{-1} \langle \asymp \rangle$

The first rule says that the value "1" is associated to a simple planar circle while the second says that to the distant union of such a circle and an arbitrary link presentation, denoted L, is associated the value of the L times $-(A^{-2}+A^{2})$. The final rule provides a recursion relation for the simplification of presentations by reducing the number of crossings. A simple calculation shows that

$$\langle \times\!\!\!\supset \rangle = -A^{-3} \langle) \rangle$$

Thus this function fails to be unchanged by the type I Reidemeister moves. Nevertheless it is easy to remedy the defect by a choice of normalization to define another Laurent polynomial by

$$X(L)(A) \equiv (-A)^{-3\omega(L)} \langle L \rangle.$$

It is easy to show that $X(L)(t^{-1/4}) = V(L)(t)$, thereby giving the promised definition of the Jones polynomial. There are two fundamental aspects of this state model construction of the Jones polynomial that are exploited in subsequent constructions of other invariants. The first is the association, to a crossing in presentation, of an expansion of the desired polynomial in terms of a linear combination polynomials of (in some way) simpler presentations of objects of the same type. The specific linear combination is chosen so as to ensure the invariance of the associated polynomial under the type II and type III Reidemeister moves or, later, their generalizations. This is manifested in rule (iii) for the state model for the Jones polynomial. The second is the calculation of an appropriate normalization so as to ensure invariance under the type I Reidemeister move.

With this introduction to state models, we shall turn to the development of invariants associated to chimerical graphs and further generalizations. At the outset it is important to note that this can usually be accomplished in two distinct manners: directly, by the creation of a "state model", analogous to the Kauffman model for the Jones polynomial, by providing for the inclusion chimerical graph vertices, or indirectly, by exploiting the knowledge of the oriented or semioriented polynomials for classical knots and links via a generalization of the averaging method which extended the definition of the linking numbers of classical links to the linking numbers of the chimerical graphs. These two approaches are presented in the description of the fundamental aspects of the polynomials by Jonish and Millett [J5], and an instructive special case of the later approach was also presented in Kauffman and Vogel, [K4]. A construction of Yamada presented later also falls into this general category. The general averaging approach has the advantage of being the quickest route to the development of the chimerical graph invariants and their properties.

An important algebraic structural simplification is obtained if one imposes an algebraic assumption upon the class of potential invariants that are to be developed. This assumption is the **connected sum axiom or localization axiom**. The axiom states that if there is a circle in the plane of the projection which transversely meets the image of a generic projection of an oriented chimerical graph in exactly 2 points, such as illustrated in Figure 5.9, we require that the invariant associated to

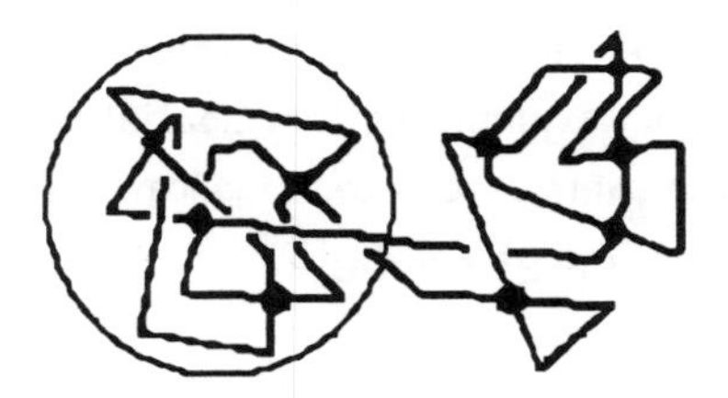

Figure 5.9 The Connected Sum of Two Graphs

the object is the product of the invariants associated to the two pieces of the graph gotten by breaking the graph at the two points and connecting the inside graph along the circle to get one graph and, similarly, connecting the outside piece along the circle to get the second graph, such as shown in Figure 5.10. This axiom can be understood as a consequence of the connected sum property of the oriented and semioriented polynomials or as an axiom of the states models proposed by Jonish-Millett [J5].

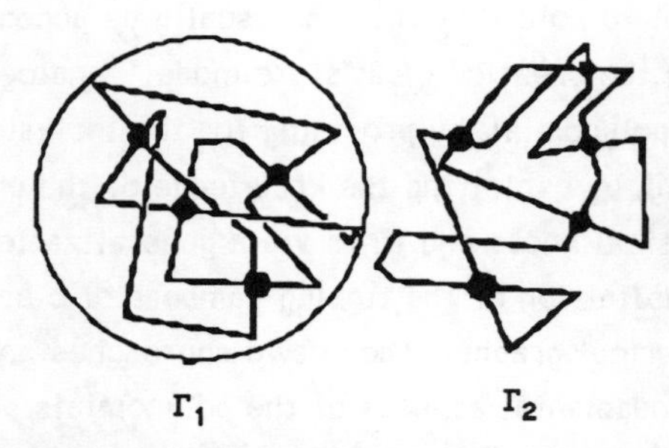

Figure 5.10 The Summands, Γ_1 and Γ_2

A second basic aspect of such theories that one would like to try to exploit is the existence of a recursion formula to facilitate calculations in the spirit of the earlier theories, c.f. [B2], [L3], and [M2], or the Kauffman state model approach. We will shall show that these oriented and semioriented polynomials, for the oriented chimerical 4-valent graphs, satisfy recursion relations which are exactly the same as those which arise for oriented knots and links. Formally the recursion relation will allow one to express the invariant of a specific placement of an oriented chimerical 4-valent graph in space in terms of the invariants of simpler placements.

There is a unique way of associating to each oriented chimerical graph in 3-dimensional Euclidian space, Γ, an algebraic function, $P_G(\Gamma)$, in the algebraic variables ℓ and m, and elementary oriented chimerical graph variables β and γ, satisfying the connected sum axiom, such that spatially equivalent oriented chimerical graphs have the same associated algebraic function and such that

(i) $P_G(U) = 1$, $P_G(\beta) = \beta$, $P_G(\gamma) = \gamma$, where

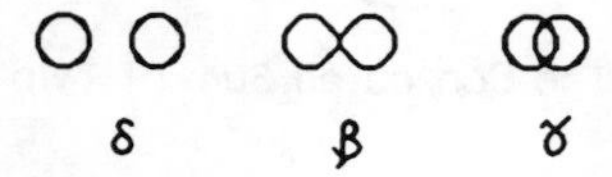

Figure 5.11 Chimerical graph variables

(ii) if Γ_+, Γ_-, and Γ_0 are any three oriented chimerical 4-valent graphs that are identical except near one point where they are as shown in Figure 5.12, then

$$\ell P_G(\Gamma_+) + \ell^{-1} P_G(\Gamma_-) + m\, P_G(\Gamma_0) = 0 \text{ and,}$$

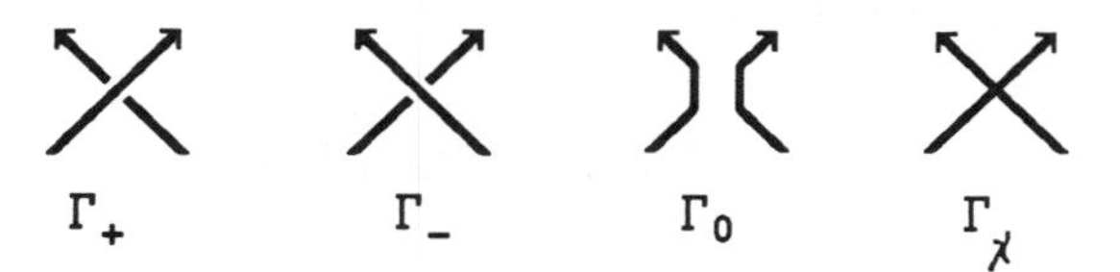

Figure 5.12. The chimerical graph state

(iii) if Γ_+, Γ_- and, Γ_χ are any three oriented chimerical graphs that are identical except near one point where they are as in Figure 5.12, then $P_G(\Gamma_\chi) = [\beta \ell - i(\beta^2-\gamma\delta)^{1/2}(\delta^2-1)^{-1/2}](\ell^{-1}+\ell)^{-1}P_G(\Gamma_+)$
$+ [\beta \ell^{-1} + i(\beta^2-\gamma\delta)^{1/2}(\delta^2-1)^{-1/2}](\ell^{-1}+\ell)^{-1}P_G(\Gamma_-)$.

The chimerical graph variables are shown in Figure 5.11 without orientations in as much as they are spatially independent of these choices when considered as chimerical graphs in S^2. The second recursion relation derives from the goal of providing an extension of the oriented polynomial invariant while the third recursion relation allows for the expression of the influence of the chimerical graph vertices in terms of the algebraic variables and the two graph variables, β and γ. Note that, as in the case of the knot and link invariants, δ can be expressed in terms of the algebraic variables. It is this later recursion relation which is the essential content of the theory. The precise relationship is determined by the requirement of invariance under spatial movements of the algebraic function to be associated to the chimerical graph. As a first step towards this result one may invoke only the first recursion relation in order to calculate the invariant in terms of elementary oriented chimerical graphs, i.e. one does not attempt to reduce the number of vertices of the graph but only simplify, as much as possible, the spatial placement of the graph by using crossing changes in the manner employed for classical knots and links.

The concept of an elementary oriented chimerical graph involves making arbitrary choices in a manner analogous to the choice of basis of a vector space. For example, exactly one of the two placements shown in Figure 5.14 must be selected to be the elementary placement. The specific choice is, for the most part, simply a question of convenience.

In addition to using the theorem to give the recursive formula for calculation one may also employ several elementary consequences of the theory so as to simplify some of the calculations. For example, it is very convenient to give a specific symbol, δ, to the invariant associated to the distant union of two unknotted circles shown as the "0" state in Figure 5.13.

Figure 5.13 Determination of δ

The recursion identity gives the equation $\ell + \ell^{-1} + m\delta = 0$ since the "+" and "-" cases are both presentations of the trivial knot. Thus $\delta = -m^{-1}(\ell^{-1} + \ell)$. An important formula for the study of chirality of a chimerical graph is $P_G(\Gamma_1)(\ell,m) = \overline{P_G(\overline{\Gamma_1})(\ell,m)}$, where $\overline{\Gamma_1}$ is the mirror image of Γ_1, i.e. reverses all the crossings in a presentation of Γ_1, and the conjugation in the algebra takes ℓ to ℓ^{-1}, i to i^{-1}, and leaves m unchanged. Specifically, if the two placements are to be topologically equivalent via the allowed chimerical spatial movements then the associated invariant must be unchanged when ℓ is replaced by ℓ^{-1} and i is replaced by i^{-1}, i.e. by complex conjugation. For example, in the classical case of knots, the figure 8 knot has such a polynomial, i.e. $(-\ell^{-2} - 1 - \ell^2) + m^2$, and is actually spatially equivalent to its mirror reflection. In order to illustrate the use of the recursion

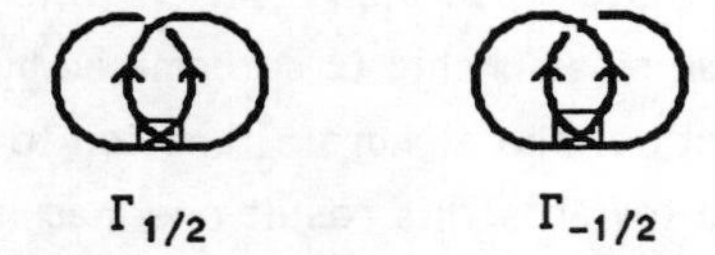

Figure 5.14.

formula and these associated properties, we shall consider the graphs, shown in Figure 5.14, and describe how the calculational methods would apply to this representative case. First one can apply property (ii) to show that

$$\ell P_G(\Gamma_{1/2}) + \ell^{-1} P_G(\Gamma_{-1/2}) + m\, P_G(\Gamma_0) = 0.$$

Since Γ_0 is β, this shows that $P_G(\Gamma_{-1/2}) = -\ell^2 P_G(\Gamma_{1/2}) - m\ell\beta$.

Thus, it is sufficient to calculate $P_G(\Gamma_{1/2})$, which is accomplished by application of property (iii) as follows:

$$P_G(\Gamma_{1/2}) = [\beta\ell - i(\beta^2-\gamma\delta)^{1/2}(\delta^2-1)^{-1/2}](\ell^{-1}+\ell)^{-1}P_G(\Gamma_+) + [\beta\ell^{-1} + i(\beta^2-\gamma\delta)^{1/2}(\delta^2-1)^{-1/2}](\ell^{-1}+\ell)^{-1}P_G(\Gamma_-),$$

where Γ_+ and Γ_- are shown in Figure 5.15. Since these are classical links,

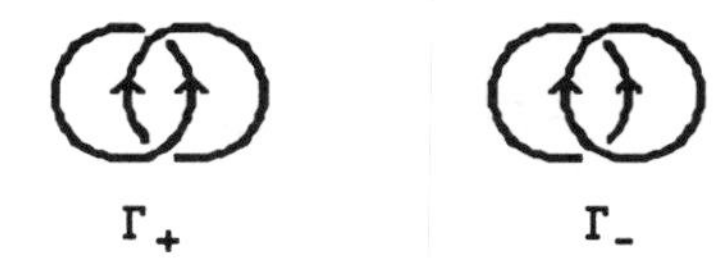

Figure 5.15

the polynomial invariant of the Γ_+ can be computed by recursion via part (ii) to find that $P_G(\Gamma_+) = (\ell^{-3}+\ell^{-1})m^{-1} - \ell^{-1}m$, while the second is equivalent to the unlink of two components and give $\delta = -(\ell^{-1}+\ell)m^{-1}$. Substitution in the formula gives

$$P_G(\Gamma_{1/2}) = -\beta(\ell^{-1}+\ell)^{-1}m + i\ell^{-1}(\beta^2-\gamma\delta)^{1/2}(\delta^2-1)^{-1/2}(\ell^{-1}+\ell)^{-1}((\ell^{-1}+\ell)\delta + m)$$
$$= \beta/\delta + i\ell^{-1}(\beta^2-\gamma\delta)^{1/2}(\delta^2-1)^{-1/2}(\delta-\delta^{-1})$$

and therefore,

$$P_G(\Gamma_{-1/2}) = -\beta(\ell^{-1}+\ell)^{-1}m - i\ell(\beta^2-\gamma\delta)^{1/2}(\delta^2-1)^{-1/2}(\ell^{-1}+\ell)^{-1}((\ell^{-1}+\ell)\delta+m)$$
$$= \beta/\delta - i\ell(\beta^2-\gamma\delta)^{1/2}(\delta^2-1)^{-1/2}(\delta-\delta^{-1}).$$

The difference between these two expressions measures the chirality therein.

The recursion formulae for the semioriented invariant require the development of some additional notation. Recall that the number of strands, or components, in a link L is denoted by c(L). Here we extend this definition to the number of circuits in the chimerical graph. In addition, we define $\langle X,Y\rangle$, X and Y mutually disjoint oriented circuits (links, in the classical case), to be the algebraic linking number of X and Y. The orientation convention that is to be employed in the recursive calculation, according to whether the crossing in L_+ involves the same or distinct strands of the link, is recalled in Figure 5.16.

There is a unique way of associating to each oriented chimerical graph in 3-dimensional Euclidian space, Γ, an algebraic function, $F_G(\Gamma)$, in the algebraic variables a and x, and elementary oriented chimerical graph variables β and γ, satisfying the connected sum axiom, such that spatially equivalent oriented chimerical graphs have the same associated algebraic function and such that

(i) $F_G(U) = 1$, $F_G(\boldsymbol{\delta}) = -\delta \equiv (a^{-1} + a)x^{-1} - 1$, $F_G(\boldsymbol{\beta}) = \beta$ and, $F_G(\boldsymbol{\gamma}) = -\gamma$, and

(ii) if Γ_+ , Γ_-, Γ_0, and Γ_∞ are planar presentations of oriented chimerical graphs in each of which there are identified small circular regions containing either a single crossing or, in the last cases, no crossing at all, and such that outside these small circular regions shown in Figure 5.16, the planar presentations are exactly the same, then the semioriented polynomial satisfies one of the formulae,

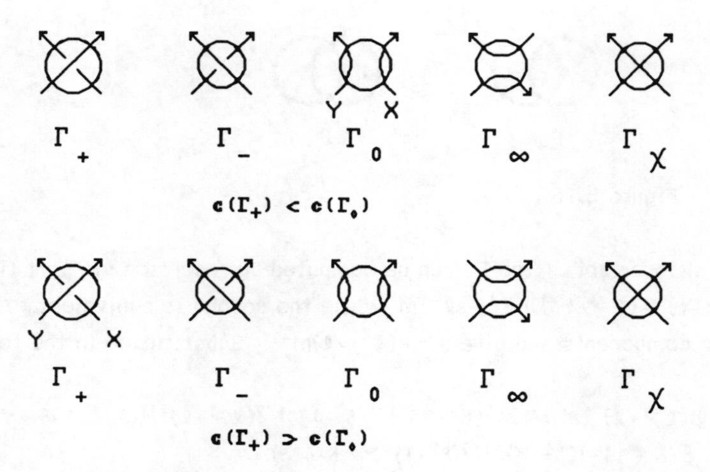

Figure 5.16

according to whether the crossing in Γ_+ involves the same or distinct components:

(I) If $c(\Gamma_+) < c(\Gamma_0)$, let $\lambda = \langle X, \Gamma_0 - X \rangle$ and

$$aF_G(\Gamma_+) + a^{-1}F_G(\Gamma_-) = x[F_G(\Gamma_0) + (ia)^{-4\lambda} F_G(\Gamma_\infty)]$$

(II) If $c(\Gamma_+) > c(\Gamma_0)$, let $\mu = \langle X, \Gamma_+ - X \rangle$ and

$$aF_G(\Gamma_+) + a^{-1}F_G(\Gamma_-) = x[F_G(\Gamma_0) - (ia)^{-4\mu + 2} F_G(\Gamma_\infty)].$$

(iii) if Γ_+, Γ_-, Γ_0 and, Γ_χ are any four oriented chimerical graphs presentations that are identical except near one point where they are as in Figure 5.16, then

$$F_G(\Gamma_\chi) = \rho(+)F_G(\Gamma_+) + \rho(-)F_G(\Gamma_-) + \rho(0)F_G(\Gamma_0), \text{ where}$$

$$\rho(+) \equiv [\alpha A/(A-B)C] = \frac{a^{-1}\beta}{x(1+\delta)} - \frac{ia^{-1}(x-a^{-1})(2\beta^2-\gamma(\delta+1))^{1/2}}{x(1+\delta)(\delta-1)^{1/2}(x^2-\delta-2)^{1/2}}$$

$$\rho(-) \equiv [-\alpha^{-1}B/(A-B)C] = \frac{a\beta}{x(1+\delta)} + \frac{ia(x-a)(2\beta^2-\gamma(\delta+1))^{1/2}}{x(1+\delta)(\delta-1)^{1/2}(x^2-\delta-2)^{1/2}}$$

$$\rho(0) \equiv [(A+B)/C] = \frac{-2\beta}{(1+\delta)} - \frac{i(a^{-1}-a)(2\beta^2-\gamma(\delta+1))^{1/2}}{(1+\delta)(\delta-1)^{1/2}(x^2-\delta-2)^{1/2}}$$

As in the development of the oriented polynomial for chimerical graphs, the first recursion relation derives from the goal of providing an extension of the semioriented polynomial invariant while the second recursion relation allows for the expression of the influence of the chimerical graph vertices in terms of the algebraic variables and the two graph variables, β and γ. There is an important difference between the form of the first recursion relation for classical knots and links and that given in the theorem for the extension to chimerical graphs. This is the occurrence of the factor $i = \sqrt{-1}$ in the "∞" term of the recursion. This factor derives from the fact that the linking number of two chimerical circuits is a multiple of 1/2 rather than being an integer as is the case for the classical knots and links. Note that, as in the case of the knot and link invariants, δ can be expressed in terms of the algebraic variables. It is this later recursion relation which is the essential content of the theory. Just as in the oriented case, the recursion identity gives the equation $a + a^{-1} = x[-\delta + 1]$ since the "+", "-", and ∞ cases are presentations of the trivial knot. Thus $\delta = -(a^{-1} + a)x^{-1} + 1$.

As with the oriented polynomial, $\overline{F_G(\Gamma_1)(a,x)} = F_G(\overline{\Gamma_1})(a,x)$, where $\overline{\Gamma_1}$ is the mirror image of Γ_1, i.e. reverses all the crossings in a generic projection of Γ_1, and the conjugation in the algebra takes a to a^{-1}, i to i^{-1}, and leaves x unchanged. To illustrate the use of the recursion formula and these associated properties, we shall again describe how the calculational methods would apply to the graphs, $\Gamma_{1/2}$ and $\Gamma_{-1/2}$, shown in Figure 5.15. First one can apply rule (ii) to show that $aF_G(\Gamma_{1/2}) + a^{-1}F_G(\Gamma_{-1/2}) = x[F_G(\Gamma_0) - a^{-4(1/2)+2} F_G(\Gamma_\infty)] = x[F_G(\Gamma_0) - F_G(\Gamma_\infty)]$ since $c(\Gamma_+) > c(\Gamma_0)$ and $\mu = \langle X,\Gamma_+ -X\rangle = 1/2$. Since Γ_0 and Γ_∞ are both β, this shows that $F_G(\Gamma_{-1/2}) = -a^2F_G(\Gamma_{1/2})$ as is implied by rule (iv). Thus, it is sufficient to calculate $F_G(\Gamma_{1/2})$, which can be accomplished by an application of Theorem 6.1 (iii) as follows: $F_G(\Gamma_{1/2}) = \rho(+)F_G(\Gamma_+) + \rho(-)F_G(\Gamma_-) + \rho(0)F_G(\Gamma_0)$, where Γ_+, Γ_-, and Γ_0 are shown in Figure 5.17. Since these are classical links,

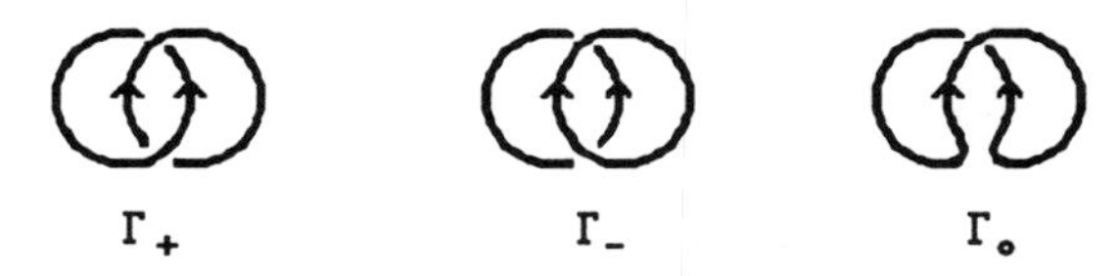

Figure 5.17

the polynomial invariant of the Γ_+ can be computed by recursion via part (ii) to find that $F_G(\Gamma_+) = -(a^{-3}+a^{-1})x^{-1}+a^{-2}+(a^{-3}+a^{-1})x$, the second is equivalent to the unlink of two components and gives $F_G(\delta) = (a^{-1} + a)x^{-1} - 1$, and the third is the trivial knot, contributing 1. Substitution in the formula gives

$$F_G(\Gamma_{1/2}) = \rho(+)[-(a^{-3}+a^{-1})x^{-1}+a^{-2}+(a^{-3}+a^{-1})x] + \rho(-)[(a^{-1} + a)x^{-1} - 1] + \rho(0)$$
$$= [(a^{-2}-1)\beta + ia^{-1}(1-\delta)^{1/2}(x^2-\delta-2)^{1/2}(-2\beta^2+\gamma(\delta+1))^{-1/2}]/(1+\delta)$$

Since $\Gamma_{-1/2}$ is the mirror reflection of $\Gamma_{1/2}$, one may employ this to complete the calculation, or use the above formula in which the quantity $[-(a^{-3}+a^{-1})x^{-1}+a^{-2}+(a^{-3}+a^{-1})x]$ is replaced by $[(a^{-1} + a)x^{-1} - 1]$ and $[(a^{-1} + a)x^{-1} - 1]$ is replaced by $[-(a+a^3)x^{-1}+a^2+(a+a^3)x]$. Expressing as much of these quantities as possible in terms of expressions which are invariant under mirror reflection and conjugation involution assists in the recognition of fundamental relationships such as chirality. Thus we employ δ, β and, γ as they represent planar graphs which are, therefore, unchanged under mirror reflection. From these calculations, one notes that the two graphs are achiral as their invariants are changed under complex conjugation and sending a to a^{-1} and that their invariants are interchanged as they are mirror images of each other, as required.

Some special cases of these chimerical graph invariants admit extensions to more general settings. One of more interesting of these is due to Yamada, [Y1]. He defines a finite one variable Laurent polynomial which is a spatial invariant of unoriented chimerical graphs up to multiplication by plus or minus a power of the variable. As noted above, in the case that the maximum degree of the vertices of the graph is three, the chimerical invariance is the same as topological invariance, i.e. the preservation of the vertex template for the vertices of degree three does not introduce an additional restriction. In addition, if one restricts to the case of regular graphs of degree 2, i.e. unoriented links, one has a specialization of the Kauffman's semioriented polynomial. Let G denote a graph having vertex set V and edge set E and let $\mu(G)$ and $\beta(G)$ denote the number of components of G and the first Betti number of G, respectively. For $F \subset E$ let $[F]$ denote the number of elements of F and let G-F denote the graph (V,E-F). Yamada defines a preliminary polynomial as follows:

$$h(G)(x,y) = \sum_{F \subset G} (-x)^{-[F]+\mu(G-F)} y^{\beta(G-F)}$$

and introduces a state model, following the method of Kauffman by which he gives a remarkably simple definition of the Jones polynomial, as follows: For any representation of a spatial graph, G, one considers the collection of functions, S,

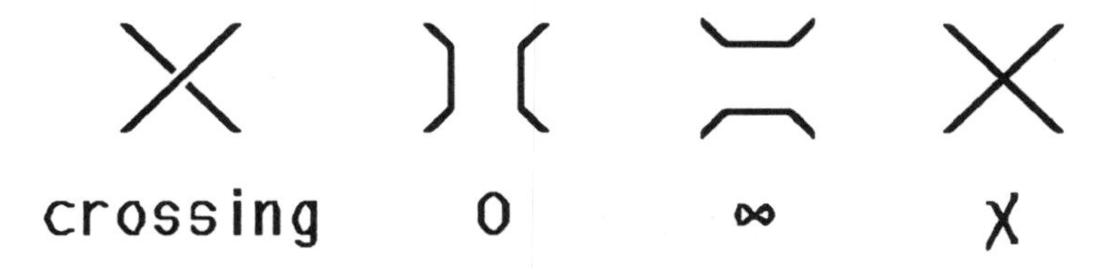

Figure 5.18

from the set of crossings in the representation into the set $\{0,\infty,\chi\}$ and for each $s \varepsilon S$ one defines the associated state of G, s(G), replacing each crossing by the configuration determined by the value of s according to Figure 5.18. Let $[s] = p - q$, where p is the number of time that s takes the value 0 and q denotes the number of times that s takes the value ∞. A one variable finite Laurent polynomial is then defined by

$$R(G)(A) = \sum_{s \varepsilon S} A^{[s]}\, h(s(G))(-1,-A^{-1}-2-A).$$

For certain graphs, such as the classical knots and links or spatial realizations of the θ_n graphs shown in Figure 5.19, one is able to remove the indeterminacy in the polynomial by an appropriate normalization as follows: Recall that for a knot presentation, K, one defines the writhe of the presentation, $\omega(K)$, to be the sum of the crossing numbers associated to the crossings in the presentation. For $\theta_n = (\{u,v\},\{e_1, \ldots ,e_n\})$ let c_{ij} denote the cycle consisting of the two vertices and the ith and jth edges. Yamada defines the writhe of a spatial realization of θ_n, θ_n, by

$$\omega(\theta_n) = \sum_{i<j} \omega(c_{ij}) \;/\; (n-1)$$

and shows that $(-A)^{-2\omega(\theta)}R(\theta)$ is chimerical invariant of θ. As above, this implies that this quantity is a topological invariant for the classical knots and links and θ-curves.

One can calculate the Yamada polynomials associated to the two graphs shown in Figure 5.3 showing that the first gives '0' while the second gives $A^{-4} + A^{-3} + A^{-2} + A^{-1} - A^2 - A^3 - A^4 - A^5$ thereby demonstrating their topological distinctness. For the trivial θ_3, shown in Figure 3.1 (a), one has the polynomial $-A^{-2} - A^{-1} - 2 - A - A^2$ while the second embedding, Figure 3.1 (b), has the polynomial $A^{-8} + A^{-7} - A^{-6} + A^{-5} + A^{-4} - A^{-3} + A^{-1} + 2A + A^2 + 2A^3 - A^5 + A^6 - 2A^7 - A^8 + A^9$ and associated

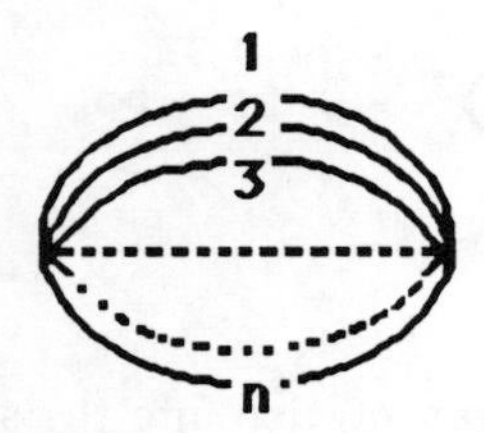

Figure 5.19

writhe equal to -3/2. As a consequence, the embedding is topologically distinct from the first and, in addition, is topologically chiral because $(-A)^{-2\omega(\theta)}R(\theta)$ of an achiral embedding, θ, must be invariant under the change of variables taking A to A^{-1}, following the model of the classical knots and links.

One may calculate that the trivial circle, as a planar graph, gives the value $\delta = A^{-1} + 1 + A$. If $G = \{V,E\}$ is a graph determined by its edges E and vertices with $v = |V|$ and $e = |E|$. Its first Betti number is given by $\beta_1 = e - v + 1$. If G is given as a planar graph one can show direct calculation that R(G) can be expressed as a polynomial in δ with integer coefficients, reflecting the planar presentation. In addition, one can show that the top two highest powers of δ which occur are $(v-\beta_0)(-1)^{\beta_1-2}\delta^{\beta_1-1} + (-1)^{\beta_1-1}\delta^{\beta_1}$. Thus, when the Yamada invariant is not of this form, one knows that the given spatial embedding is not equivalent to a planar embedding even though the associated polynomial may be invariant under the change of variables taking A to A^{-1}. For another approach to planarity testing see Kobayashi, [K8].

Further analysis of the Yamada invariant shows that it is an extension of the semioriented polynomial invariant of Kauffman [K1]. For example, if one restricts attention to classical knots and links, then

$$(-1)^{c(L)-1}A^{4\lambda(L)}(A^{-1}+1+A)F_G(L)(iA^2, i(A^{-1}-A)) = A^{-2t(L)}R(L)(A),$$

where $\lambda(L)$ denotes the total linking number of L, i.e. the sum of the linking numbers of all pairs of distinct components of L. Thus, in a fundamental way, the Yamada polynomial is an extension of the semioriented polynomial to the case of 3 valent graphs and topological equivalence or, for the case of vertices of valence greater than 3, spatial equivalences in which a rigid template structure, analogous to that employed in the chimerical graph case, is preserved. The price of this extension is the current absence of a method to give a normalization which can ensure invariance under the type I Reidemeister moves except in the cases mentioned above. In the Yamada case the resulting polynomial is invariant up to product of a factor $\pm A^n$. Sufficient structure remains in the Yamada invariant to successively detect the chirality as well as the non-planarity of a spatial embedding in a large number of cases.

It is possible to extend the Yamada theory, via an appropriate normalization to account for the potential change in the factor $\pm A^n$, so as to define a spatial invariant of regular 3-valent graphs. This is accomplished, Millett [M1], by defining an enhanced structure associated to a fixed embedding of the graph in an oriented surface so as to determine a reduced writhe that can be associated to the spatial embedding. The resulting theory can be applied to detect knotting, linking and, supercoiling intrinsic to the spatial placement of the Mobius ladders and analogous molecular graphs associated to DNA.

The early observation of Jones, [B1,J1,J3], of similarities between the formal structure of the knot polynomial invariants and soluble models of two dimensional statistical mechanics, i.e. the Temperley-Lieb algebras, has stimulated a significant body of research by mathematicians and physicists attempting to understand and exploit these interconnections. This has also lead to the proposal of intrinsically three dimensional gauge theory models for both the knot invariants and for invariants associated to three dimensional manifolds by Witten, [W5-7]. This interaction between mathematics and physics has also given rise to spatial invariants which can be used to detect chirality and which one might imagine would lead to a deeper understanding of the molecular graph models themselves. Without going into the theoretical details of this method we may nevertheless describe the essential aspects of the theory as it pertains to spatial embeddings of θ_3. The standard planar embedding is assigned a value which we shall denote by $[\theta_3]$ to which the invariant associated to another spatial placement is related. In this case one is required, by the physical constraints, to consider the Chern-Simons SU(3) gauge theory giving rise to an extension of the oriented polynomial with $\ell = -i\, q^{1/6}$ and $m = -i(q^{-1/2} - q^{1/2})$. For this model one makes, as is required in the previous state models, a normalization so as to insure invariance under the type I Reidemeister moves, i.e. the introduction of the factor $q^{4\omega(\Gamma)/3}$. Thus, if Γ is an oriented 3 valent graph (the edges incident at a vertex have the same orientation), then

(i) $q^{1/6}P_G(\Gamma_+) - q^{-1/6}P_G(\Gamma_-) - (q^{-1/2} - q^{1/2})P_G(\Gamma_0) = 0$, and

(ii)

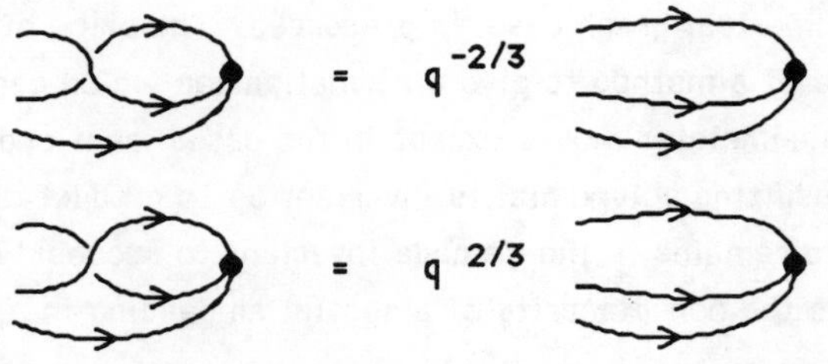

$$= q^{-2/3}$$

$$= q^{2/3}$$

Using these relations one is able to calculate the spatial invariant associated to an embedding of Θ_3 such a shown in Figure 3.1 (b). This gives

$$P_G(\Gamma) = (q^{-2} - 2\,q^{-1} + 1 - q^{2/3} + 2\,q^{5/3})\,[\Theta_3].$$

Since this polynomial is altered by the change of variables taking q to q^{-1} we have another proof of the chirality of the embedding.

The connections between the new polynomial invariants of knotting and linking and fundamental aspects of integrable models in statistical mechanics and three dimensional generally covariant gauge theories, of which the Chern-Simons theory is an example, contain many mysteries which are the subject of intense study. The success of these efforts could well lead to new perspectives and deeper understanding of chirality, its detection and quantification, and to new geometrical and topological connections between integrable models, knot theory and the properties of molecular graphs.

REFERENCES

[A] J. W. Alexander, Topological invariants of knots and links, Trans. Amer. Math. Soc. 20 (1928), 275-306.

[B1] R. J. Baxter, Exactly solved models in statistical mechanics, Academic Press, London, 1982.

[B2] R. D. Brandt, W. B. R. Lickorish, and K. C. Millett, A polynomial invariant for nonoriented knots and links, Invent. Math. 84, (1986), 563-573.

[C] J. H. Conway, An enumeration of knots and links and some of their algebraic properties, Computational problems in abstract algebra, (John Leech,ed.), Pergamon Press, Oxford and New York 1969, 329-358.

[Co] N. R. Cozzaralli, J. H. White, and K. C. Millet, Description of the topological entanglements of DNA catenanes and knots by a powerful method involving strand passage and recombination, J. Mol. Biol. (1987) 197, 585-603.

[D] C. Dietrich-Buchecker, J.-P. Sauvage, A synthetic molecular trefoil knot, Angew. Chem. Int. Ed. Engl. 28 (1989), 189-192.

[F1] E. Flapan, Rigid and non-rigid achirality, Pacific Journal of Math. 129(1987), 57-66

[F2] E. Flapan, Symmetries of knotted molecular graphs, Discrete Applied Math, Vol. 19 (1988), 157-166.

[F3] E. Flapan, Symmetries of Mobius ladders, Mathematische Annalen, 283 (1989), 271-283.

[F4] P. Freyd, D. Yetter; J. Hoste; W. B. R. Lickorish, K. Millett; A. Ocneanu, A new polynomial invariant for knots and links, Bulletin (New Series) of the American Mathematical Society, Vol 12, No. 2, April 1985, 239-246.

[F5] H. L. Frisch and E. Wasserman, Chemical Topology, J. Am. Chem. Soc. 83 (1961), 3789-3795.

[J1] V. F. R. Jones, A polynomial invariant for knots via von Neumann algebras, Bull. Am. Math Soc. 12(1985),103-111.

[J2] V. F. R. Jones, Hecke algebra representations of braid groups and link polynomials, Ann. of Math. 126 (1987). 335 -388.

[J3] V. F. R. Jones, On knot invariants related to some statistical mechanical models, preprint 1988.

[J4] D. Jonish, K. C. Millett, Extrinsic topological chirality indices of molecular graphs (with Jonish), Graph Theory and Topology in Chemistry 51(1987), 82-90.

[J5] D. Jonish, K. C. Millett, Isotopy invariants of graphs, Lab. de Math. Marseille, n^{0}89-22, July 1989

[K1] L. Kauffman, Invariants of Graphs in three space, Trans. Amer. Math. Soc., vol. 311 (1989), 679-710.

[K2] L. Kauffman, An invariant of regular isotopy, to appear in Trans. Amer. Math. Soc.

[K3] L. Kauffman, State models and the Jones polynomial, Topology, 26 (1987), 395-407.

[K4] L. Kauffman and P. Vogel, Link polynomials and graphical calculus, preprint 1987.

[K5] S. Kinoshita, Elementary ideals in knot theory, Kwansei Gakuin Univ. Annual Studies, 35.

[K6] S. Kinoshita, On elemenatary ideals of polyhedra in the 3-shpere, Pac. Jour. of Math. 42 (1972), 89-98.

[K6] S. Kinoshita, Alexander polynomial as isotopy invariants I, Osaka Math. J. 10(1958), 263-271.

[K7] S. Kinoshita, Alexander polynomial as isotopy invariants II, Osaka Math. J. 11(1959), 91-94.

[K8] K. Kobayashi, Reduced degree of Yamada polynomial and planarity of graphs, preprint 1987.

[L1] W. B. R. Lickorish, The panorama of polynomials for knots, links, and skeins, Proc. Artin's Braid Group Conference, Santa Cruz (1986).

[L2] W. B. R. Lickorish, Polynomials for links, Bull. London Math. Soc. 20(1988),558-588.

[L3] W. B. R. Lickorish and K. C. Millett, The new polynomials for knots and links, Mathematics Magazine 61(1988), 3-23.

[L4] W. B. R. Lickorish and K. C. Millett, A polynomial invariant for oriented links, Topology 26(1987), 107-141.

[M1] K. C. Millett, An invariant of 3-valent Spatial Graphs, preprint 1990.

[M2] K. C. Millett, Stereotopological indices for a family of chemical graphs, J. Comp. Chem. 8 (1987), 536-550.

[M3] K. C. Millett, Configuration census, topological chirality and the new combinatorial invariants, The Proceedings of the International Symposium on Applications of Mathematical Concepts to Chemistry, Croatica Chemica Acta., Vol. 59 (3) (1986), 669-684.

[N] S. Negami, Polynomial invariants of graphs, Trans. Amer. Math. Soc. 299 (1987), 601-622.

[P] C. D. Papakyriokopolous, On Dehn's lemma and the asphericity of knots, Ann. of Math. 66 (1957), 1-26.

[S1] M. Scharlemann, A. Thompson, Detecting unknotted graphs in 3-space, preprint 1989

[S2] G. Schill, Catenanes, Rotaxanes, and Knots, Academic Press, (Org. Chem. Mono. Ser., No. 22), 1971.

[S3] J. Simon, Topological chirality of certain molecules, Topology Vol. 25 (2) (1986), 229-235.

[S4] J. Simon, Molecular graphs as topological objects in space, J. Comp. Chem. 8 (1987), 718-726.

[S5] J. Simon, K. Wolcott, Minimally knotted graphs in S^3, preprint 1989

[S6] V. I. Sokolov, Topological ideas in stereochemistry, Russian Chemical Reviews 42(6), (1973), 452-463.

[S7] D. W. Sumners, Knots, macromolecules and chemical dynamics, Graph Theory and Topology in Chemistry, Elsevier (1987), 3-22.

[S8] S. Suzuki, On linear graphs is 3-space, Osaka J. Math. 7 (1970), 375-396.

[W1] D. M. Walba, Stereochemical topology, Proceedings of Symposium on Chemical Applications of Topology and Graph Theory, University of Georgia, 1983, R. B. King, Ed., Elsevier Pub., 1983.

[W2] D. M. Walba, R. M. Richards, and R. C. Haltiwanger, Total synthesis of the first molecular Möbius strip, J. Am. Chem. Soc, 104 (1982), 3219-3221.

[W3] D. M. Walba, Topological Stereochemistry, Tetrahedron Vol. 41(16) (1985), 3161-3212.

[W4] E. Wasserman, Chemical topology, Scientific American 207(5) (1962), 94-102.

[W5] E. Witten, Quantum field theory and the Jones polynomial, preprint 1988

[W6] E. Witten, Gauge theories and integrable lattice models, preprint February 1989

[W7] E. Witten, Gauge theories, vertex models, and quantum groups, preprint May 1989

[Y1] S. Yamada, An invariant of spatial graphs, preprint 1987.

[Y2] S. Yamamoto, Knots in spatial embeddings of the complete graph on four vertices, preprint 1988.

TOPOLOGICAL TECHNIQUES TO DETECT CHIRALITY

Erica Flapan
Department of Mathematics
Pomona College
Claremont, CA 91711

Recent progress in topological stereochemistry has led to the synthesis of molecules whose embeddings in 3–spaces are topologically non–trivial, for example, the molecular Möbius strip which was synthesized in 1982 by Walba, Richards and Haltiwanger [WRH]. Synthesis of a knotted molecule has been sought for some time and has finally now been achieved by Dietrich–Buchecker and Sauvage [DS]. Also, recent work with DNA has resulted in various knots being synthesized and even used to help understand recombinant mechanisms [WDC].

As knots and other topologically interesting molecules are now under investigation by synthetic chemists and molecular biologists, it may be useful for chemists to become familiar with recent results from the field of mathematics known as topology, which can be used to detect chirality for knots and graphs. Before we begin, however, it is important to note that topological achirality does not have precisely the same meaning as chemical achirality. Since the chemical achirality of a molecule depends on physical conditions, it is mathematically somewhat intractable. So instead we introduce the notion of the topological achirality of a molecular graph which depends only on how the graph is embedded in space. That is, it assumes complete flexibility with no geometric constraints on bond lengths or angles. In particular, we define a molecule to be <u>topologically achiral</u> if its molecular bond graph can be deformed in 3–dimensional space to its mirror image. Although topological achirality is easier to detect mathematically than chemical achirality, it may not always provide the most useful information for the chemist. In contrast, a molecule which is topologically chiral cannot be deformed to its mirror image by realizable molecular motions. Hence, topological chirality does imply chemical chirality.

Topologists in the subfield of knot theory have been interested in determining which knots and links have mirror image symmetry since the mid–nineteenth century [Lis], [Tai], [Has]. Historically, in the topological literature, a knot or link which can be deformed to its mirror image is said to be <u>amphicheiral</u> (There was no mathematical term for a knot or link which cannot be deformed to its mirror image). However, very recently, as a result

P. G. Mezey (ed.), New Developments in Molecular Chirality, 209–239.

of increased communication with chemists, some topologists are using the terms topological achirality and topological chirality.

This chapter is divided into three sections. The first section summarizes results from knot theory on determining the topological chirality of knots and links. In the second section we discuss the concept of rigid achirality, which is a mathematically characterizable property that is more closely related to chemical achirality than topological achirality. In particular, we show that there exist topologically achiral structures which have no topologically accessible symmetry presentation, that is, although such a structure can be deformed to its mirror image, it cannot be deformed to a rigidly non–dissymmetric conformation. The third section focuses on results about the topological chirality of graphs. Here we discuss how the embedding of a graph may or may not affect its chirality. Specifically, we present examples of graphs which are intrinsically chiral. That is, which are topologically chiral no matter how they are embedded in space. For the most part, throughout this chapter we present Theorems without proofs, because the proofs are too technically complex to include. However, we have chosen to present a few of the less technical proofs in order to illustrate the type of ideas which form the basis for these results.

Section 1 : Topological Chirality of Knots & Links

We begin by specifying that by a knot we mean a curve in three dimensional space which begins and ends at the same point and does not intersect itself. A knot is said to be the unknot if it can be deformed (without breaking it or pushing it through itself) so that it lies in a plane, otherwise it is non–trivial. Figure 1 illustrates an unknot and a non–trivial knot. The non–trivial knot in Figure 1 is known as the trefoil knot. It is the simplest knot and the first knot which was proved to be topologically chiral [De]. Shortly, we shall present a more modern proof of the chirality of the trefoil knot.

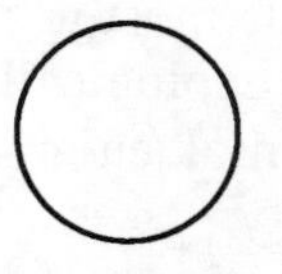
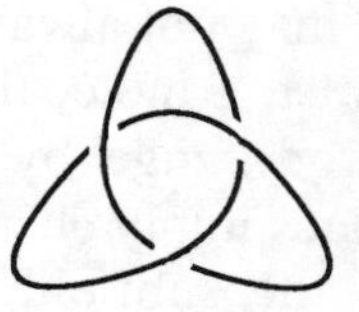

Figure 1

A link is any collection of one or more disjoint knots known as the components of L. In general, we illustrate knots and links (as in Figure 1) by projecting them into a plane and drawing a broken arc to indicate a lower string which passes under an upper string. Such a planar diagram is called a projection of the knot or link. Given a projection of a link, a projection of its mirror image can be obtained by changing all of its overcrossings to undercrossings and vice versa. The trefoil and its mirror image are illustrated by the projections in Figure 2.

Figure 2

We can conclude that a knot is topologically achiral if we can illustrate the deformation of it to its mirror image by some manipulation of a projection of the knot to a projection of its mirror image. Reidemeister proved [Rei] that one knot (or link) can be deformed into another if and only if a projection of the first can be transformed into a projection of the second by a deformation within the plane together with the following three moves.

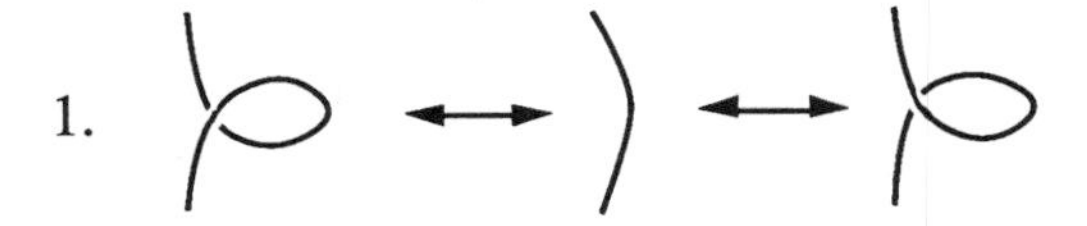

(i.e. add or remove a kink)

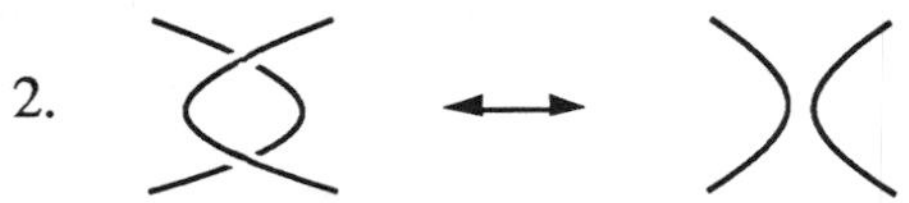

(i.e. add or remove two adjacent undercrossings or overcrossings)

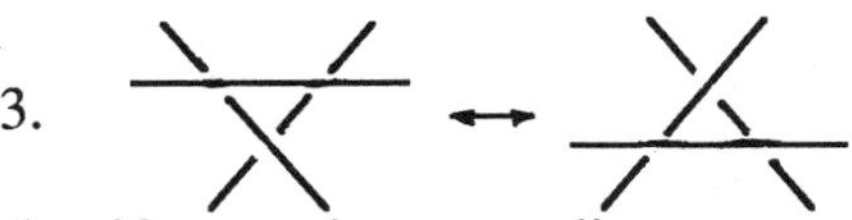

(i.e. if an arc has two adjacent overcrossings which are also each adjacent to another common crossing, then the arc can be passed over that common crossing.The move is analogous for undercrossings)

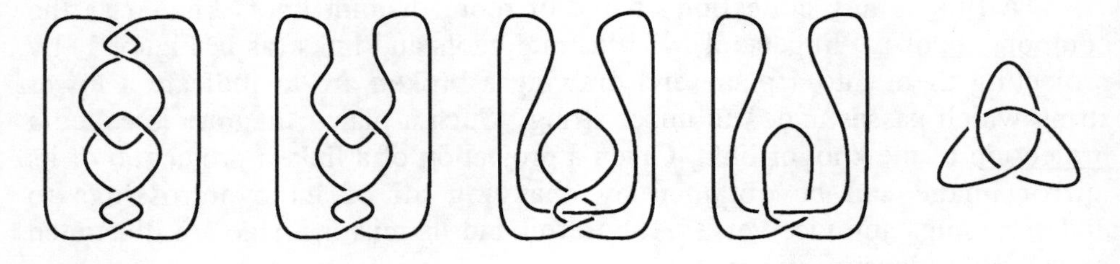

Figure 3

These three types of moves are called <u>Reidemeister moves</u>. Figure 3 illustrates how to go from a knot projection to a simpler one by using deformation within the plane together with the Reidemeister moves.

Although Reidemeister's Theorem tells us that any deformation can be accomplished by using these moves, it does not tell us how many such moves are needed or in what order they should be performed. Thus these moves do not tell us whether or not a knot is topologically achiral. However, the Reidemeister moves are used to establish certain topological invariants which are an important tool for detecting chirality. The invariants which we are interested in are polynomials which are associated to links and which do not change even when the link is deformed. Thus, if a knot, K can be deformed to its mirror image, then the polynomial associated to K and the one associated to the mirror image of K, should be the same. If they are different then we know that K is topologically chiral.

There are several important link polynomials [Kau], [LM], however they are similar in flavor, and so we present only one. (Our presentation here is based on that of [LM]). This polynomial invariant is known as the P–polynomial and was discovered simultaneously by five sets of authors [FYHLMO] and [PT]. Given a link L, with one or more components, we orient L by assigning an arrow to each component to indicate a preferred direction. To the oriented link L we associate a two variable Laurent polynomial P(L) in the variables ℓ and m. A <u>Laurent polynomial</u> is a polynomial which can contain both positive and negative powers of the variables. What is important about P(L) is that if L* can be obtained from L by a deformation then P(L*) = P(L).

Given a projection of the oriented link L, we define P(L) recursively by the following two axioms.

1. P(unknot) =1
2. If L_+, L_-, and L_o are three oriented link projections which are identical except near a crossing where they differ by a positive, negative, or null

crossing respectively (as is illustrated in Figure 4). Then
$\ell P(L_+) + \ell^{-1}P(L_-) + m\ P(L_o) = 0$.

Figure 4

What is meant by saying that P(L) is defined recursively is that, given a link, we can compute its polynomial in terms of the polynomials of simpler links, which in turn were computed in terms of links which are simpler still, and so on until we get to one or more unknots whose polynomials are each known to be 1.

As an illustration, we compute several examples. Let L consist of two oriented circles which are unknotted and unlinked. Then L is just the link designated by L_o in Figure 5.

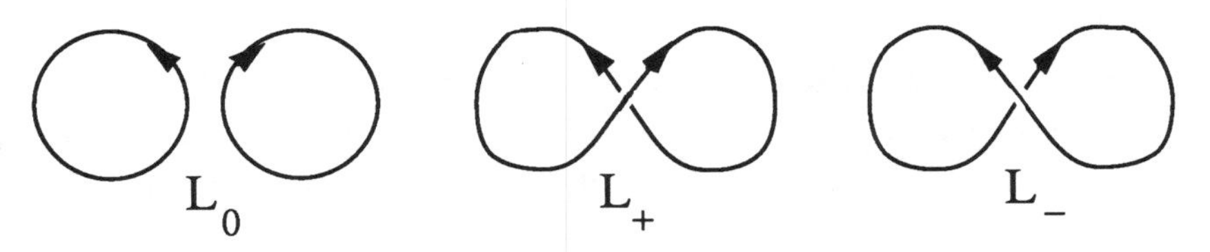

Figure 5

By creating a positive crossing or a negative crossing we get L_+ or L_- respectively. By the second axiom in the definition of P(L) we have
$\ell\ P(L_+) + \ell^{-1}\ P(L_-) + m\ \ P(L_o) =0$. Since in our case, $L=L_o$, we have
$m\ P(L) = -\ell P(L_+) - \ell^{-1}\ P(L_-)$. But, in fact L_+ and L_- are both unknots, so by Axiom 1, $P(L_+) =1= P(L_-)$. Thus $P(L) = -m^{-1}(\ell + \ell^{-1})$.

Now suppose we want to compute $P(L^*)$ where L^* denotes the oriented linked rings . We select a crossing that we are going to change, say the upper crossing. We observe that the upper crossing in L^* is a negative crossing, so we will designate L^* by L_-. Then we change the upper crossing to a positive crossing or a null crossing to create L_+ or L_0, as is illustrated in Figure 6. Note that we must be careful that when we change a crossing we do so in such a way that the new link is coherently oriented.

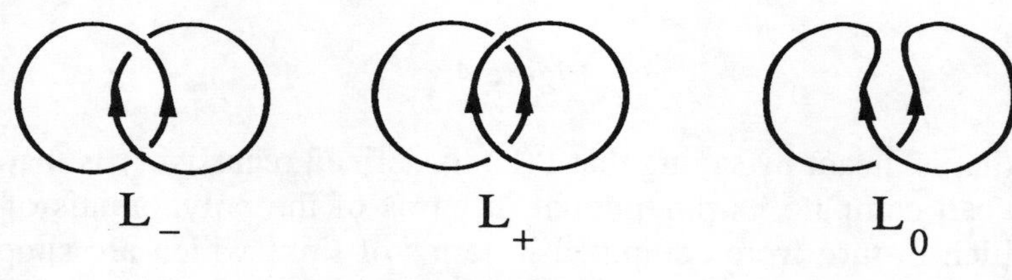

Figure 6

Now L_+ is the link whose polynomial we just computed, and L_0 is the unknot. So $P(L_+) = -m^{-1}(\ell + \ell^{-1})$ and $P(L_0) = 1$. Thus $P(L^*) = P(L_-) = -\ell(mP(L_0) = \ell\, P(L_+)) = -\ell(m - \ell m^{-1}(\ell + \ell^{-1})) = -\ell m = \ell^3 m^{-1} + \ell m^{-1}$.

As a final example, let K denote the oriented trefoil knot . We select the upper right hand crossing to work with. It is again a negative crossing, so let $L_- = K$ and create L_+ and L_0 as illustrated in Figure 7.

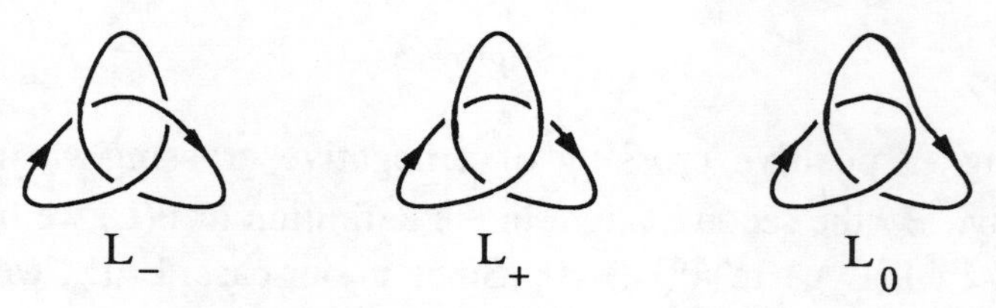

Figure 7

By deforming L_0 slightly we get the link L^* whose polynomial we computed to be $P(L^*) = -\ell m = \ell^3 m^{-1} + \ell m^{-1}$. Since L_+ is again the unknot, we obtain $P(K) = P(L_-) = -\ell(m(-\ell m + \ell^3 m^{-1} + \ell m^{-1}) + \ell) = \ell^2 m^2 - \ell^4 - \ell^2 - \ell^2 = -\ell^4 + \ell^2 m^2 - 2\ell^2$.

We can often use the P–polynomial to detect chirality for oriented links, according to the following Theorem.

Theorem 1 : Let L be an oriented link with polynomial P(L). Let $\overline{P}(L)$ denote P(L) with ℓ and ℓ^{-1} interchanged. If P(L) does not equal $\overline{P}(L)$ then L is topologically chiral.

Proof: Let L* denote the mirror image of L. Then L* is obtained by reversing all of the crossings of L. Thus every positive crossing  of L, becomes a negative crossing  of L* and every negative crossing of L becomes a positive crossing of L*. So the roles of positive and negative crossings in L* are the reverse of what they are in L. Since the defining equation is $\ell P(L_+) + \ell^{-1}P(L_-) + m\, P(L_0) = 0$, the polynomial of L* is just P(L) with the roles of ℓ and ℓ^{-1} interchanged. Hence P(L*) =$\overline{P}(L)$. If L is topologically achiral then L can be deformed to L*. So P(L) = P(L*). Thus if P(L) does not equal $\overline{P}(L)$ then L is topologically chiral.

We can use Theorem 1 to conclude that the trefoil knot K is topologically chiral, since $\overline{P}(K) = -\ell^{-4} + \ell^{-2}m^2 - 2\ell^{-2}$ does not equal P(K).

We remark here that while a link must be oriented to compute its polynomial, for knots the choice of orientation has no effect on the polynomial. This is because changing the orientation of a knot has the effect of reversing the direction of both arrows occurring at any crossing. So a positive crossing  becomes  , which by rotating 180 degrees we see is still a positive crossing. Similarly, a negative crossing remains a negative crossing.

Hence we conclude from Theorem 1, that the following corollary is true.

Corollary 1 : Regardless of choice of orientation, a knot K is topologically chiral if $\overline{P}(K)$ does not equal P(K).

In particular, the unoriented trefoil knot is topologically chiral. Analogously, for a link L, if we reverse the orientation of all of the components of L, it has no effect on P(L). But if we change the orientation of some components of L, and leave other components with their original orientation, it may very likely change the value of P(L).

Certain types of knots are more easily analyzed than others. One particular class of knots which is very well understood is the collection of "torus knots." Consider an ordinary doughnut or bagel. The surface of such a doughnut is called a torus. Any knot which can be drawn on such a surface is known as a torus knot. For example, Figure 8 illustrates two different torus knots.

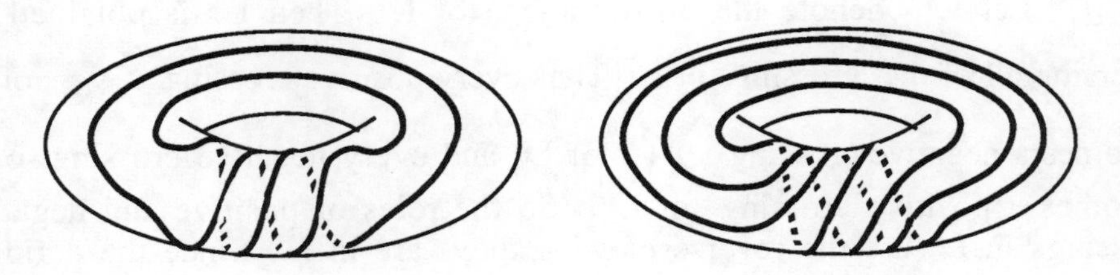

Figure 8

Note that the knot illustrated on the left side of Figure 8 is just the trefoil knot. Torus knots are characterized by two numbers, how many times the knot winds around the hole of the doughnut and how many times it wraps around the doughnut itself. For example, the trefoil knot goes twice around the hole, but three times around the doughnut itself. For this reason the trefoil is also known as a torus knot of type 2, 3. For any two integers p and q whose greatest common divisor is 1, there is a non–self–intersecting curve Tp,q which goes p times around the doughnut hole and q times around the doughnut. Thus there are infinitely many different torus knots.

We saw how to prove that the trefoil knot is topologically chiral. In fact, in 1923, Schreier [Schr] proved the following Theorem using group theory.

Theorem 2 : All torus knots are topologically chiral.

Although we do not present the proof here, the chirality of torus knots can also be proved using polynomials [Jon] as we did for the trefoil knot.

Recall that the P–polynomial is one example of a topological invariant of links. That is, given a link L, P(L) will remain the same regardless of how the link is deformed. Polynomials are not the only topological invariants of links. There are many numbers associated with a link, which will remain constant when the link is deformed. The simplest such link invariant is the number of components in the link. For example, a link made of two rings cannot be deformed into a link of three rings. While the number of components of a link is easy to count, it does not give us very much information, since there are infinitely many different links with the same number of components. However, it is possible to use more than one invariant at a time. For example, for a link L with a given number of components we can then ask what is the minimum number of crossing points needed in a projection of L. For example, the trefoil knot has minimum crossing number three. Tables of knots (see for example [Rol]) are generally organized according to minimum crossing number. However, the minimum crossing number is often difficult to compute. Given a projection of a knot, it is not always obvious whether or not it could be deformed so as to have fewer crossings. Of course, the minimum crossing number is not a sufficient invariant to classify all knots. In fact, a knot and its mirror image will always have the same minimum crossing number. So the crossing number itself will not detect chirality. However, there is a very old important conjecture which still remains unsolved [Tai], which is as follows:

Tait conjecture : Any knot whose minimum crossing number is odd must be topologically chiral.

While this conjecture has not, as of yet, been proven, there is also no known example of a topologically achiral knot which has an odd minimum crossing number. On the other hand, not every knot with even minimum crossing number is topologically achiral. For example, the knot 6_1 , which is illustrated in Figure 9, has minimum crossing number six (in fact, the reason it is called 6_1, is that in the knot tables it is the first knot listed with minimum crossing number six).
Yet $P(6_1) = (-\ell^{-2} + \ell^2 + \ell^4) + (1-\ell^2)m^2$ does not equal $\overline{P}(6_1)$, hence 6_1 is topologically chiral.

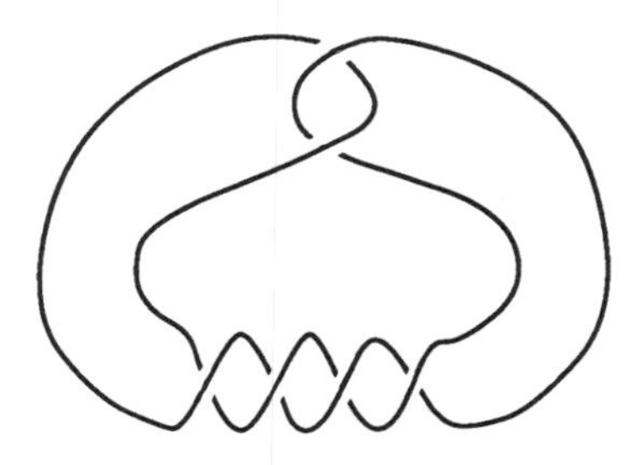

Figure 9

While the above mentioned conjecture regarding the minimum crossing number has not yet been solved for arbitrary knots, it has recently been solved for a large class of knots [Mur], as we shall explain below. Given a link L, a projection of L is said to be alternating if each component goes alternately over and under at successive crossing points. So, for example, the projection of 6_1 illustrated in Figure 9 is an alternating projection. In contrast, the knot projection in Figure 10 is not alternating.

Figure 10

Any link which has an alternating projection is called an alternating link. The alternating links are a large and important class of links which are often easier to work with than arbitrary links. For example, for an arbitrary knot, it is difficult to determine whether a given projection has the minimum number of crossings. However, it was recently proved [Mur], [Thi] using polynomials, that if the projection of a link is a so called "reduced" alternating projection then it has the minimum number of crossings. An example of an alternating projection which is not reduced is given in Figure 11. In general, a crossing is said to be removable if the projection looks like (or its mirror image), where each ball contains a tangled arc and possibly other components of the link. So the knot projection in Figure 11 contains a removable crossing in the center.

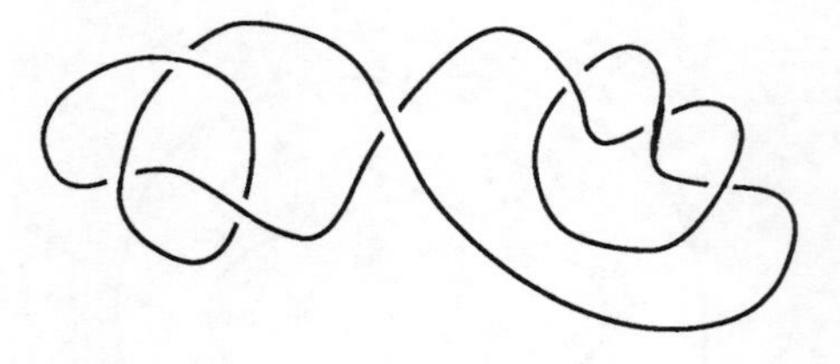

Figure 11

A removable crossing is easy to recognize, and can be eliminated by grabbing one of the balls and untwisting it. An alternating link projection which contains no removable crossings is said to be reduced. Since it is not hard to determine if a projection is a reduced alternating projection, this gives us an easy way to determine that such a projection has a minimal number of crossings.

Returning now to the question of chirality, we have the following important Theorem, recently proved by Murasugi [Mur].

Theorem 3 : Let K be a knot with a reduced alternating projection which has an odd number of crossings. Then K is topologically chiral.

This Theorem proves that Tait's conjecture is true for alternating knots. In particular, we can use Theorem 3 to immediately detect the chirality of many knots. For example, we can conclude by inspecting its projection that the knot illustrated in Figure 12 is topologically chiral.

Figure 12

While Theorem 3 is very useful for detecting chirality of knots with a reduced alternating projection which has an odd number of crossings, we need to use another method for those which have an even number of crossings. For this purpose we introduce the notion of the "writhe" of a knot projection. Recall, from our discussion of the P–polynomials, that if K is an oriented knot projection then any crossing of K which looks like ╳ is called a positive crossing, and any crossing of K which looks like ╳ is called a negative crossing. To obtain the writhe of the projection we add +1 for every positive crossing and –1 for every negative crossing. For example, we have computed the writhe for the two oriented knot projections shown in Figure 13.

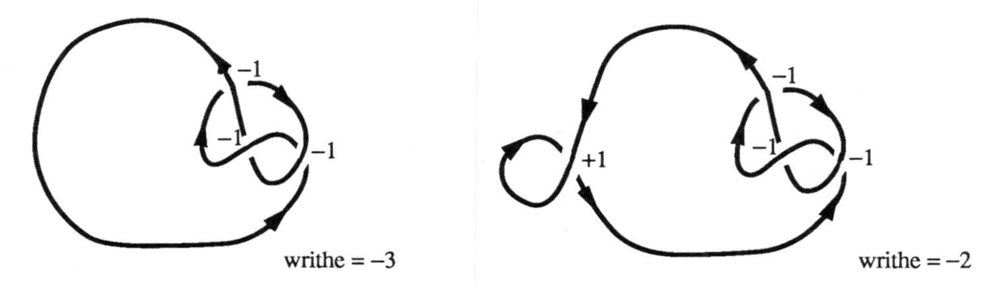

Figure 13

As the reader can easily observe, the two knot projections illustrated in Figure 13 can be deformed into the same knot. Hence, the writhe is not an invariant of the knot. In fact, given a knot we can always add an appropriate number of small kinks to create any writhe that we want. However, Thistlethwaite [Thi] has recently proved that given an alternating knot, any oriented reduced alternating projection of it has the same writhe. Thus for an alternating knot K, we let w(K) denote the writhe of an oriented reduced alternating projection of K. Note that, as with knot polynomials, while we need to orient K in order to compute the writhe of a reduced alternating projection, the choice of orientation itself has no effect on the number of positive or negative crossings. Thus either orientation will give the same w(K). From Thistlethwaite's result, we prove the following Theorem.

Theorem 4 : Let K be an alternating knot. If w(K) does not equal 0 then K is topologically chiral.

Proof : Consider the mirror image K* of an oriented reduced alternating projection of K. As we observed in the proof of Theorem 1, every positive crossing of K corresponds to a negative crossing of K*, and every negative crossing of K corresponds to a positive crossing of K*. Thus w(K) = –w(K*). Suppose K is topologically achiral. Then K can be deformed to K*. Hence K* is also a reduced alternating projection of K. So by Thistlethwaite's Theorem, w(K) = w(K*). Hence if K is topologically achiral then w(K) = 0. Given that w(K) does not equal 0 we conclude that K is topologically chiral.

For alternating knots, whose reduced alternating projections have an even number of crossings we can often use Theorem 4 to detect chirality. (Recall, that if the reduced alternating projection has an odd number of crossings we conclude immediately by Theorem 3 that the knot is topologically chiral.) For example, the knot projection illustrated in Figure 14 is easily seen to be reduced alternating and to have writhe 4 Hence, without having to compute its polynomial, we know that the knot is topologically chiral.

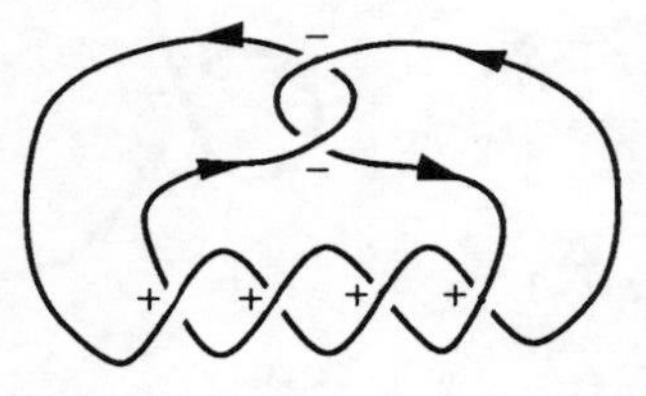

Figure 14

The reader must be careful to remember that the converse of Theorem 4 is not true. That is, there exist alternating knots K, such that w(K) =0 and K is nonetheless topologically chiral. For example, the knot which is shown in Figure 15 has P–polynomial $-\ell^4m^4 + 3\ell^4m^2 - 2\ell^4 + \ell^2m^6 - 4\ell^2m^4 + 5\ell^2m^2 - \ell^2 - m^4 - 4m^2 + 5 - \ell^{-2}m^4 - 3\ell^{-2}m^2 + 3\ell^{-2}$ so it is topologically chiral. Yet we see that the reduced alternating projection which is given in Figure 15 has writhe 0. Hence even for alternating knots we may have to use more than one method to detect topological chirality, and as of now there is no sure–fire method of detecting chirality either for alternating or non–alternating knots.

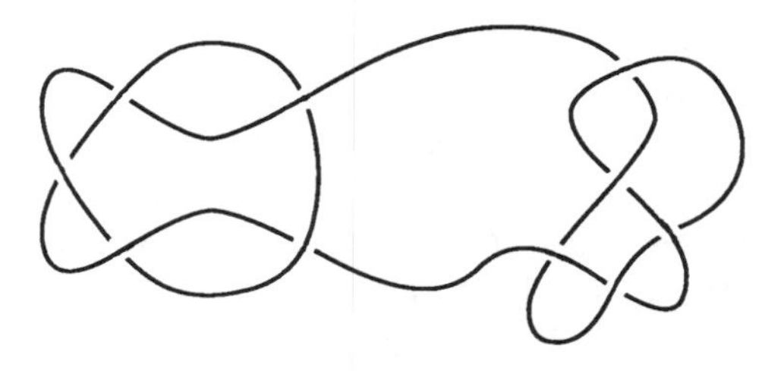

Figure 15

Section 2 : Rigid Achirality

We have seen several methods of detecting topological chirality, which for molecular structures will imply chemical chirality. However, structures which are seen to be topologically achiral may nonetheless be chemically chiral due to molecular rigidity. Here, we follow Walba [Wal] and introduce the notion of rigid achirality as follows. Rather than allowing any deformation of the molecular bond graph to its mirror image, we first deform the graph into an optimal position and then require that it remain rigid. This optimal position is said to be an accessible symmetry presentation. If the structure can be deformed to a symmetry presentation which can then be rotated through an angle of $2\pi p/q$, where p and q are integers, to obtain its mirror image, then the structure is said to be rigidly achiral. (The restriction that we rotate only through such rational angles is not unreasonable since a molecular graph is only a model of reality.)

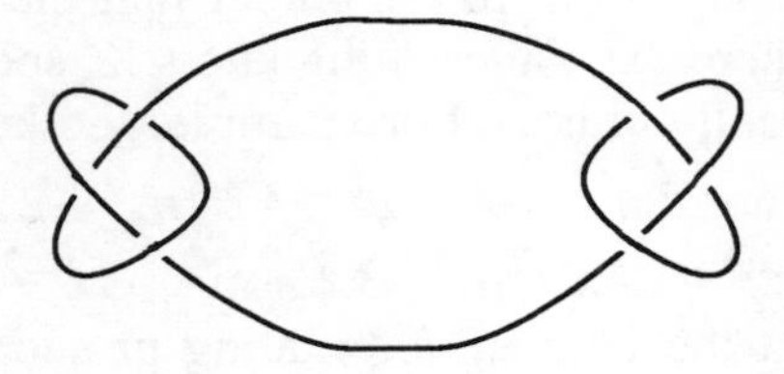

Figure 16

The knot, seen in Figure 16 has, in fact, been drawn in a symmetry presentation. Thus this knot is rigidly achiral. Any graph which is rigidly achiral is necessarily topologically achiral. Since a rotation is a particular type of deformation, any graph which can be deformed to a symmetry presentation can certainly be deformed to its mirror image. We are interested in the reverse question. That is, is every molecular bond graph which is topologically achiral necessarily rigidly achiral? (This question was asked in [Wal_1] and in [Wal_2]). Mislow [Mis] gave examples of disubstituted biphenyls that are achiral yet have no <u>chemically</u> accessible symmetry presentations. Walba raised the question of whether such a phenomenon can occur in completely flexible structures, with no geometric constraints on bond lengths. That is, we replace <u>chemically</u> accessible with <u>topologically</u> accessible.

We show that there are knots as well as other graphs which are topologically achiral but not rigidly achiral, and thus have no symmetry presentation. These may be regarded as hypothetical molecular bond graphs, since it is not yet known whether these knots could actually be synthesized.

In looking for topologically achiral knots which are not rigidly achiral it is necessary to be cautious because the symmetry presentation of a rigidly achiral knot is often hard to find. In fact, given a knot which is known to be rigidly achiral, there is no known algorithm to deform it to a symmetry presentation. For example, the figure eight knot is topologically achiral, as can be seen from the deformation indicated in Figure 17.

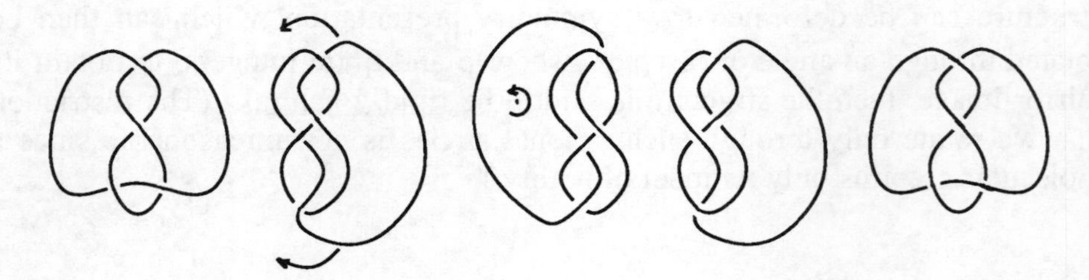

Figure 17

On the other hand, a symmetry presentation for the figure eight knot does not resemble any stage of the deformation. In fact, the figure eight knot must be deformed so that its projection has some additional crossings in order to achieve a symmetry presentation. A symmetry presentation, with S_4 symmetry, for the figure eight knot is illustrated in Figure 18.

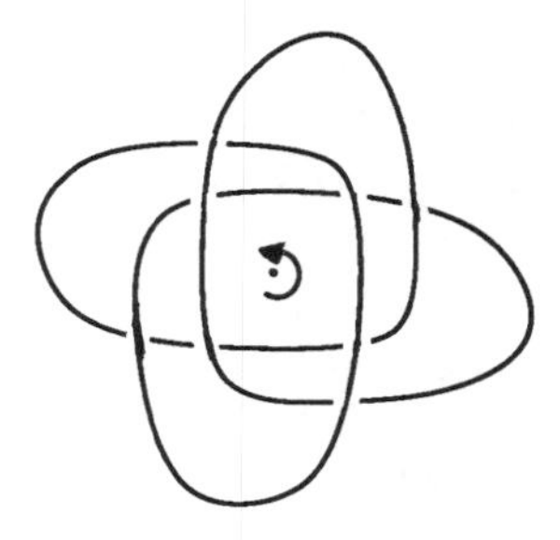

Figure 18

Note that while the reader might expect the figure eight knot to have an S_2 symmetry, such symmetries are actually not that common for knots. In fact, among the knots that can be drawn with less than 11 crossings, only the two illustrated in Figure 19 have symmetry presentations with an S_2 symmetry (see [HK]). Note that we illustrate these knots in their symmetry presentations (which have more than 10 crossings), rather than in their minimum crossing projections.

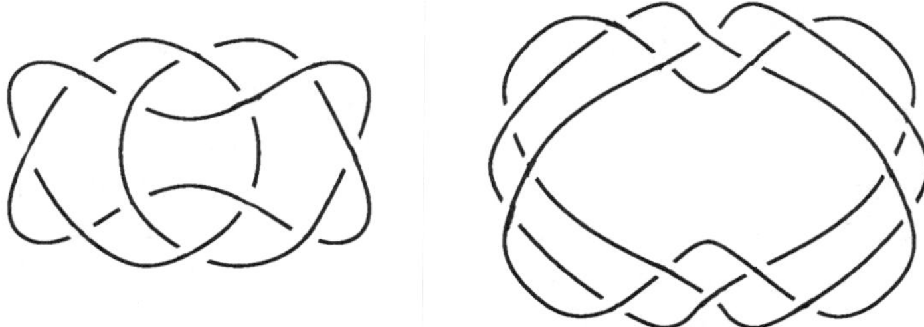

Figure 19

It turns out that our deformations actually contain somewhat more information than we have mentioned so far. If we begin by orienting our knots by drawing an arrow to indicate a preferred direction, we can also take note of whether a given deformation preserves the direction of the arrow or reverses it. Looking back at Figures 17, 18 and 19, the reader will see that the orientation of the knot in Figure 17 has been reversed by the deformation, while the orientations of the knots in Figures 18 and 19 are preserved by the rotation which takes the knot to its mirror image. In general, if we can deform an oriented knot to its mirror image in such a way that the orientation of the knot is preserved then we say the knot is positive achiral. On the other hand, if there is a deformation of the oriented knot to its mirror image which reverses the orientation of the knot, then we say the knot is negative achiral. Note that formally we should say "topologically positive achiral" and "topologically negative achiral," but for the sake of simplicity we suppress the word "topologically." A topologically achiral knot, can be positive achiral, negative achiral or both. For example, Figures 17 and 18 indicate that the figure eight knot is both positive and negative achiral.

Before we obtain the desired knots which are topologically achiral but not rigidly achiral, we need to introduce a few more definitions. Let G denote either a knot, a link or a graph which is embedded in 3–space. Let h represent a rigid motion of space which is either a reflection, a rational rotation (i.e. rotation through an angle of $2\pi p/q$ where p and q are integers), or a combination of the two. If h sends G to itself we write h(G) = G, and we say that h is a symmetry of G. Note that the word "symmetry" is used in mathematical literature with various different meanings. Many authors would refer to the symmetries which we describe here as "rigid symmetries." However, in order to keep the terminology as simple as possible, we simply use the word "symmetry" in this situation.

If h represents a reflection of 3–space, possibly together with a rational rotation, then we say that h is an orientation reversing symmetry. On the other hand if h is simply a rational rotation, we say that h is orientation preserving. This terminology refers to the fact that if we choose a right hand or left hand to indicate an orientation for 3–space, then h may reverse or preserve this orientation. Although we mentioned above that it is often hard to find a symmetry presentation for a knot, we shall assume here that any knot which is rigidly achiral has been put in a symmetry presentation. In this case, a knot K is rigidly achiral if and only if K has an orientation reversing symmetry h. If, in addition, the knot K is given an orientation, then we can distinguish between cases where the symmetry h preserves or reverses the orientation of K. In the case where h preserves the orientation of K, then K is said to be rigidly positive achiral. On the other hand, if h reverses the orientation of K, then K is said to be rigidly negative achiral.

Suppose h is simply a reflection of 3–space. Then we say h has order two since if we perform h twice (denoted by h^2) every point is returned to its original place. Suppose now that h is a reflection together with a rotation through an angle of $2\pi p/q$. By cancelling any common factors if necessary, we can assume that the greatest common divisor of p and q is 1. Thus q is the smallest number such that if we do the rotation q times every point will return to its original position. If q is even, then the order of h is q since h^q is the identity function. If q is odd, then the order of h is 2q since h is a reflection composed with a rotation. In any case the order of h is the smallest number such that if we iterate h that many times every point will return to its original place.

Figure 20 shows an example of a symmetry presentation of a knot having two different orientation reversing symmetries which are both of order two. Figure 20a is meant to illustrate the orientation reversing symmetry h_1, which is the composition of a 180 degrees rotation about the axis indicated, followed by a reflection through the plane of the paper. The orientation reversing symmetry h_2, indicated in Figure 20b is simply reflection through a plane which is perpendicular to the plane of the paper. Observe that h_1 preserves the orientation of the knot, while h_2 reverses it. Hence Figure 20 illustrates a knot which is both rigidly positive achiral, and rigidly negative achiral.

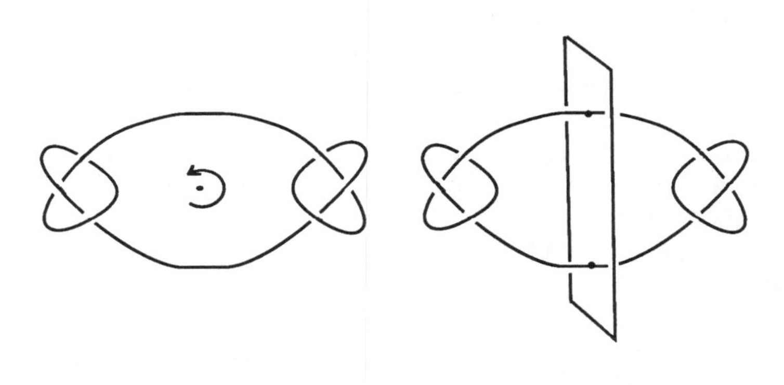

Figure 20a Figure 20b

In general, it is much easier to understand the symmetries of a knot, than to understand arbitrary deformations of a knot. One tool that is often helpful in analyzing symmetries is the "fixed point set." Given a symmetry h, the fixed point set of h, written fix(h), consists of all those points x in 3–space such that h(x) =x. For the examples in Figure 20, fix(h_1) is one point whereas fix(h_2) is a plane. In fact, these two examples illustrate all of the possibilities for the fixed point set of an orientation reversing symmetry. That is, if the symmetry is a rotation together with a reflection then it fixes precisely one point, while a reflection alone fixes a plane of points. In contrast, if h is an orientation preserving symmetry, (i.e. a rotation) then fix(h) is a line.

Now suppose K is a non–trivial knot which is rigidly achiral. Then let h be an orientation reversing symmetry of K. We consider the two possibilities for h separately. First suppose h is just a reflection. Then fix(h) is a plane separating 3–space into two half–spaces which are interchanged by the reflection. If K were entirely contained in one half–space, then h(K) would be contained in the other half–space. But we are assuming h takes K to itself, that is h(K) =K, so K cannot be entirely contained in one half–space. Also since K is not the unknot, K cannot lie entirely in the plane of fix(h). Hence K lies partially in each of the two half–spaces, and K intersects fix(h) as it goes from one half–space to the other.

Let A be an arc of K which is entirely contained in one of the half–spaces. Then the endpoints of A lie in fix(h). So h reflects A to an arc B lying in the other half–space but with the same endpoints as A. Hence taking the arcs A and B together we obtain a loop. Now since A is contained in K , B= h(A) is also contained in K=h(K). So the loop A∪B is contained in K. But K is a knot, and so consists of just one loop. In other words K is precisely the loop A∪B. Since K is a non–trivial knot, at least one of the arcs A or B must be knotted. But the arc B is the mirror image of the arc A. Thus if one arc is knotted, then both arcs must be knotted. So, in fact, K must look similar to the knot depicted in Figure 20. In other words, K consists of two knotted arcs which are separated by a plane.

What we have just discovered in this case about our knot K, is that it is a very special type of knot, known as a <u>composite knot</u>. In contrast, any non–trivial knot which cannot be separated by a plane into two knotted arcs, is known as a prime knot. The knots in Figure 19, as well as the trefoil and figure eight knot are examples of prime knots.

Thus if K is a non–trivial knot having a symmetry which is a reflection then K must be a composite knot. In addition, if we indicate a orientation for K by drawing an arrow on the arc A, then after the reflection we will see an arrow going in the opposite direction on the arc B. So the reflection h also reverses the orientation of the knot K. Thus the knot is rigidly negative achiral.

Let us now consider the case where h is a reflection together with a rotation. We want to show that in this case h preserves the orientation of K. We know that fix(h) is one point. So pick a point x on K, such that h(x) does not equal x. Let I denote one arc on K between x and h(x), and let J denote the other arc on K between x and h(x). Thus K is the union of I and J. Orient K by drawing an arrow on I pointing towards h(x). Now h(I) is an arc on K of the same length as I, but with endpoints h(x) and $h^2(x)$, and the arrow on h(I) is going towards $h^2(x)$. If h preserves the orientation of K, then the situation is as in Figure 21a), whereas if h reverses the orientation of K, we have the situation illustrated in 21b). That is, if h reverses the orientation of K then the arc h(I) is the same as the arc I but facing in the opposite direction. In this case, h fixes the midpoint of the arc I.

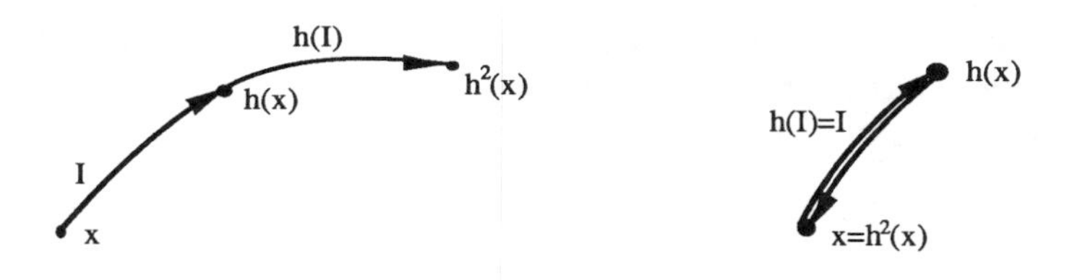

Figure 21

By a similar argument for the arc J, we see that if h reverses the orientation of K then h also sends the arc J to itself reversing its orientation. So in this case h also fixes the midpoint of the arc J. Thus if h reverses the orientation of K, then there are two points on K which are fixed by h. But we have seen above that when h is a rotation together with a reflection then h fixes precisely one point. Hence, we conclude that in this case h must preserve the orientation of K.

From the above discussion, Theorem 5 follows. We summarize the proof below for the sake of clarity.

Theorem 5 : No prime knot is rigidly negative achiral.

Proof : Let K be a prime knot, with an orientation reversing symmetry h. Suppose h is just a reflection, then as shown above K is composite. Since K is assumed to be prime, h cannot be just a reflection. Thus h is a reflection together with a rotation. In this case, we have seen that h preserves the orientation of K. Hence there is no orientation reversing symmetry which reverses the orientation of K.

As a result of this Theorem we can now easily find examples of knots in 3–space which are topologically achiral but not rigidly negative achiral. The simplest such example is the figure eight knot. We have already seen in Figure 17 that the figure eight knot can be deformed to its mirror image in such a way that its orientation is reversed. But since it is a prime knot it cannot be rigidly negative achiral. However, the figure eight knot is rigidly positive achiral, as we saw in Figure 18.

Nonetheless, Theorem 5 also suggests how to find examples of knots in 3–space which are topologically achiral but neither rigidly negative achiral nor rigidly positive achiral, and hence have no symmetry presentation. If there is a knot which is prime, topologically negative achiral, but not positive achiral, by Theorem 5, such a knot cannot be rigidly negative achiral; and since it is not positive achiral it is clearly not rigidly positive achiral. Since the knot is negative achiral it will be topologically achiral. Thus any such knot is topologically but not rigidly achiral. The knot illustrated in Figure 22 is the simplest such knot. In the standard knot tables (see, for example, [Rol]) this knot is known as 8_{17} because it is the seventeenth knot which has minimum crossing number eight. In Figure 22 we have drawn this knot as it usually appears in the tables.

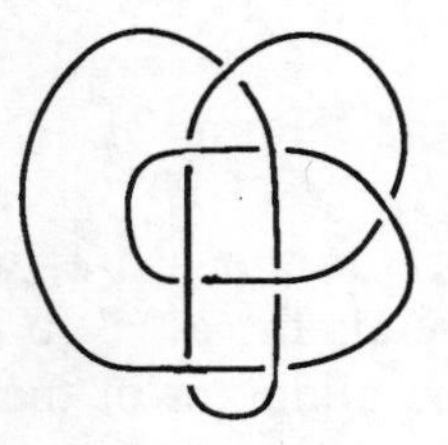

Figure 22

In order to see how the knot 8_{17} can be deformed to its mirror image it is easier if we do not use the standard projection shown in Figure 22. Instead we first deform it so that it appears as in Figure 23, where we then show how to deform it to its mirror image. Since this knot is topologically achiral but not rigidly achiral this deformation is necessarily a chiral pathway for racemization.

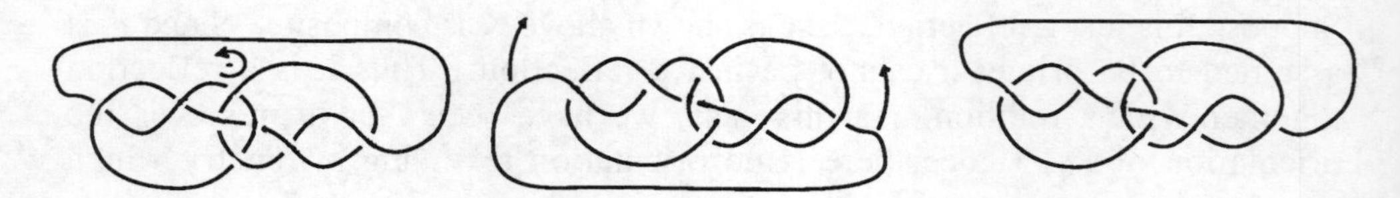

Figure 23

There are many other prime negative achiral knots which are not positive achiral, and hence will yield similar examples. However, the knot 8_{17} is the simplest such example. It is also possible to construct knots which are topologically positive achiral but not rigidly achiral (see [Fla_1]). But that construction is substantially more difficult than the construction of topologically negative achiral knots which are not rigidly achiral.

In contrast to knots, it is not hard to find graphs which are topologically achiral but not rigidly achiral. Let K be a prime achiral knot, for example the figure eight knot. Let C be an ordinary unknotted circle which meets K at one point. Let the graph G = K∪C be as illustrated in Figure 24. Since K can be deformed to its mirror image and C is its own mirror image, one simply performs the deformation of K shown in Figure 17, then slides C back into place at the end to get a deformation of G to its mirror image. Hence G is topologically achiral. However, suppose h were an orientation reversing symmetry of G. Then since K is knotted and C is the unknot, h would have to take K to itself, and h would have to take C to itself. Hence h would fix the point where C is attached to K. Thus h must also fix the point on the circle C which is directly opposite the point where C is attached to K. Now recall that a reflection fixes a plane of points, whereas a reflection together with a rotation fixes only one point of 3–space. Since h fixes at least two points we conclude that h is a reflection. By ignoring C we see that h is a symmetry of K which is a reflection. But, by the argument before Theorem 5, we saw that this implies that K is a composite knot. However we chose our knot K to be the figure eight knot, which is a prime knot. So we conclude that there is no orientation reversing symmetry of G. In other words, our graph G is topologically achiral but not rigidly achiral.

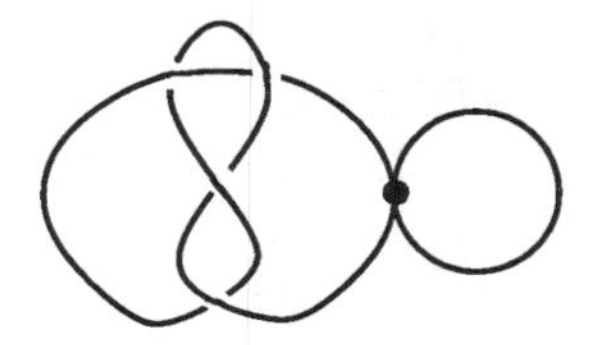

Figure 24

Observe that for graphs we do not distinguish between positive and negative achirality because, unlike for a knot, there is no natural way to give a graph an orientation.

Section 3 : Chirality of Graphs

As the reader may have observed in the example above, because of the internal structure that graphs have, it is often easier to analyze the topological achirality of graphs than that of knots. In fact, for graphs there are two levels of chirality that can be distinguished. As above, a graph embedded in 3–space is topologically chiral if it cannot be deformed to its mirror image. In contrast, a graph is intrinsically chiral if no matter how it is embedded in 3–space it can never be deformed to its mirror image. A knot can always be reembedded in 3–space as a planar circle, so knots are never intrinsically chiral. It is not immediately obvious that there exist graphs which are intrinsically chiral. However, we shall see examples of such graphs.

We begin our discussion by looking at the "molecular Möbius ladder" which was synthesized by Walba et al [WRH]. This is a molecule shaped like a ladder with three rungs which was made to join itself end–to–end with one half twist. See Figure 25.

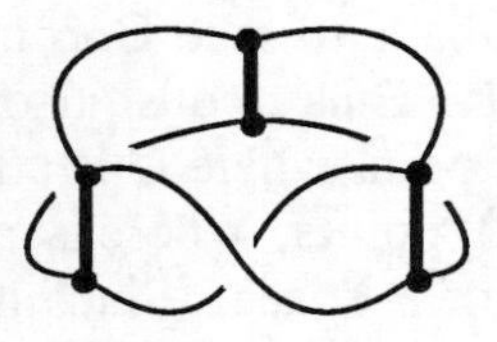

Figure 25

Walba [Wal] observed that this Möbius ladder can be represented by a graph with six vertices, where the six edges along the sides of the ladder represent the polyethyleneoxy chains and the three rungs of the ladder represent the C=C bonds. Walba noted that there was chemical evidence to suggest that this molecule was chemically chiral, however chemical achirality could not be completely ruled out until Simon [Sim] proved that its associated embedded graph was topologically chiral.

More generally, let M_n denote the graph illustrated in Figure 26, with $n \geq 3$ where the rungs of the ladder are denoted by $\alpha_1, ..., \alpha_n$ and the sides of the ladder together form the loop K. Observe that the graph M_3 is just the bipartite graph $K_{3,3}$ which is one of Kuratowski's non–planar graphs. For all $n \geq 3$, M_n contains $K_{3,3}$ and hence is itself non–planar

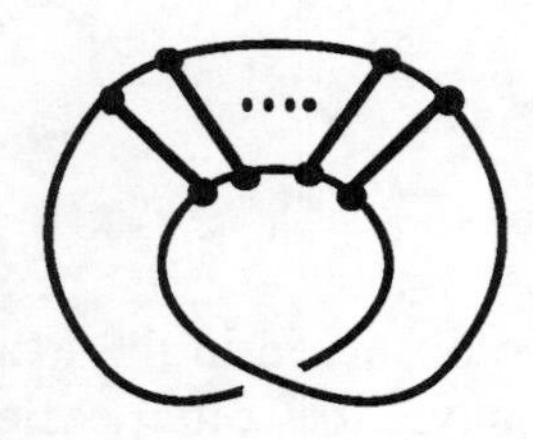

Figure 26

Specifically what Simon showed is that for the embedding of M_n

illustrated in Figure 26, for any n ≥3, there is no deformation of 3–space taking M_n to its mirror image, where K also goes to its mirror image. Since the loop K represents the polyethyleneoxy chains, and the rungs are the C = C bonds, it makes sense to require that K goes to its mirror image, rather than allowing some of the edges in K to play the role of rungs. We note however, that Simon [Sim] has also shown that if n ≥ 4 then every deformation of 3–space which sends the Möbius ladder to its mirror image will, in particular, send the loop to its mirror image. Thus only in the case n =3 does the hypothesis, that K also go to its mirror image, make any difference. In fact, for the sake of simplicity, from now on if we say M_n is deformed to its mirror image we shall mean in such a way that K also goes to the mirror image of K.

Simon's results naturally led to the question of whether or not other embeddings of the graph M_n could be topologically achiral. In particular, is it possible to reembed M_n in 3–space in such a way that it can be deformed to its mirror image or will M_n be topologically chiral no matter how it is embedded? Here, rather than thinking of M_n as the embedded graph illustrated in Figure 26, we think of M_n as any abstract graph consisting of a 2n–gon K together with disjoint edges $\alpha_1, ..., \alpha_n$ joining opposite pairs of vertices for n ≥3. For example, a different embedding of M_4 is illustrated in Figure 27.

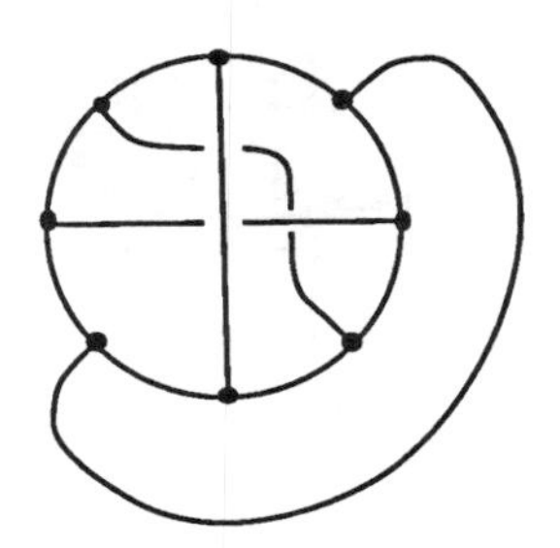

Figure 27

In [Fla_2] we show that the answer to the above question depends on whether the number of rungs is odd or even. In particular, we prove the following Theorem.

Theorem 6 : Any Möbius ladder with an odd number of rungs is intrinsically chiral.

In other words, for any n ≥ 3 which is odd, and for any embedding of M_n in 3–space there is no deformation of M_n to its mirror image. The proof of this Theorem is too technical to explain here. However, we note that both this proof and the proof of Simon's original Theorem [Sim] use the idea of "branched covering spaces." This concept translates the problem of detecting the topological chirality of a graph, to detecting the chirality of a certain link.

In contrast with the Möbius ladders with an odd number of rungs, the Möbius ladders with an even number of rungs can be either topologically chiral or topologically achiral, depending on how they are embedded in 3–space. Simon has shown that the standard embedding of M_n, which is shown in Figure 26, is topologically chiral. We illustrate, in Figure 28 an embedded Möbius ladder M_4 with four rungs, which is topologically achiral. Let P denote the plane containing the unknotted circle K. A rotation of K by 90 degrees about an axis perpendicular to the plane P, will take M_4 to its mirror image through the plane P. By adding another rung above P, and one below P which is perpendicular to the first, we get a similar achiral embedding of M_6. In an analogous fashion we can find an achiral embedding of M_n for any n which is even.

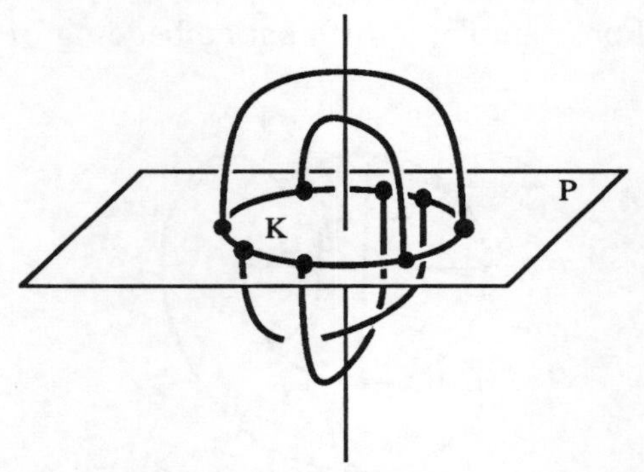

Figure 28

The reader will note that all of the achiral Möbius ladders that we have exhibited above are rigidly achiral and, in fact, have an S_4 symmetry. We suspect there are examples of embeddings of Möbius ladders, with an even number of rungs, which are topologically achiral but not rigidly achiral. But, at the present, we know of no such examples. However, the question we address here is whether every rigidly achiral embedding of a Möbius ladder, in fact, has an S_4 symmetry. We will answer the question in the affirmative by proving Theorem 7. However, the proof of Theorem 7 is rather involved, so in order to shorten it somewhat we omit the proof of the following Lemma which we will use.

Lemma : Let M_n be a Möbius ladder which is embedded in 3–space with loop K. If h is an orientation reversing symmetry of M_n, and h(K) =K, then h preserves the orientation of K.

The proof of this Lemma is similar to the proof of Theorem 5 which is given in section 2, but is a bit more complicated. The interested reader can find the proof in [Fla_2]. We shall now state and prove Theorem 7, but note first that since the proof is rather longer than the others we have been giving, the reader should feel free to skip it. The idea of the proof is to consider different possibilities for the order of h, and see what we can conclude about the fixed point set of h.

Theorem 7 : Let M_n be a Möbius ladder which is embedded in 3–space. Suppose h is an orientation reversing symmetry of M_n. Then the order of h is four.

Proof : First note that by Theorem 6, we know that the number of rungs n is even. By the definition of a Möbius ladder $n \geq 3$, thus in fact $n \geq 4$. Now by Simon's paper [Sim] since $h(M_n) = M_n$ and $n \geq 4$ we know that also h(K) = K. So by ignoring the rungs we see that h is also an orientation reversing symmetry of K.

Suppose that h is a reflection, then we saw in section 2 that K is composite (or the unknot) and h reverses the orientation of K. However, by the Lemma any orientation reversing symmetry of M_n taking K to itself, must preserve the orientation of K. Thus h cannot be just a reflection. So we conclude that h is a reflection together with a rotation.

Now suppose that the order of h is two. Then h rotates each point of K to a point exactly half–way around K. Since each of the endpoints of each rung are diametrically opposite points of K, we see that h takes each rung and flips it over. In other words, if α is a rung with an arrow to indicate a preferred direction, then $h(\alpha) = \alpha$ with the direction of the arrow reversed. In particular, the midpoint of each rung is fixed by h. But since h is a reflection together with a rotation, we know that fix(h) is just one point, whereas M_n has at least four rungs. Therefore the order of h is not two.

Note that the order of h is an even number because h is orientation reversing. So if p is the order of h then there is some number $r > 1$ such that $p = 2r$. Let $f = h^r$. That is, f is the motion we obtain if we repeat h over and over, a total of r times. Then f is a symmetry of M_n, but whether f is orientation preserving or orientation reversing depends on whether r is even or odd. Raise both sides of the equation $f = h^r$ to the power two and we get $f^2 = h^{2r} = h^p$. But p is the order of h. So h^p takes every point back to its original position. Hence f^2 also takes every point back to its original position. Therefore the order of f is two.

Suppose that f is an orientation reversing symmetry of M_n. Then we could repeat the first three paragraphs of this proof with f in place of h. Hence we would conclude that the order of f is not two. But since we know that the order of f is two, in fact f could not have been orientation reversing. Therefore f is an orientation preserving symmetry of M_n (i.e. a rotation).

But recall that $f = h^r$. If r were an odd number, then since h is a reflection together with a rotation, h^r would be a reflection possibly with a rotation. Since we saw above that f is orientation preserving, the number r must be even. Therefore, there is a number q such that $r = 2q$.

Now let $g = h^2$. That is, g is the rotation we get by performing h twice. By taking both sides of the equation to the q power we get $g^q = h^{2q}$. But $r = 2q$, so $g^q = h^r = f$. Thus if we perform the rotation g a total of q times we will get the rotation f. So f and g must have the same axis of rotation. Let A denote this axis. Then the fixed point set of each rotation is the axis A.

Recall that the order of f is two. So f rotates every point on K to its diametrically opposite point. In particular f takes every rung and flips it over, fixing one point in the center. So the axis A intersects each rung α of M_n. But A is also the fixed point set of g. So g fixes the center point of each rung. Hence g takes each rung and flips it over. So g^2 takes every point back to its original position. Hence the order of g is two. Recall $g = h^2$. So $g^2 = h^4$. Since g^2 takes each point back to its original position, h^4 must also take each point back to its original position. Thus we conclude that the order of h is four. So the proof of Theorem 7 is complete.

It follows from Theorem 7 that the symmetries of Möbius ladders illustrated in Figure 28 are typical of the type of orientation reversing symmetries that a Möbius ladder can have. That is, while there are different embeddings which are also rigidly achiral they will always have S_4 symmetries. This Theorem as well as Theorem 6 (i.e. that every Möbius ladder with an odd number of rungs is intrinsically chiral) tell us that there is often a great deal of information in a graph which will rule out achirality or certain types of symmetries, regardless of how that graph is embedded in space.

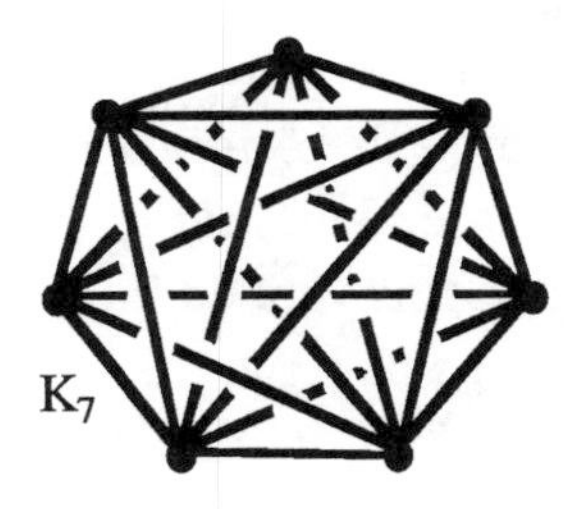

Figure 29

As a final example of this phenomenon, we look at the complete graphs. The complete graph on n vertices is defined as a set of n vertices together with an edge between every pair of vertices. We denote the complete graph on n vertices by K_n. The question which we address here is, for which n is the graph K_n intrinsically chiral and for which n does there exist a topologically achiral embedding of K_n? In [FW] we prove that the answer to this question depends on what remainder we get when we divide n by the number four. The possible remainders are 0, 1, 2, and 3. If n can be divided by four with no remainder then $n = 4p$ for some number p. If the remainder is 1 then $n = 4p + 1$ for some p, if the remainder is 2 then $n = 4p + 2$, and if the remainder is 3 then $n = 4p + 3$. In [FW] we prove the following Theorem.

<u>Theorem 8</u> : If $n = 4p + 3$ for $p \geq 1$, then K_n is intrinsically chiral.

Therefore, for example, no matter how K_7 is embedded in space it will always be topologically chiral. Similarly K_{11}, K_{15}, K_{19} and so on are intrinsically chiral graphs. The proof of Theorem 8 uses [Fla_2] as well as [CG]. We require $p \geq 1$ since, for $p=0$, the graph K_{4p+3} is a triangle which is planar and hence achirally embeddable. In contrast with those n which are of the form $4p + 3$, for all other n there exist embeddings of K_n which are topologically achiral. In fact, in these cases, there exists a symmetry presentation of K_n which has an S_4 symmetry. The general idea of the constructions is to start with $n = 4p$ and to embed K_{4p} so that it has S_4 symmetry, where all of the vertices are in the plane of reflection. Then to get a topologically achiral embedding of K_{4p+1}, we start with the symmetry presentation for K_{4p} and add one vertex at the point where the plane of reflection meets the axis of rotation. Now all $4p + 1$ vertices are in the plane of reflection. So we can add edges within the plane from the new vertex straight out to each of the original vertices. This embedding of K_{4p+1} will still have the S_4 symmetry which the original embedding of K_{4p} had. As an illustration Figure 30 shows an embedding of K_4 with an S_4 symmetry, and Figure 31 shows the corresponding embedding of K_5 with an S_4 symmetry. We can draw similar embeddings of K_8 and K_9, (or K_{12} and K_{13}, etc.), but the graphs have so many edges that the symmetries are more difficult to see.

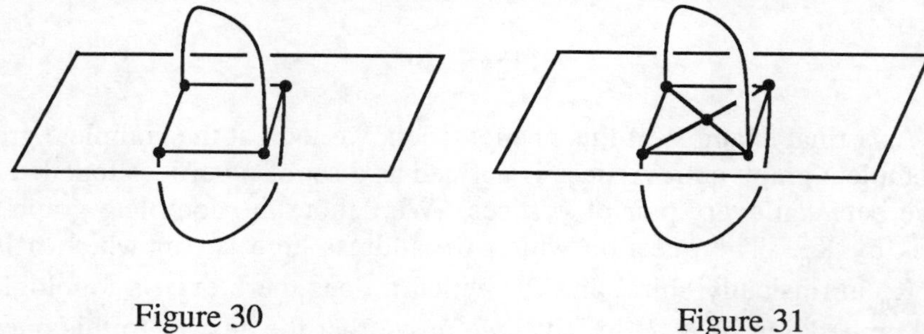

Figure 30 Figure 31

To obtain a topologically achiral embedding of K_{4p+2} again we start with the symmetry presentation for K_{4p}. Then we add two vertices on the axis of rotation, but on opposite sides of the plane of reflection, in such a way that these vertices will be interchanged by the S_4 symmetry. Then add straight edges from the two new vertices to each of the 4p vertices in the plane of reflection, and add the segment of the axis of rotation between these two vertices. Now this embedding of K_{4p+2} has an S_4 symmetry. For example, see the K_6 which is illustrated in Figure 32.

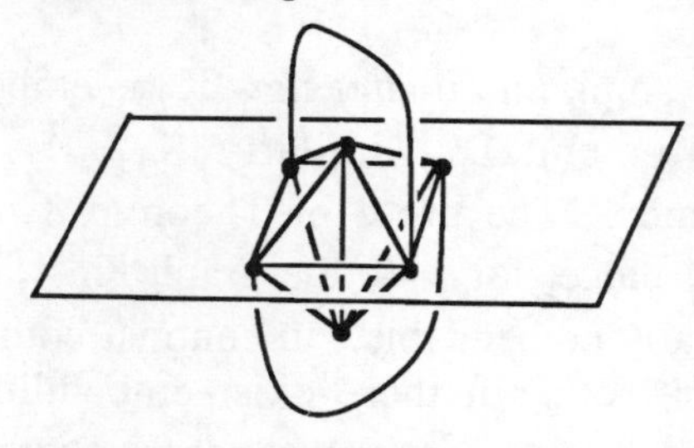

Figure 32

For further discussion of chirality and achirality of complete graphs, see [FW]. Since the intrinsic chirality of a graph implies that all embeddings of it are topologically chiral, it is quite a strong and hence important property. Although, presently, there are not very many methods available to detect topological chirality or intrinsic chirality of arbitrary graphs, topologists have only recently begun studying chirality of graphs. Nonetheless, we expect in the future that some of the techniques that have been used to detect topological chirality of knots will prove useful in detecting topological chirality of graphs, and that the methods that enabled us to detect intrinsic chirality for Möbius ladders and complete graphs will be extended to include much larger classes of graphs.

References

[CG] J.H. Conway, C.McA. Gordon, "Knots and links in spatial graphs," J. Graph Theory 7 (1983) 445–453.

[De] M. Dehn, "Die beiden Kleeblattschlingen," Math. Ann. 75 (1914) 402–413.

[DS] C. Dietrich–Buchecker, J.–P. Sauvage,"A synthetic molecular trefoil knot," Angew. Chem. Int. Ed. Engl. 28 (1989) 189–192.

[Fla_1] E. Flapan, "Rigid and non–rigid achirality," Pacific Journal of Math., Vol. 129, No. 1, (1987) 57–66.

[Fla_2] E. Flapan, "Symmetries of Möbius ladders," Math. Ann., 283, (1989) 271–283.

[FW] E. Flapan and N. Weaver, "Intrinsic chirality of complete graphs," Preprint, Mathematics Department, Pomona College, Claremont, CA 91711.

[Has] M.G. Haseman, "On knots, with a census of the amphicheirals with twelve crossings," Trans. Roy. Soc. Edinburgh 52 (1918), 235–255.

[HOMFLY] P. Freyd, D. Yetter, J. Hoste, W. Lickorish, K. Millett, and A. Ocneanu, "A new polynomial invariant of knots and links," Bull. Amer. Math. Soc. 12 (1985) 239–246.

[HK] R. Hartley and A. Kawauchi, "Polynomials of amphicheiral knots," Math. Ann., 243, (1979) 63–70.

[Jon] V.F.R. Jones, "Hecke algebra representations of braid groups and link polynomials," Ann. of Math. 126 (1987) 335–388.

[Kau] L. Kauffman, "New invariants in the theory of links," Amer. Math. Monthly, Vol. 95, No. 3, (1988) 195–242.

[LM] W.R.B. Lickorish and K.C. Millett, "The new polynomial invariants of knots and links," Mathematics Magazine, Vol. 61, No. 1, (1988) 3–23.

[Lis] J.B. Listing, *Vorstudien zur Topologie*, Göttingen Studien, (1847).

[Mis] K. Mislow, *Introduction to Stereochemistry*, (Bejamin, Reading, MA, 1965).

[Mur] K. Murasugi, "Jones polynomials and classical conjectures in knot theory," Topology, Vol. 26, No. 2 , (1987) 187–194.

[PT] J.H. Przytycki and P. Traczyk, "Invariants of links of Conway type," Kobe J. Math. 4 (1987) 115–139.

[Rei] K. Reidemeister, *Knotentheorie*, Chelsea Publishing Company, New York, (1948).

[Rol] D. Rolfsen, *Knots and links*, Publish or Perish press, (1976).

[Sim] J. Simon, "Topological chirality of certain molecules," Topology, 25, (1986) 229–235.

[Tai] P.G. Tait, "On knots I, II, III," Scientific Papers Vol. I, Cambridge University Press, London, (1898), 273–347.

[Thi] M. Thistlethwaite, "Kauffman's polynomial and alternating links," Topology, Vol. 27, No. 3, (1988) 311–318.

[Wal_1] D. Walba, "Stereochemical Topology," in *Chemical Applications of Topology and Graph Theory*, R.B. King (ed.) Studies in Physical and Theoretical Chemistry, Vol. 28 (Elsevier, Amsterdam, 1983) 17–32.

[Wal_2] D. Walba, "Topological stereochemistry," Tetrahedron 41 (1985) 3161–3212.

[WDC] S.A. Wasserman, J.M. Dungan, and N.R. Cozzarelli, "Discovery of a predicted DNA knot substantiates a model for site–specific recombination," Science 229 (July 12, 1985) 171–174.

[WRH] D. Walba, R. Richards, and R.C. Haltiwanger, "Total synthesis of the first molecular Möbius strip," J. Am. Chem. Soc. 104 (1982) 3219–3221.

Chiral and achiral square-cell configurations; the degree of chirality

Frank Harary
Department of Computer Science
New Mexico State University
Las Cruces, NM 88003, USA,

Paul G. Mezey
Department of Chemistry and
Department of Mathematics
University of Saskatchewan,
Saskatoon, Canada, S7N 0W0

ABSTRACT

The shape features, in particular, the chirality properties of the patterns of molecules adsorbed on various surfaces, can be modeled by square cell configurations in the plane. Square-cell configurations can be represented by planar graphs. Various families of planar graphs are chiral in two dimensions. A drawing of a tree in the plane, for example, the letter F, is called a plane tree. Chiral plane trees were counted by Harary and Robinson; this result was extended to the enumeration of chiral alkanes in 3-space by Robinson, Harary and Balaban. The shape of a Jordan curve in the plane can be modeled by square cell configurations, which we call animals. Many of the animals show two-dimensional chirality. This suggests the dichotomy of chiral and achiral animals for the shape characterization of Jordan curves, such as molecular curves, and cross-sections of molecular contour surfaces. Chirality of animals can be analysed in terms of the animal codes introduced earlier, leading to the concept of the degree of chirality of Jordan curves and to the conjecture that almost all animals are chiral.

P. G. Mezey (ed.), New Developments in Molecular Chirality, 241–256.

Introduction.

The analysis of shapes of planar patterns is of importance in a variety of fields, ranging from the chemistry of molecular adsorption to the study of photographs of galaxies. For example, the study of the shape of letters is relevant in pattern recognition, e.g., for computerized reading of mail addresses. We show in Figure 1 one of the standard presentations of the 26 capital letters of the roman alphabet, as in [1]. When taking one of the possible most symmetric representations for each, we find that only eight: F, G, J, N, P, R, S, Z are chiral. Similarly, as is easily seen from Figure 2, when taking one of the most symmetric representations for each of the ten numerals 0 to 9, we find that only four: 2, 5, 6 and 9 are chiral.

In a classic study of chirality in chemistry, V. Prelog *et al.* [2] illustrated the phenomenon of handedness or chirality, the property of an object having a mirror image different from itself, using the roman alphabet as the set of objects. These objects, as well as objects of more complex shapes [3-5], can be studied using the tools of graph theory.

In the examples of molecules adsorbed on a surface, or letters and numerals, as in the rest of this report, we are concerned with the special, two-dimensional case of chirality, where the objects and all their allowed motions are restricted to the two-dimensional plane. A geometric configuration Γ in the plane is called *achiral* if the mirror image $\Gamma^{\Diamond}$ of Γ can be rotated and translated in the plane so that it can be made to coincide with the original object Γ. For example, the mirror image of letter E is $E^{\Diamond} = \exists$ which when rotated 180 degrees, coincides with E. On the other hand the letter R is chiral in the plane, since no rotation and translation in the plane can lead to coincidence with its mirror image, and the fact that its mirror image is a letter in the cyrillic alphabet is irrelevant. Note, however, that if one regards the letter R as an object in the three-dimensional space, then R is not chiral, since three-dimensional rotations and translations can bring coincidence with its mirror image; if this mirror image (the cyrillic "ya") is painted on a glass window, it reads R from the other side of the window. By turning the window around (a three-dimensional motion), the mirror image is superimposable on the original letter R. Whereas most

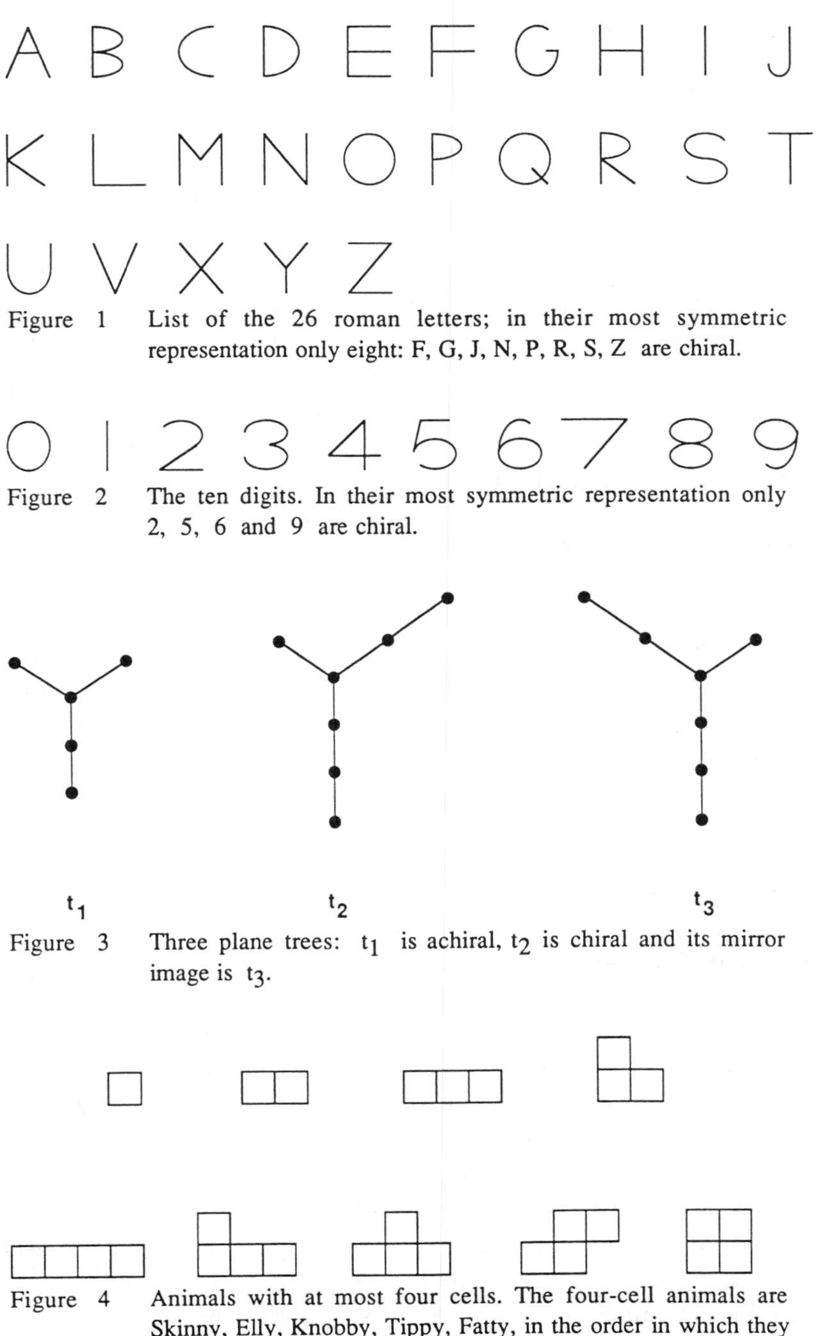

Figure 1 List of the 26 roman letters; in their most symmetric representation only eight: F, G, J, N, P, R, S, Z are chiral.

Figure 2 The ten digits. In their most symmetric representation only 2, 5, 6 and 9 are chiral.

Figure 3 Three plane trees: t_1 is achiral, t_2 is chiral and its mirror image is t_3.

Figure 4 Animals with at most four cells. The four-cell animals are Skinny, Elly, Knobby, Tippy, Fatty, in the order in which they appear in the Figure.

chemical applications of the concepts of chirality are three-dimensional, see, e.g., the pioneering work of Cahn, Ingold and Prelog [2], nevertheless, two-dimensional chirality is of importance in detailed shape studies of molecular curves, such as cross-sections of molecular contour surfaces [3-5].

Some of the shape properties of letters and numerals can be studied using the tools of graph theory. Many of these symbols can be represented by graphs from a special family: plane trees. For example, the letter F, when drawn in the plane, is a plane tree. In Figure 3 three plane trees are shown: t_1 is achiral whereas t_2 and t_3 are chiral mirror images of each other. Counting formulas for chiral plane trees were derived by Harary and Robinson [6]. Based on the results of [6], the century-old problem of van't Hoff, concerning the enumeration of chiral and achiral alkanes (specifically, the open chain, saturated hydrocarbons in euclidean 3-space), has been solved by Robinson, Harary and Balaban [7].

Two-dimensional chirality of closed curves and areas enclosed by them in the plane is another problem with many applications in chemistry. For example, besides the cross-sections of molecular contour surfaces [3-5], the pattern of adsorbed molecules on metal surfaces, important in catalysis [8], can also be analysed by the shape of a Jordan curve enclosing the molecules on the planar metal surface. The interiors of Jordan curves can be modeled by square cell configurations, which we call animals. Planar chirality of animals is an important aspect of shape characterization of such curves.

Square-cell animals.

The (square-cell) animals [9] with at most four cells are shown in Figure 4; those with five cells in Figure 5. The latter are called "pentaminoes " by Golomb [10], who uses the word "polyominoes" (obtained by the extension of "dominoes") for our word, animals. The names of the five 4-cell animals [11], in the order in which they appear in Figure 4 are: Skinny, Elly, Knobby, Tippy, and Fatty. A very readable account of Harary's achievement and avoidance games involving animals was presented by Martin Gardner [11]. A numerical code for shape characterization of animals was proposed by Harary and Mezey [12], using graph theoretical methods [13], developed earlier for the representation of animals [14-16]. Statistical properties of various families of animals embedded in a square lattice are of importance in thermodynamic models [17,18].

Some graph theoretical concepts and tools used in this report are briefly reviewed below.

A *mesh* $M_{m,n}$ is defined [16] as the cartesian product $P_m \times P_n$ of two nontrivial paths P_m and P_n. When $m = n$, then the notation M_n may be used for the mesh $P_n \times P_n$. The *interior* of a given drawing of a graph G in the plane is the union of all open point sets enclosed by the cycles of G. With reference to Jordan curves of the plane of M_n, a Jordan cycle C of mesh M_n is a cycle that is a subgraph of M_n and has a vertex degree of two for all of its nodes. A subgraph A of a mesh M_n is called an *animal* if it contains all the nodes and edges of mesh M_n that fall on a Jordan cycle C of M_n or within the interior of C. Each 4-cycle C_4 contained in animal A is called a *cell* c of A. The Jordan cycle C is the *perimeter* of A that, evidently, contains all edges of A which are on exactly one of its cells. As a point set, the perimeter C of A is a single Jordan curve, denoted by J(A).

The *circumscribed mesh* M(A) *of* A is the unique smallest mesh $M_{m,n}$ containing animal A as a subgraph, where $m \le n$ without loss of generality. Animal A can be represented as a rectangular $(m-1) \times (n-1)$ matrix $R(A) = [r_{ij}]$ of elements $r_{ij} = 1$ if the i,j-cell of mesh $M_{m,n}$ is a cell of A, while $r_{ij} = 0$ otherwise [15]. We do not distinguish between animals which can be obtained from each other by rotations in the plane, but in this report chiral animals that are mirror images of each other are distinguished. Hence each animal A may have at most four different matrix representations, obtained by successive 90 degree rotations of the animal A together with its mesh M(A). For example, the four matrix representations of Elly of Figure 4 are

$$R_1 = \begin{vmatrix} 1 & 0 & 0 \\ 1 & 1 & 1 \end{vmatrix}, \quad R_2 = \begin{vmatrix} 0 & 1 \\ 0 & 1 \\ 1 & 1 \end{vmatrix}, \quad R_3 = \begin{vmatrix} 1 & 1 & 1 \\ 0 & 0 & 1 \end{vmatrix}, \quad R_4 = \begin{vmatrix} 1 & 1 \\ 1 & 0 \\ 1 & 0 \end{vmatrix}, \quad (1)$$

whereas the transposes of these matrices, R'_1, R'_2, R'_3, and R'_4 are the four representations of the chiral mirror image of Elly. In the case of Elly, all eight of these matrices are different. Note that for the chiral Tippy there exist only two matrix representations, $R_1 = R_3$, and $R_2 = R_4$, and, evidently, the same applies for its mirror image, since $R'_1 = R'_3$, and $R'_2 = R'_4$. For the achiral Fatty all eight matrices are the same.

In general, a necessary and sufficient condition for achirality of an animal A is the existence of an i,j $(1 \le i, j \le 4)$ index pair such that for the

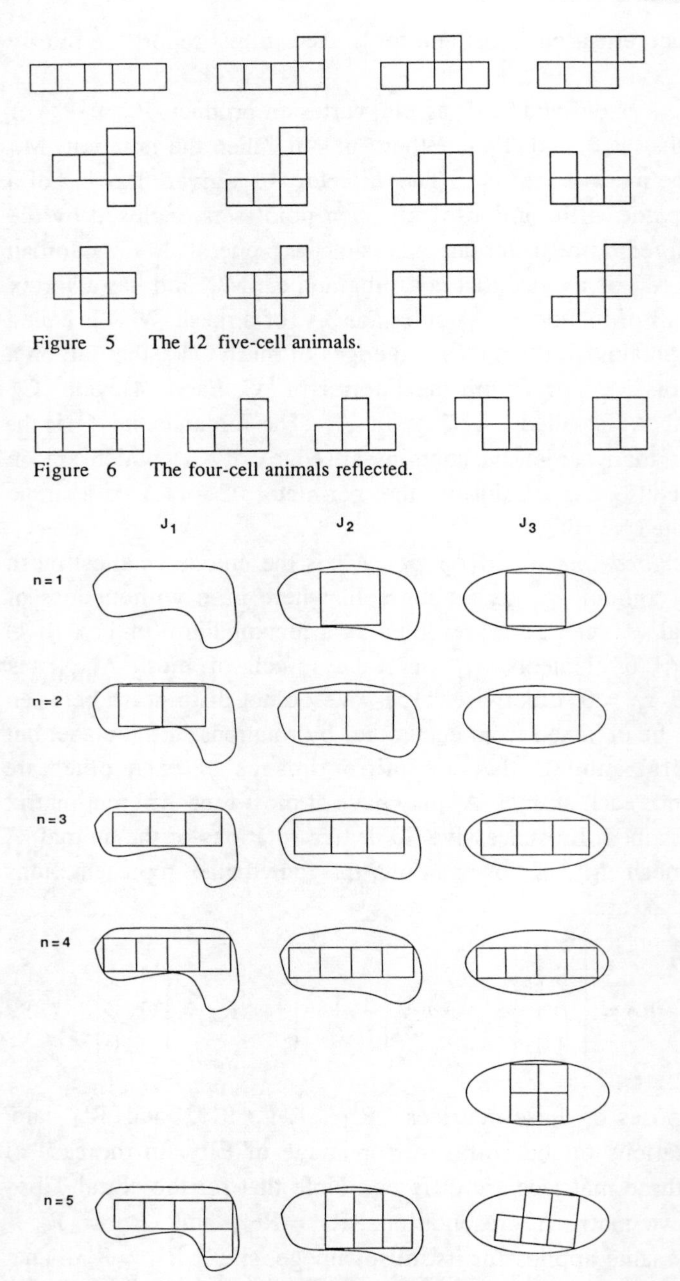

Figure 5 The 12 five-cell animals.

Figure 6 The four-cell animals reflected.

Figure 7 The n-cell interior filling animals of three Jordan curves J_1, J_2 and J_3, for n=1,...,5. Curves J_1 and J_2 are chiral whereas the ellipse J_3 is achiral.

corresponding matrix representations

$$R_i(A) = R'_j(A) \tag{2}$$

holds. If no such index pair i, j exists, then A is chiral.

For any matrix representation $R_i(A)$ of an animal A of precisely n cells the sum of eigenvalues of the product matrix $R_i(A)\, R'_i(A)$ is equal to the number of cells:

$$\text{trace} (R_i(A)\, R'_i(A)) = n \tag{3}$$

The matrix representations define a code $c(A)$ for an animal A [12]. For each $(p-1) \times (r-1)$ binary matrix $R_i(A)$ (where either $p=m$ and $r=n$ or $p=n$ and $r=m$), the binary number $b(R_i)$ is the concatenation of the rows $1,\ldots,p-1$ of $R_i(A)$. If the mirror image $A^\diamond$ of animal A is taken with respect to a northwest-southeast reflection line within a given representation, then $R_i(A^\diamond) = R'_i(A)$, $i=1,\ldots,4$, holds for the corresponding matrices. The notation $R_{i+4}(A) = R'_i(A)$, $i=1,\ldots,4$, was used in an earlier study, where only the intrinsic shape features of animals were considered, and animals that are mirror images of each other were not distinguished [12]. The common code $c(A)$ of an animal A and its mirror image $A^\diamond$ was defined as

$$c(A) = \max \{ b(R_i), b(R'_i), i=1,2,\ldots,4\} . \tag{4}$$

Code $c(A)$ itself does not completely determine the animal pair A and $A^\diamond$, as the values of m and n are also needed, even if one choses the convention that $m \leq n$. A determining numerical invariant $i(A)$ of any animal A (and its mirror image $A^\diamond$) is the ordered triple, in which $c = c(A)$:

$$i(A) = (c,m,n) . \tag{5}$$

Note that c as a binary number contains mn digits, hence c and m determine n, and in short form this invariant can be given as the ordered pair $i(A)=(c,m)$. In actual applications, however, the triple form (5) is more convenient for the reconstruction of the animal A.

In the present report an animal A and its mirror image $A^\diamond$ are distinguished, and we propose a code $d(A)$ and determining numerical invariant $j(A)$ to reflect this. By analogy with (4), the code $d(A)$ is defined as

$$d(A) = \max \{ b(R_i), \ i=1,2,...,4 \} . \qquad (6)$$

This code $d(A)$ itself is insufficient to determine the animal A. For example, d(Elly) = d(Knobby) = 111010, even though c(Elly) = 111100 ≠ c(Knobby) = 111010. For a reconstruction of the animal A from its code $d(A)$ one must also know

(i) the values of m and n of its mesh $M_{m,n}$,

(ii) which mesh, the standard mesh $M_{m,n}$ $(m \leq n)$ or the transposed mesh $M_{n,m}$, belongs to the actual matrix R_i of maximum $b(R_i)$.

Both (i) and (ii) are known if $d(A)$ and the length of the row of the matrix selected in eq. (6) are given. We define the number $r(A)$ as follows: $r(A)-1$ is the number of columns of matrix $R_i(A)$ which realizes the code of A by having its binary number equal to the code, $b(R_i) = d(A)$. Then, a determining numerical invariant $j(A)$ of any animal A is the ordered pair (d,r), in which $d = d(A)$ and $r=r(A)$:

$$j(A) = (d,r) \qquad (7)$$

This numerical invariant, involving code $d(A)$ defined by eq. (6), distinguishes animals that are chiral mirror images.

A necessary and sufficient condition for achirality of an animal A can be expressed in terms of the invariants $j(A)$ and $j(A^\lozenge)$ of A and its mirror image $A^\lozenge$, respectively:

$$j(A) = j(A^\lozenge) \qquad (8)$$

A conjecture on chiral and achiral animals.

All animals of three cells or less are evidently achiral. We show in Figure 6 the mirror images of the five 4-cell animals in the order in which they appear within Figure 4.

Clearly, Skinny, Knobby and Fatty are achiral. By contrast, Elly and

Tippy are chiral, since neither can be made congruent to its mirror image by rotation and translation confined to the plane. Hence these two are the smallest chiral animals.

The number of n-cell animals of different intrinsic shape is denoted by a_n. When intrinsic shape is considered, then chiral mirror images are not distinguished. The number of chiral animals among the a_n n-cell animals is denoted by χ_n. Among the 12 animals of five cells in Figure 5, it is immediately verified that exactly six are chiral.

TABLE 1
The numbers a_n of animals and χ_n of chiral animals for $n \leq 5$

n	1	2	3	4	5
a_n	1	1	2	5	12
χ_n	0	0	0	2	6

From the data of Table 1 we find that $\chi_4/a_4 = 2/5$ and $\chi_5/a_5 = 1/2$. One expects that for larger animals, i.e., for animals with more cells, chirality is more the rule than the exception and we propose the following conjecture.

Conjecture. Almost all animals are chiral, i.e.,

$$\lim_{n \to \infty} \chi_n/a_n = 1 \tag{9}$$

Inscribed animals and interior filling animals of Jordan curves and planar domains.

The perimeter $J(A)$ of an animal A is a Jordan curve. This suggests a possible characterization of shapes of Jordan curves by the shapes of inscribed animals [12]. Here we shall study the chirality properties of both Jordan curves J and plane domains $D=Int(J)$ enclosed by them in terms of the chirality properties of inscribed animals.

Whether an animal A fits within the interior D of a given Jordan curve J, depends on the size of the cells of the animal. Evidently, by chosing a small enough size s for the length of the side of the square cells, any finite animal can fit within D. When fitting an animal A within D, J and A may be rotated with respect to one another, that is, the relative orientation of J and mesh $M(A)$ is not fixed. Note that the forthcoming analysis can be adapted to the case with orientation constraints, however, in this report no such constraints will be considered.

For a given Jordan curve J and size s there exists a countable family $F(J,s)$ of animals which fit within D. Members $A_i(J,s)$ of this family $F(J,s)$ are the *inscribed animals*, with reference to J and s. If J and the cell size s are given then there exists a maximum number n of cells for inscribed animals. A subfamily $F(J,s,n)$ of $F(J,s)$ contains all animals $A_i(J,s,n)$ for which the number of cells is the maximum possible value n. Without orientation constraints, this number $n=n(J,s)$ depends only on the given Jordan curve J and cell size s. Evidently, the perimeters of these $A_i(J,s,n)$ animals are also Jordan curves which provide approximations to the original Jordan curve J, and the smaller the size s the better the approximation. A small change of cell size s does not necessarily change n. For the given J, take the maximal interval $s(J,n) = [s(n,1), s(n,2))$ within which the n value is invariant, and generate the union $F(J,n)$ of all $F(J,s,n)$ animal families for this interval:

$$F(J,n) = \bigcup_{s(J,n)} F(J,s,n) \qquad (10)$$

The family $F(J,n)$ contains all n-cell animals $A_i(J,n)$ which fit within the interior of Jordan curve J, with the maximum number n of cells for the given range $s(J,n)$ of cell size s. These animals are the n-cell *interior filling animals* of the Jordan curve J.

Stated differently, the n-cell animal $A_i(J,n)$ inscribed in Jordan curve J is an interior filling animal of J if and only if no animal of the same cell size s and more than n cells can be inscribed in J. In particular, none of the animals $A_i(J,n)$ can be enlarged by a cell and still fit within the interior of J, as long as $s \in s(J,n)$.

The relative size of D=Int(J) and the cells c is implied by the maximum number n of cells which fit within domain D. The continuum of size range s(J,n) is replaced by a discrete descriptor, integer n. In particular, in the F(J,n) and $A_i(J,n)$ notations the cell size information is not given directly.

Some of the n-cell interior filling animals of three Jordan curves, J_1, J_2 and J_3 are shown in Figures 7 and 8, for cases $1 \leq n \leq 5$, and $6 \leq n \leq 9$, respectively. In Figure 9 some additional interior filling animals of the Jordan curve J_2 are listed for $10 \leq n \leq 16$.

The degree of chirality of Jordan curves and planar domains.

The following description of two-dimensional chirality properties of Jordan curves J and planar domains D=Int(J) is motivated by an analogous treatment of molecular similarity [19], based on some of the fundamental processes of visual perception of shape features of objects.

Consider the shape characterization of a planar domain D at various levels of resolution. At very low resolution, e.g., from a great distance, D appears as a single dot, hence it appears achiral. At a slightly better resolution, it may appear as a small, achiral disk. Somewhat higher resolution is required to observe details of its shape, and to detect the presence or the lack of two-dimensional chirality of D. A domain D_1 may exhibit chirality at relatively low resolution, while some other chiral domain D_2 manifests its chirality only at a much higher resolution. One may wish to introduce a measure to express how detailed shape description is needed to detect chirality of a given planar domain D. We shall use the chirality properties of interior filling animals of Jordan curves to introduce the concept of the *degree of chirality* of Jordan curves and planar domains D=Int(J). The generalization of this concept to three-dimensional objects is straightforward, using the analogous method involving polycubes.

The chirality properties of the interior filling animals $A_i(J,n)$ for each

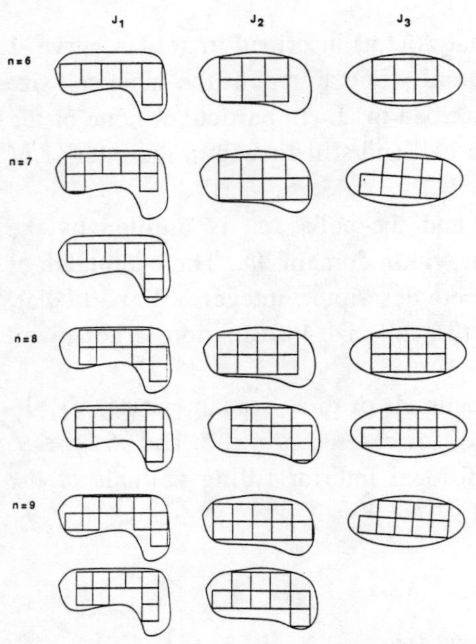

Figure 8 Some of the n-cell interior filling animals of three Jordan curves J_1 and J_2, for n=6,...,9.

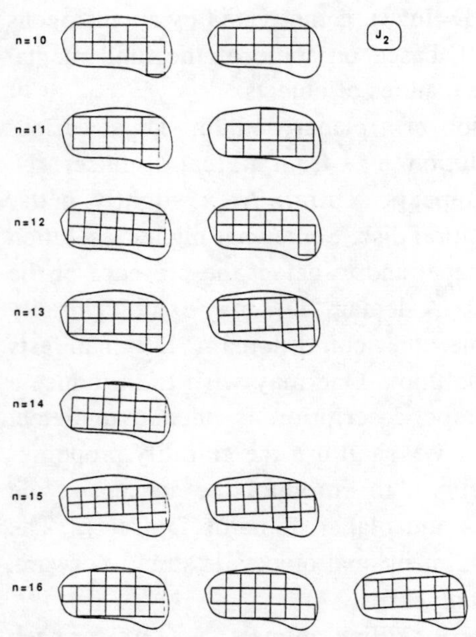

Figure 9 Some of the n-cell interior filling animals of the Jordan curve J_2, for n=10,...,16.

value n provide information on the chirality of the Jordan curve J at various levels of resolution. Each level of resolution is defined indirectly by n, that in turn depends on the cell size s. A Jordan curve J can be approximated by the perimeters $P_i(J,n)$ of n-cell interior filling animals $A_i(J,n)$ of increasing n. Even if J is chiral, some of the perimeters $P_i(J,n)$ may well be achiral for low values of n. It is natural to consider the smallest number n_1 for which there exists a chiral interior filling animal $A_i(J,n)$. However, as our examples in Figures 7-9 show, this number is not a good measure of the degree of chirality. Instead, for each Jordan curve J we shall consider a chirality index, defined as a critical cell number $n_o(J)$, above which all interior filling animals $A_i(J,n)$ are chiral.

We say that J is chiral at and above cell number n_c if each $A_i(J,n)$ is chiral if $n \geq n_c$. The *chirality index* $n_o(J)$ is the smallest n_c value above which all interior filling animals $A_i(J,n)$ are chiral,

$$n_o(J) = \begin{cases} \min\{n_c : A_i(J,n) \text{ is chiral if } n \geq n_c\}, & \text{if the minimum exists} \\ \infty & \text{otherwise.} \end{cases} \qquad (11)$$

Since Elly and Tippy are the smallest chiral animals, the minimum possible value for chirality index is $n_o(J)=4$. In order to provide a measure of chirality that is 1 for "very chiral" curves and 0 for achiral ones, we propose the following choice for the degree of chirality of Jordan curves:

Definition. The degree of chirality $\chi(J)$ of a Jordan curve J is

$$\chi(J) = 1 / (n_o(J)-3) \qquad (12)$$

It is intuitively evident that $\chi(J) > 0$ if and only if J is chiral. The extreme value $\chi(J) = 1$ corresponds to the case of most pronounced chirality, whereas $\chi(J) = 0$ corresponds to an achiral Jordan curve J.

Examples of chirality degrees.

In Figures 7 and 8 some of the smallest interior filling animals of three Jordan curves, J_1, J_2, and J_3 are shown. Both J_1 and J_2 are chiral, whereas the ellipse J_3 is achiral. For an observer the chirality of J_1 is more evident than that of J_2, a feature that we shall quantify using the concept of the degree of chirality introduced above.

On level n=1, curves J_2 and J_3 have the only possible interior filling animal of a single cell $A(J_2,1)= A(J_3,1)$. This animal is not an interior filling animal for curve J_1, since at the cell size s that allows the single cell animal to fit within J_1, the unique two-cell animal also fits within J_1. Hence, no $A(J_1,1)$ animal exists.

Up to level n=3, all three Jordan curves have identical shape features but at level n=4, J_3 deviates from the other two. At level n=5, all three curves, including the achiral ellipse J_3, have chiral interior filling animals, that justifies our comment above concerning the unsuitability of n of the smallest chiral interior filling animal as a descriptor of the degree of chirality. In particular, Figure 8 shows achiral animals filling J_3 at the n=6 and the n=8 levels. Evidently the achiral ellipse J_3 has achiral interior filling n-cell animals $A(J_3,n)$ of n exceeding any finite threshold n_t, consequently, for J_3 the chirality index $n_o(J_3) = \infty$, and $\chi(J_3) = 0$. These are precisely the values one expects for an achiral curve J.

The chirality of curve J_1 manifests itself at a much lower level n than that of J_2. Although at level n=7, J_1 still has an achiral interior filling animal, from level $n \geq 8$ all $A_i(J_1,n)$ animals are chiral. Consequently, $n_o(J_1) = 8$, and $\chi(J_1) = 1/5$. By contrast, curve J_2 has an achiral interior filling animal $A_i(J_2,n)$ even at the level n=14, and one finds that $n_o(J_2) = 15$ and $\chi(J_2) = 1/12$. The degree of chirality $\chi(J)$ correctly reflects the fact that both curves are chiral: $\chi(J_1) > 0$, and $\chi(J_2) > 0$. However, in agreement with a qualitative, visual description that the chirality of J_1 is much more evident than that of J_2, the degree of chirality $\chi(J)$ shows major difference for the two curves. According to this measure, the chirality of J_1 is more than twice as pronounced as that of curve J_2:

$$\chi(J_1) / \chi(J_2) = (1/5)/(1/12) = 2.4$$

Conclusions.

Square cell configurations (animals) are used to characterize the two-dimensional chirality properties of planar domains D enclosing them. A conjecture is given on the predominance of chiral n-cell animals for large n. The concept of *degree of chirality* is introduced for Jordan curves and their interiors D=Int(J). These concepts are applicable for the shape characterization of curves defined as cross-sections of molecular contour surfaces, and to molecular aggregates as well as to patterns of adsorbed molecules on metallic surfaces in studies of catalysis. The shape characterization technique and the concept of degree of chirality can be extended easily to three-dimensional objects such as formal molecular bodies and molecular boundary surfaces, using polycubes as the three-dimensional analogues of square cell configurations.

Acknowledgement. This work has been supported by both strategic and operating research grants from the Natural Sciences and Engineering Research Council of Canada.

References.

1 F. Harary, Typographs. Visible Language **7**, 199 (1973).

2 R.S. Cahn, C.K. Ingold and V. Prelog, Angew. Chem., Internat. Ed. Engl., **5**, 385 (1966).

3 P.G. Mezey, The shape of molecular charge distributions: group theory without symmetry. J. Comput. Chem., **8**, 462 (1987).

4 P.G. Mezey, Shape group studies of molecular similarity: shape groups and shape graphs of molecular contour surfaces. J. Math. Chem., **2**, 299 (1988).

5 P.G. Mezey, Potential Energy Hypersurfaces. Elsevier, Amsterdam, 1987.

6 F. Harary and R.W. Robinson, The number of achiral trees. J. Reine Angew. Math. **278**, 322 (1975).

7 R.W. Robinson, F. Harary and A.T. Balaban, The numbers of chiral and achiral alkanes and monosubstituted alkanes. Tetrahedron **32**, 355 (1976).

8 G. A. Somorjai, Modern concepts in surface science and heterogeneous catalysis. J. Phys. Chem., **94**, 1013 (1990).

9 F. Harary, The cell growth problem and its attempted solutions. Beitrage zur Graphentheorie. Teubner, Leipzig, 49 (1968).

10 S. Golomb, Polyominoes. Scribner's, New York (1965).

11 M. Gardner, Mathematical games in which players of ticktacktoe are taught to hunt for bigger game. Scientific American **240**, 18 (April 1979).

12 F. Harary and P.G. Mezey, Similarity and complexity of the shapes of square-cell configurations. (to appear).

13 F. Harary, Graph Theory. Addison-Wesley, Reading (1969).

14. F. Harary and E. M. Palmer, Graphical Enumeration, Academic Press, New York, 1973.

15. G. Exoo and F. Harary, Animal-trees and caterpillars are equivalent. Indian Nat. Acad. Sci. Letters, **10**, 67 (1987).

16. F. Harary and M. Lewinter, Spanning subgraphs of a hypercube III: meshes. Internat. J. Computer Math., **25**, 1 (1988).

17. C. Soteros and S.G. Whittington, J. Phys. A, **21**, 2187 (1988).

18. N. Madras, C. Soteros and S.G. Whittington, J. Phys. A, **21**, 4617 (1988).

19. P.G. Mezey, J. Math. Chem., **6**, xxx (1991).

A global approach to molecular chirality

Paul G. Mezey
Department of Chemistry and
Department of Mathematics
University of Saskatchewan,
Saskatoon, Canada, S7N 0W0

ABSTRACT

The three-dimensional arrangements of molecular fragments that may or may not lead to molecular chirality are subject to quantum mechanical uncertainty and to a more significant uncertainty due to the non-rigid nature of molecules. Molecular chirality is an energy dependent property: if sufficient energy is available, then no molecule can retain its chiral properties since rearrangement and decomposition reactions may take place at a significant rate. In a rigorous model, it is impossible to separate the problems of chirality from energetic considerations and from explicitly taking into account the deformability of molecules. This underlines the importance of the global chirality properties of a whole range of formal nuclear arrangements. In a global approach to the problems of molecular chirality, it appears worthwhile to consider the nuclear configuration space approach where all possible arrangements of a given stoichiometric family of nuclei are considered. Such a stoichiometric family includes all isomers, intermediates and decomposition products of all molecules and their transition structures with a given atomic composition, as well as all distorted conformations of the above. The global approach leads to the recognition of several rules on the presence of stable chiral nuclear configurations within families of nuclear arrangements, on the facility of configuration inversion and racemization processes, and on chirality changes due to electronic excitation.

P. G. Mezey (ed.), New Developments in Molecular Chirality, 257–289.

Introduction.

Molecular chirality is an energy dependent property. Molecules that are chiral at low temperatures may lose their chirality at higher temperatures where sufficient energy is available to overcome the activation energy barriers of conformational changes, leading to configuration inversion or to achiral conformations. In the extreme case, at high enough temperatures, any chiral molecule may deform into a planar shape, or simply decompose into achiral fragments. The conformational freedom and, in general, the range of deformations available to a molecule is a function of its energy. Generally, the more energy available, the larger the deformations that are possible, some of which may be classified as chemical reactions. A minor bond stretching may be regarded as a mere molecular deformation, whereas a major change in a internuclear distance may be regarded as a dissociation reaction. The important, common feature is that both deformations (conformational changes) and decomposition reactions (such as dissociation processes) involve changes of the nuclear arrangements. In fact, conformational changes and chemical reactions differ from one another only in the *extent* of the associated deformations of the nuclear arrangements. All ofthese changes in the nuclear arrangements can be studied in a formal *nuclear configuration space*.

In the traditional approach to molecular chirality, the analysis has been focussed on individual nuclear arrangements [1-3], and many, subsequent developments are based on this approach [4-11]. In some cases, chirality has been associated with formal chiral centers in a molecule, with reference to the local site symmetry [8]. In other cases, the overall symmetry and nuclear arangement of the molecular system as a whole is the dominant factor determining chirality [10-15].

We shall distinguish the concepts of *configuration* and *nuclear configuration*. The former is the usual concept referring to the spatial arrangement of substituents around a chiral (or achiral) center, whereas the latter is specifying the mutual arrangements of all nuclei in a given molecular system. Most of the problems discussed in this chapter are formulated in terms of nuclear configurations.

Let us consider a given collection of N atoms. In general, many different molecules, molecule pairs, or sets of several molecules can be constructed from these atoms, and all of these arrangements belong to the same *stoichiometric family* of species [16,17]. A stoichiometric family includes all isomers, intermediates and decomposition products of all molecules and their transition structures with a given atomic composition, as well as all their distorted conformations. Note that in this context a pair of molecules with the given total stoichiometry, for example the pair of molecules A and B of N_A and N_B nuclei, respectively, where $N_A + N_B = N$, is regarded as one formal species.

Each nuclear arrangement of the N atoms can be characterized by 3N cartesian coordinates. Alternatively, if one disregards rigid translation and rigid rotation of the arrangement, then the internal (relative) nuclear configuration for $N \geq 3$ nuclei can be described by 3N-6 internal coordinates. The collection of all possible nuclear arrangements is regarded to form a nuclear configuration space, where each internal (relative) nuclear configuration is represented as a *point* of the space. A nuclear configuration space belongs to the entire stoichiometric family, not just to a selected molecule. It is possible to introduce (usually local) coordinate systems, as well as a global distance function into the nuclear configuration space [16,18,19]. The distance between two points K and K' of the space is a measure of dissimilarity between the corresponding two nuclear configurations. This distance may be defined as a global *metric* for the space, turning the nuclear configuration space into a formal *metric space*, denoted by M.

The identity of a chemical species is not affected by most minor deformations of the nuclear configuration. Consequently, a stable or unstable chemical species of the given stoichiometric family can be represented by a subset of the nuclear configuration space M. Similarly, a transition structure for a molecular process may exhibit deformability along internal coordinates that do not contribute to the progress of the reaction (for example, those orthogonal to a formal reaction coordinate). Consequently, a transition structure also corresponds to a subset of the nuclear configuration space M. Typically, however, this latter subset has a dimension lower than that of a stable species, since even small deformations along the reaction coordinate change the nuclear configuration of the transition structure to a distorted product or a distorted reactant configuration, thereby changing chemical identity. In contrast, for a stable species minor deformations along *all* internal coordinates are possible without affecting chemical identity.

One useful tool for the analysis of molecular deformations, conformational changes, chemical reactions, stability problems and many other aspects of chemistry, is the potential energy hypersurface. This represents an energy function E(K) where an energy value is assigned to each nuclear configuration K of the space M. (For some relevant results on potential surfaces and reaction paths, see references [16-33]).

Whether or not a deformation is significant enough to change chemical identity depends on the given potential energy hypersurface. The potential energy surfaces differ for each electronic state. Hence, chemical identity and the range of deformations that preserves it are also functions of the electronic state. For a given electronic state, the subsets of the nuclear configuration space M that represent chemical species can be chosen as the *catchment regions* of the corresponding potential energy hypersurface [16,17,31-33].

Informally, a catchment region is a subset of M, representing the range of deformations that preserve chemical identity of a chemical species of a given electronic state [31]. If the local internal coordinates are derived from the mass-weighted cartesian coordinates of the nuclei, then catchment regions can be defined in terms of *steepest descent paths*. These paths represent formal, infinitely slow relaxations of the various deformed nuclear configurations. A catchment region of a potential energy hypersurface is that subset of nuclear configurations from where the relaxation paths lead to a common stationary point of energy (critical point of the potential energy hypersurface). The mathematical concept of catchment region is based on the early work of Cayley and Maxwell on geographical watersheds [34,35]. The definition [31] applies to stable chemical species as well as transition structures, where the latter typically correspond to lower dimensional catchment regions.

At a critical point of a potential energy surface the energy gradient vanishes. However, the second energy derivatives, that are the elements of the Hessian matrix of the critical point, are not in general zero. The eigenvalues of the Hessian matrix (the local canonical curvatures of the potential energy hypersurface) are used to characterize the critical point. For an energy minimum, all eigenvalues are positive; for a simple saddle point of a transition structure, precisely one eigenvalue is negative. In general, the index λ of the critical point $K(\lambda,i)$, defined as the number of negative eigenvalues of the Hessian matrix $H(K(\lambda,i))$, is used for characterization, while index i is a serial index. For example, the catchment region C(0,i) of the minimum point K(0,i) represents the i-th stable molecular species of the given stoichiometry and electronic state that is associated with the given potential energy hypersurface. The steepest descent paths from all points of the catchment region C(0,i) lead to the minimum point K(0,i) and the dimension of C(0,i) is 3N-6. Similarly, a transition structure is represented by a catchment region C(1,j) of a saddle point K(1,j) of critical point index λ=1. The set C(1,j) is the collection of all nuclear configurations from where a formal, infinitely slow relaxation (steepest descent path on the potential energy hypersurface) leads to the saddle point K(1,j). With the rare exception of some degenerate cases [31], the dimension of C(1,j) is 3N-7.

Catchment regions C(λ,i) of critical points K(λ,i) of higher indices, $\lambda > 1$, and of dimensions lower than 3N-7, are of lesser direct chemical significance than those with $\lambda = 0$ (minima) and $\lambda = 1$ (simple saddle points). However, for sake of simplicity in the terminology, catchment regions of higher critical point indices are also referred to as formal chemical species.

There is a one to one correspondence between critical points and catchment regions. The catchment regions of each potential energy hypersurface, as defined in [31], generate a rigorous partitioning of the entire nuclear configuration space M, that has advantages over earlier treatments (see, e.g. [36]) of subsets of a potential surface. The catchment region partitioning provides a global approach to the analysis of all chemical species of the given stoichiometric family, as well as to the study of their conformational changes and interconversion reactions. A review of the background of the catchment region approach and alternative approaches can be found in references [16,17].

Chirality properties of a given stable molecule may be studied in the context of all possible nuclear configurations occurring in its catchment region C(0,i), as well as all nuclear configurations in the catchment regions that are the neighbors of C(0,i) within the nuclear configuration space M. Alternatively, one may view the problem from a global perspective by studying those subsets of the nuclear configuration space M where chiral and achiral nuclear configurations are located. In the following sections of this chapter, we shall discuss molecular chirality problems within both of these two approaches, utilizing, as the primary tool, some recent results on global point symmetry analysis [16,17].

Subsets of chirality and achirality of a nuclear configuration space M.

For the given stoichiometric family of chemical species of N atoms, the nuclear configuration space M of dimension n=3N-6 can be subdivided into subsets according to the point symmetry elements of nuclear configurations. In particular, two types of symmetry elements are of special concern to us: reflection planes σ and improper rotations S_{2n}, where among the latter the point of inversion $i= S_2$ of n=1 is the most important.

Let us denote by G_σ and G_i the collections of all those points K of the nuclear configuration space M where the three-dimensional point symmetry elements of nuclear configurations, a reflection plane σ and an improper rotation S_{2n}, respectively, are present:

$$G_\sigma = \{K:\ K \in M,\ \text{nuclear configuration } K \text{ has } \sigma\}, \quad (1)$$

$$G_i = \{K:\ K \in M,\ \text{nuclear configuration } K \text{ has } S_{2n}\}. \quad (2)$$

The union of these two subsets,

$$G_\alpha = G_\sigma \cup G_i, \quad (3)$$

contains precisely the achiral nuclear configurations of the space M, that is, *all the achiral nuclear arrangements* of the entire stoichiometric family of the given N nuclei. The relative complement of set G_α,

$$G_\chi = M \setminus G_\alpha , \quad (4)$$

contains all the chiral nuclear configurations of space M, that is, *all chiral nuclear arrangements* of the given stoichiometric family.

A nuclear configuration K may have several reflection plane symmetry elements σ but at most one symmetry element i. If some internal motions of the molecule preserve a σ or an S_{2n} symmetry element, then these motions are confined to the respective G_σ or G_i subset of M. These motions are *achiral motions,* since achirality is preserved along these changes of nuclear configurations. However, for molecules containing more than three atoms there are always internal motions which destroy these symmetry elements. Consequently, for $N\geq4$, $n\geq6$, the dimensions of subsets G_σ and G_i are less than n, and G_α is a subset of measure zero within the space M. Informally, set G_α is "thin" within the nuclear configuration space M. Generally, the greater the dimension n, the greater is the number of those independent deformations (normal modes in the special case of energy minima) that destroy achirality.

For a general polyatomic stoichiometric family of $N\geq4$, the subset G_χ of chiral nuclear configurations is disconnected. That is, from a given chiral nuclear configuration it is impossible to reach all other chiral nuclear configurations without passing through some achiral ones. The set G_α of achiral nuclear configurations subdivides the set G_χ of all chiral nuclear configurations into its maximum connected components $G_{\chi,k}$,

$$G_\chi = \bigcup_k G_{\chi,k}, \qquad G_{\chi,k} \cap G_{\chi,k'} = \emptyset, \qquad \text{if } k\neq k', \tag{5}$$

where set $\emptyset$ is the empty set.

In the following sections, we shall derive conditions for the presence of critical point nuclear configurations $K(\lambda,i)$ within the $G_{\chi,k}$ subsets of chiral nuclear configurations. In addition, we shall compare these chirality domains $G_{\chi,k}$ with the catchment regions $C(\lambda,i)$ of the potential energy hypersurfaces $E(K)$ of ground and various excited electronic states of the stoichiometric family of chemical species represented by the nuclear configuration space M.

A critical point theorem for chiral nuclear configurations.

A complete inventory of all stable molecular species and all transition structures of the given electronic state of the stoichiometric family is available if all critical points of the corresponding potential energy hypersurface are known (with the exception of some rare, potential defying species [37]). Whereas chirality properties by themselves are insufficient to give a complete solution of this problem, there exist some relevant results. First we shall prove the following theorem:

Theorem 1. There exists at least one critical point nuclear configuration $K(\lambda,i)$ either within or on the boundary of each chirality domain $G_{\chi,k}$ of the nuclear configuration space M of each stoichiometric family, for any electronic state and any total electronic charge.

In order to prove this theorem, we first observe that each chirality domain $G_{\chi,k}$ is an open subset of the nuclear configuration space M. If a nuclear configuration K does not possess any of the symmetry elements σ and S_{2n}, then it is possible to choose a small enough configuration space distance d such that no nuclear configuration K', deviating from K by less than d, possesses symmetry elements σ or S_{2n} either. Hence, all boundary points of an arbitrary chirality domain $G_{\chi,k}$ must belong to the set G_α of achiral nuclear configurations. The *boundary set* $\Delta G_{\chi,k}$ of chirality domain $G_{\chi,k}$ is defined as the intersection

$$\Delta G_{\chi,k} = G_\alpha \cap \mathrm{clos}[G_{\chi,k}] , \tag{6}$$

where $\mathrm{clos}[G_{\chi,k}]$ is the closure of chirality domain $G_{\chi,k}$.

Since $G_{\chi,k}$ is a maximum connected component of the set of all chiral nuclear configurations, the boundary $\Delta G_{\chi,k}$ must separate the chirality domain $G_{\chi,k}$ from all other chiral nuclear configurations of the nuclear configuration space M. As a consequence of the definition (6), each nuclear configuration K' of the boundary $\Delta G_{\chi,k}$ must possess at least one of the point symmetry elements σ or S_{2n}. Furthermore, for any nuclear configuration K from chirality domain $G_{\chi,k}$, no symmetry element σ or S_{2n} may be present. Consequently, by taking

$$M_1 = \mathrm{clos}[G_{\chi,k}] , \tag{7}$$

and

$$M_2 = M \setminus M_1 , \tag{8}$$

as two subsets of M, and the boundary

$$B = \Delta G_{\chi,k} , \tag{9}$$

the conditions of the vertical point symmetry theorem of nuclear configuration spaces (ref.[17], see also Appendix) are fulfilled. Hence, there must exist at least one critical point nuclear configuration $K(\lambda,i)$ either on the boundary $\Delta G_{\chi,k}$ or within the chirality domain $G_{\chi,k}$ of the nuclear configuration space M of the given stoichiometric family of N nuclei, for the potential energy hypersurface of each electronic state and any total electronic charge. Q.E.D.

Note that the above theorem does not specify the type (e.g., minimum or saddle point) of the critical point. It simply states that there must exist *some* critical point within or on the boundary of the chirality domain $G_{\chi,k}$. The theorem can detect the presence of at least one critical point, but it cannot provide their precise number; there may be several citical points, possibly of different types, within the chirality domain $G_{\chi,k}$. The result is independent of the electronic state and of the electronic charge. Hence, it applies to *all potential energy hypersurfaces* defined over the nuclear configuration space M of the given stoichiometric family of N nuclei.

Three catchment region chirality theorems and related results.

The first two results we shall discuss in this section are two catchment region chirality theorems:

Theorem 2. If the nuclear configuration of the critical point $K(\lambda,i)$ of a catchment region $C(\lambda,i)$ is chiral, then all nuclear configurations of the catchment region $C(\lambda,i)$ are chiral. That is, if

$$K(\lambda,i) \in G_{\chi}, \tag{10}$$

then

$$C(\lambda,i) \subset G_{\chi} \tag{11}$$

follows.

The proof of Theorem 2 follows from the catchment region point symmetry theorem (ref.[17], see also the Appendix) that states that the critical point nuclear configuration must have all the point symmetry elements of all nuclear configurations present within the catchment region. If the critical point lacks the symmetry elements σ and S_{2n}, then these symmetry elements cannot be present for any nuclear configuration K of the entire catchment region $C(\lambda,i)$. Hence, relation (11) follows. Q.E.D.

Theorem 3. If there exists an achiral nuclear configuration K anywhere within a catchment region $C(\lambda,i)$, then the critical point $K(\lambda,i)$ of $C(\lambda,i)$ is also achiral. That is, if

$$K \in C(\lambda,i) \tag{12}$$

and

$$K \in G_\alpha \tag{13}$$

then

$$K(\lambda,i) \in G_\alpha \tag{14}$$

follows.

The proof of this result also follows from the catchment region point symmetry theorem (ref.[17], see also the Appendix). If a point symmetry element σ or S_{2n} is present for any nuclear configuration K of the catchement region $C(\lambda,i)$, then the same point symmetry element must also be present at the critical point $K(\lambda,i)$. Consequently, the nuclear configuration of the critical point $K(\lambda,i)$ must be achiral. Q.E.D.

Here we present a related result on chirality properties of catchment regions: the achirality of nuclear configurations in a small neighborhood of the critical point implies achirality of all nuclear configurations within the entire catchment region.

Theorem 4. All nuclear configurations K of a catchment region $C(\lambda,i)$ are achiral if and only if all nuclear configurations K' are achiral within any neighborhood A of the critical point $K(\lambda,i)$ of $C(\lambda,i)$, where the entire set A is contained and open within the corresponding catchment region $C(\lambda,i)$. That is,

$$C(\lambda,i) \subset G_\alpha \tag{15}$$

if and only if for any set A open within $C(\lambda,i)$, such that

$$K(\lambda,i) \in A \subset C(\lambda,i) , \tag{16}$$

the condition

$$A \subset G_\alpha \tag{17}$$

also holds.

The proof of necessity of condition (17) is trivial, since relations (15) and (16) evidently imply relation (17).

In order to prove that condition (17) is sufficient, let us choose an arbitrary, non-critical point K of the catchment region $C(\lambda,i)$,

$$K \in C(\lambda,i) , \tag{18}$$

and a subset A fulfilling condition (14) and open within $C(\lambda,i)$. On the one hand, the subset A is open, hence the steepest descent path P from the point K of the catchment region $C(\lambda,i)$ must pass through a non-critical point K' of subset A before the path reaches the critical point $K(\lambda,i)$,

$$K' \in P , \tag{19}$$

$$K' \in A \subset C(\lambda,i) , \tag{20}$$

$$K' \neq K(\lambda,i) . \tag{21}$$

Since according to (17)

$$K' \in A \subset G_\alpha , \tag{22}$$

either a symmetry element σ or a symmetry element S_{2n} must be present at the nuclear configuration represented by point K'. On the other hand, all point symmetry elements are preserved along paths of steepest descent as long as no critical point is encountered [25]. Consequently, all nuclear configurations along the entire K to K' segment of the steepest descent path, including the starting point K of the path, must have either a reflection plane σ or an improper rotation S_{2n} symmetry element. Hence, point K must belong to the set G_α of achiral nuclear configurations,

$$K \in G_\alpha . \tag{23}$$

Since point K of the catchment region has been chosen arbitrarily, the entire catchment region $C(\lambda,i)$ must belong to the set G_α of achiral nuclear configurations. This proves (15). Q.E.D.

The above result has some computational significance. If a critical point $K(\lambda,i)$ of a potential energy hypersurface has been located, then it is sufficient to test an arbitrarily small, open neighborhood A of this point for achirality. If all nuclear configurations in A are achiral, then the entire catchment region, representing the given chemical species, contains only achiral nuclear configurations. In fact, it is sufficient to test only the boundary points of A for achirality.

Also note that this result is easily generalized for any point symmetry element R, by replacing σ or i by R, and replacing G_α by the subset G_R of all nuclear configurations having point symmetry element R. Then,

$$C(\lambda,i) \subset G_R \tag{24}$$

if and only if for any set A open within $C(\lambda,i)$, such that

$$K(\lambda,i) \in A \subset C(\lambda,i) , \tag{25}$$

the condition

$$A \subset G_R \tag{26}$$

also holds.

Chiral and achiral deformations and the cases of strong and weak achirality.

Following the method described earlier (ref.[16], p367]), one may analyse those cases where the critical point itself corresponds to an achiral nuclear configuration, but the test of achirality of Theorem 4 fails at some points of subset A. The paths of steepest descent can be reversed and followed "back" from the critical point $K(\lambda,i)$, through the open neighborhood A, toward the boundary of the catchment region $C(\lambda,i)$. By this technique, the chirality domain decomposition of the catchment region $C(\lambda,i)$ may be obtained as a "blown up", and possibly deformed, version of the chirality domain decomposition of the open set A itself: the two decompositions are topologically equivalent. Hence, if one is interested in the pattern of chirality domains in $C(\lambda,i)$, then it is sufficient to study the pattern within an arbitrarily small subset A around the critical point $K(\lambda,i)$.

In most cases, there are more degrees of freedom for deformations that destroy a symmetry element than those preserving the symmetry element. Hence, if

$$K(\lambda,i) \in G_\alpha , \tag{27}$$

$$K(\lambda,i) \in A , \tag{28}$$

but

$$A \not\subset G_\alpha , \tag{29}$$

then for a typical catchment region $C(\lambda,i)$ the subset

$$C(\lambda,i)_\alpha = C(\lambda,i) \cap G_\alpha \tag{30}$$

has a measure of zero within $C(\lambda,i)$ (that is, the set $C(\lambda,i)_\alpha$ is "infinitely thin" within $C(\lambda,i)$). In contrast, the relative complement,

$$C(\lambda,i)_\chi = C(\lambda,i) \cap G_\chi = C(\lambda,i) \setminus C(\lambda,i)_\alpha \tag{31}$$

has the same dimension as that of the catchment region $C(\lambda,i)$, with the exception of some degenerate cases.

Subsets $C(\lambda,i)_\alpha$ and $C(\lambda,i)_\chi$, respectively, contain all achiral and all chiral nuclear configurations of the catchment region $C(\lambda,i)$.

Let us define two subsets of the open set A as follows:

$$A_\alpha = A \cap G_\alpha \tag{32}$$

and

$$A_\chi = A \cap G_\chi \, . \tag{33}$$

The above chirality domain decompositions of A and $C(\lambda,i)$ are interrelated. Any achiral nuclear configuration K_α of $C(\lambda,i)$ (that necessarily falls within $C(\lambda,i)_\alpha$) is reachable from a point K'_α of A_α by following a reversed steepest descent path (a steepest ascent path). Similarly, any chiral nuclear configuration K_χ of $C(\lambda,i)$ (that necessarily falls within $C(\lambda,i)_\chi$) is reachable from a point K'_χ of A_χ by following a reversed steepest descent path. These very paths are suitable to establish a homeomorphism (topological equivalence, refs.[38,39]) between the chirality domain decompositions of subset A and the catchment region $C(\lambda,i)$.

The deformations within a catchment region $C(\lambda,i)$ can also be classified by chirality, as

(i) chirality preserving,
(ii) achirality preserving, and
(iii) chirality-achirality mixing deformations.

The deformation paths in these three classes are denoted by P_χ , P_α , and $P_{\chi\alpha}$, respectively.

Whereas any two achiral nuclear configurations K_α and K'_α of the catchment region $C(\lambda,i)$ can be deformed into one another by an achirality preserving path P_α within $C(\lambda,i)$, two chiral nuclear configurations, K_χ and K'_χ of $C(\lambda,i)$ are not necessarily deformable into one another by a chirality preserving path P_χ , confined to the catchment region $C(\lambda,i)$. That is to say, subset $C(\lambda,i)_\alpha$ is necessarily pathwise connected, whereas for $C(\lambda,i)_\chi$ this is not necessarily true. The maximum connected components of $C(\lambda,i)_\chi$ are denoted by $C(\lambda,i)_{\alpha,j}$.

Note , however, that by adding the critical point $K(\lambda,i)$, one can join all of these maximum connected components. That is, the set obtained as the union

$$C(\lambda,i)'_\chi = \{K(\lambda,i)\} \cup C(\lambda,i)_\chi \tag{34}$$

is pathwise connected.

A general path may be composed from chirality preserving segments connected by achiral nuclear configurations. Such paths represent chirality-achirality mixing deformations.

A chemical species exhibits achiral properties if the critical point configuration $C(\lambda,i)$ within its catchment region $C(\lambda,i)$ is achiral. There are two possibilities:

(i) all the nuclear configurations within $C(\lambda,i)$ are achiral

(ii) $C(\lambda,i)$ contains both achiral and chiral nuclear configurations.

Trivial examples for case (i) are the catchment regions of all triatomic molecules, or the five-dimensional catchment region $C(1,i)$ of the planar transition structure of a substituted ammonia, for example, NHDX. We assume that the stable NHDX species is chiral, but all nuclear configurations within the transition structure catchment region, including, of course, the transition structure critical point $K(1,i)$ for the planar inversion process, are certainly achiral. Molecular species belonging to family (i) are achiral under all deformations preserving chemical identity, and we shall refer to case (i) as *strong achirality.* The transition structure of NHDX is strongly achiral.

An example for case (ii) is the ammonia molecule, NH_3, with an achiral minimum point $K(0,i)$ in a catchment region $C(0,i)$ that contains both chiral and achiral distorted nuclear configurations. Clearly, most distortions of NH_3 lead to nuclear configurations which do not possess either one of the point symmetry elements σ or S_{2n}. Some (in fact most) deformations that preserve chemical identity of molecular species from family (ii) do not preserve achirality, and we shall refer to case (ii) as *weak achirality.* The ammonia molecule is weakly achiral.

Paths of chirality-achirality mixing deformations, if not confined to a single catchment region, are responsible for racemization and configuration inversion processes. These processes typically involve three catchment regions. Two of them, $C(0,i)$ and $C(0,i^*)$, are catchment regions of stable species with chiral nuclear configurations at their respective minimum points $K(0,i)$ and $K(0,i^*)$. The third one is a strongly achiral transition structure catchment region $C(1,j)$ between $C(0,i)$ and $C(0,i^*)$. Pairs of nuclear configurations of $C(0,i)$ and $C(0,i^*)$ are mirror images of each other.

Whether or not a molecular species is strongly achiral or weakly achiral may have some interesting consequences on chirality changes involving electronic excitations. Some aspects of these problems will be described in the next section.

Chirality relations for electronic excitations.

Consider a weakly achiral molecular species, a stable molecule in its ground electronic state, represented by the catchment region C(0,i). If an electronic excitation occurs and an excited state species, an excimer, is formed, then a nuclear rearrangement may also occur and the new species may exhibit new chirality features. The excimer is represented by a catchment region C'(0,i') of the excited state potential energy hypersurface. Let us assume that the excimer is chiral. The overall process may be modelled by two formal steps:

(a) an electronic excitation and

(b) a nuclear rearrangement.

Within a semiclassical model there are three nuclear configurations of special interest: the equilibrium nuclear configurations K(0,i) and K'(0,i') on the ground and excited electronic state potential energy hypersurfaces, and the formal nuclear configuration K" from which the electronic excitation occurs on the ground state surface. In the simplest model, one may consider a simple vertical process for electronic excitation. The resulting nuclear configuration on the excited electronic state potential surface, at the moment of excitation, is assumed to be identical to K". A subsequent relaxation on the excited state potential surface leads to the equilibrium nuclear configuration K'(0,i') of the excimer.

Let us first assume that the actual nuclear configuration K" of the weakly achiral ground state molecule involved in the electronic excitation is achiral. That is, the point K" falls within the $C(0,i)_\alpha$ part of the catchment region C(0,i),

$$K'' \in C(0,i)_\alpha \subset G_\alpha . \quad (35)$$

We consider the possibility of the following process:

(1a) vertical electronic excitation from an achiral nuclear configuration K",

(1b) relaxation of the nuclear configuration K" to the chiral equilibrium nuclear configuration K"(0,i') on the excited state potential surface.

We shall show that, according to Theorem 2, this is not an allowed process.

In step (1a) the vertical electronic excitation alone does not alter the instantaneous nuclear configuration K". Consequently, the relaxation process (1b) on the excited state potential energy hypersurface is supposed to start at an achiral nuclear configuration, $K'' \in G_\alpha$. However, the equilibrium nuclear configuration of the excimer is chiral, and, according to Theorem 2, if the nuclear configuration of the critical point is chiral, then all nuclear configurations in the catchment region must be chiral;

$$C'(0,i') \subset G_\chi \, . \tag{36}$$

Consequently, the achiral nuclear configuration K" cannot belong to the excimer catchment region,

$$K'' \notin C'(0,i') \, , \tag{37}$$

and no formal vibrationless relaxation path of step (1b) may lead from K" to K'(0,i'). Hence, according to Theorem 2, the overall process (1a)-(1b) is prohibited.

Let us now assume that the actual nuclear configuration K" of the weakly achiral ground state molecule involved in the electronic excitation is chiral. That is, K" is a point within the $C(0,i)_\chi$ part of the catchment region C(0,i),

$$K'' \in C(0,i)_\chi \subset G_\chi \, . \tag{38}$$

We consider the possibility of the following process:

(2a) vertical electronic excitation from a chiral nuclear configuration K",

(2b) relaxation of the nuclear configuration K" to the chiral equilibrium nuclear configuration K"(0,i') on the excited state potential surface.

As before, the vertical electronic excitation alone does not alter the instantaneous nuclear configuration K". Hence, the relaxation process on the excited state potential energy hypersurface starts at a chiral nuclear configuration,

$$K'' \in G_\chi . \tag{39}$$

This requirement is not in conflict with the condition

$$K'' \in C'(0,i'). \tag{40}$$

Since the excimer C'(0,i') is chiral,

$$C'(0,i') \subset G_\chi , \tag{41}$$

there exists a chiral deformation path P_χ , (a path of formal, infinitely slow, vibrationless relaxation) that leads from K" to the minimum point K'(0,i') of the excimer C'(0,i'). This is an allowed process, permitted by all four theorems.

If the ground state molecular species is chiral, then

$$C(0,i) \subset G_\chi , \tag{42}$$

and the excimer C'(0,i'), obtained from C(0,i) by an elementary process (a)-(b), cannot be strongly achiral, since then one would have

$$C'(0,i') \subset G_\alpha . \tag{43}$$

This would imply that C(0,i) and C'(0,i)' cannot have any common nuclear configurations,

$$C(0,i) \cap C'(0,i') = \emptyset . \tag{44}$$

In turn, this would contradict the requirement that K" is a nuclear configuration of both species,

$$K'' \in C(0,i) \cap C'(0,i') . \tag{45}$$

Fuzzy chirality and syntopy.

Chirality, as a special aspect of symmetry, has been introduced within a geometrical context. However, actual chemical species are topological, rather than geometrical entities [40], and it is natural to investigate chirality problems within a topological framework. The relations between symmetry and three-dimensional molecular topology has been the subject of several studies [16,17,41-46]. Among these, the syntopy model, based on a fuzzy set approach to the concept of symmetry resemblance [42,43], has interesting applications to chirality problems.

Fuzzy sets have been proposed for the mathematical description of the "grade of belonging" of objects to various classes, if the classification criteria cannot be defined in absolute terms [47-51]. In ordinary classification schemes, an object is either a member or not a member of a class, indicated by the 1 or 0 value of the membership function of ordinary set theory. In fuzzy set theory, the membership function, representing belonging, can take noninteger, intermediate values. Fuzzy set methods have been applied to a variety of fields of the natural sciences. The description of quantum mechanical uncertainty [52-56], a fuzzy set generalization of the chemical concept of catchment regions [33], and a fuzzy set description of symmetry resemblance within the syntopy model [42,43] are examples.

The membership function $\mu_k(K,\varepsilon)$ of the fuzzy syntopy model [43] assigns a value from the [0,1] interval to each nuclear configuration K. This value describes the symmetry resemblance of K to a nuclear configuration K' that has the precise point symmetry of the kth point symmetry group. The $\mu_k(K,\varepsilon)$ value is dependent on an energy criterion ε. This criterion describes the accessibility of nuclear configuration K' from K, if the maximum energy variation allowed is ε. The accessibility may vary from potential surface to potential surface. Hence, the energy dependent fuzzy membership function $\mu_k(K,\varepsilon)$, in accordance with the actual symmetry resemblance manifested by physical experiments, may be different for each electronic state. By taking these membership functions, one may define fuzzy sets of the nuclear configuration space M, leading to the syntopy concept, a generalization of point symmetry. For precise definitions and further details see ref.[43].

By analogy with the fuzzy syntopy model, we may develop a fuzzy set model for achirality, suitable for describing cases one may call "almost achiral". By replacing the subset G_k of the kth point symmetry group with the union $G_\alpha = G_\sigma \cup G_i$ of eq. (3), and, accordingly, replacing the fuzzy membership function $\mu_k(K,\varepsilon)$ with the corresponding fuzzy membership function $\mu_\alpha(K,\varepsilon)$, one obtains a set $S_\alpha(\varepsilon)$ of *fuzzy achirality* within the nuclear configuration space M :

$$S_\alpha(\varepsilon) = \{\ K:\ \mu_\alpha(K,\varepsilon) > 0\ \}. \tag{46}$$

The accessibility of any formal nuclear configuration K from any given achiral nuclear configuration K' depends on the energy variation and not on the M space distance along a transformation path in M. Consequently, the membership function $\mu_\alpha(K,\varepsilon)$ and the set $S_\alpha(\varepsilon)$ of fuzzy achirality are dependent on the energy parameter ε. If a greater energy threshold ε is chosen, then greater variations of the nuclear configurations are allowed. Consequently, *more* chiral nuclear configurations may show a *greater degree* of resemblance in their physical and chemical behavior to achiral nuclear configurations.

By definition of $\mu_\alpha(K,\varepsilon)$ (for the original definition of $\mu_k(K,\varepsilon)$, see ref.[43]), the value of this membership function is one for achiral nuclear configurations, and zero for those chiral configurations from which no achiral nuclear configuration is reachable if the deformation paths are restricted to an energy range ε. The function $\mu_\alpha(K,\varepsilon)$ takes intermediate values for chiral nuclear configurations from which some achiral nuclear configurations are reachable under threshold ε of energy variation. A greater $\mu_\alpha(K,\varepsilon)$ value indicates a greater resemblance of K to an achiral nuclear configuration K'. The symmetry resemblance to an achiral nuclear configuration, as expressed by the membership function $\mu_\alpha(K,\varepsilon)$, changes *continuously* with the change of nuclear configuration K, if displacements are given in terms of the metric of M. Reachability depends not only on the energy threshold ε but also on the actual potential energy hypersurface, that is, on the electronic state. Hence, the fuzzy achirality concept is not purely geometrical. It also involves deformability of various nuclear configurations that may well be different for different electronic states.

In the special case of $\varepsilon = 0$, the fuzzy achirality set $S_\alpha(\varepsilon)$ becomes identical to the set G_α of achiral nuclear configurations of space M :

$$S_\alpha(0) = G_\alpha\ . \tag{47}$$

In somewhat oversimplified terms, if no energy is available for changes of nuclear configurations, then the fuzzy set of achirality $S_\alpha(\varepsilon)$ becomes the "crisp" set G_α, with ordinary "yes or no" membership for the nuclear configurations K. In this case, the concept of fuzzy achirality resemblance is replaced with that of strict achirality.

Chiral and achiral charge distributions and molecular surfaces.

It is convenient to represent molecular electronic charge densities by isodensity contour surfaces G(a), where G(a) is the set of those points of the 3D space where the electronic density is equal to some chosen threshold value a . A formal molecular body B(a) can be defined as the collection of all points where the electronic density is equal to or greater than the value a . If the specification of the nuclear configuration K is also needed then the notations G(K,a) and B(K,a) can be used.

Electron density of a molecule is dependent on the nuclear configuration, and it is natural to interrelate symmetry properties of electron densities and those of nuclear configurations. The direct characterization of chirality properties of electron densities, however, represents a somewhat different problem than that of nuclear configurations. For nuclear arrangements, the point symmetry concept is relevant. However, for a three-dimensional continuum of the electronic density function some of the continuum aspects of symmetry are also of importance. This leads to various generalizations, symmorphy [15,57], for example. Similar considerations apply to formal molecular surfaces [58], defined, for example, as isodensity contours or contours of electrostatic potentials.

In general, one may consider a partitioning of the nuclear configuration space M in terms of properties of electron density functions or those of formal molecular surfaces, expressed as functions of nuclear configuration K [59]. If symmetry is the property of interest, then one can generate a partitioning of M into domains defined in terms of symmetry elements and symmetry groups of the 3D or 2D objects (electronic density functions or molecular contour surfaces) associated with each nuclear configuration [60].

Symmetry rules analogous to the catchment region point symmetry theorem and to the family of vertical point symmetry theorems are applicable to these objects. However, these rules are somewhat weaker then the original point symmetry theorems [16,17]. For example, symmetries of excited electronic state charge densities are not necessarily the same as that of the nuclear configuration, and accidental symmetry of electronic density contours may occur even if the nuclear arrangement does not have the corresponding symmetry element [60]. However, the accidental symmetries of electron densities or contour surfaces occur (if at all) at nuclear configurations which form a measure zero subset of the nuclear configuration space M, and the rules are reliable in most instances.

For example, the appearance of new point symmetry elements of the nuclear configuration at a critical point cannot lower the symmetry of an isodensity contour G(K,a). Hence, by analogy with the catchment region point symmetry theorem [16,17], one expects few exceptions to the following rule:

The density contour surface G(K,a) at the critical point has the highest symmetry within the catchment region.

Based on the above rule, the following additional rule has been proposed [60]:

Subject to the symmetry selection rule of a given one-step electronic excitation or de-excitation process, the isodensity contour $G_2(K(\lambda,i),a')$ of the product electronic state equilibrium nuclear configuration $K(\lambda,i)$, with contour density a', has all the symmetry elements of the isodensity contour $G_1(K,a)$ of the reactant electronic state initial (possibly non-equilibrium) nuclear configuration K, with contour density a, where the two density threshold values a' and a may be different.

One interesting consequence of the above result of ref.[60] is the following rule:

If the electronic transition is allowed and the excitation or de-excitation itself does not lead to loss of symmetry, then, in a one-step electronic transition, an initial achiral density contour surface cannot lead to a chiral product density contour surface in the final electronic state.

Whereas, for actual electronic densities, accidental symmetry (that is, symmetry not implied by the point symmetry of the nuclear arrangement) is a rare event, such symmetries may be induced easily in simple models of molecular surfaces obtained as fused sphere Van der Waals surfaces [61]. Accidental symmetry may occur if two different nuclei are assigned the same Van der Waals radius.

Polycubes and the degree of chirality concept for molecular bodies.

The shapes of molecular contour surfaces, for example, isodensity contours G(a), and the formal molecular bodies B(a) enclosed by them, can be studied by various topological techniques (see, for example, refs.[62-71]). One approach that is particularly suitable for the analysis of molecular chirality is based on the generation of polycubes enclosed by the contour surface G(a), following a technique proposed for the quantification of similarity of molecular shapes [71]. A polycube P_n is a connected arrangement of n impenetrable cubes of uniform size where only three types of contacts between cubes are allowed: common face, common edge or common vertex. For n>1 each cube must have a face contact with another cube. It is often advantageous to consider polycubes embedded in a cubic lattice. The mesh $M(P_n)$ of a polycube P_n is the smallest rectangular block of the cubic lattice that contains P_n.

We shall insist on three further restrictions:

(i) for any edge contact between two cubes C and C' of P_n there must be also a face contact between them, or there must exist a cube C'' having face contact with both C and C',

(ii) for any vertex contact between two cubes C and C' of P_n there must be also an edge contact between them, or there must exist two cubes C'' and C''' with face contact to each other and C'' having face contact to C and C''' having face contact to C',

(iii) P_n is topologically equivalent to a solid ball. This constraint refers to the most common case where B(a) is topologically a solid ball. However, toroidal or more complicated topologies are also possible for a formal molecular body. In general, a topological equivalence is required between the B(a) bodies and the polycubes used for their characterization.

In the discussion below, we shall assume that the polycubes fulfill conditions (i)-(iii).

The surface $G(P_n)$ of a polycube P_n is the point set union of all those faces of the cubes of P_n that are on precisely one cube. We shall use the surfaces of polycubes to approximate molecular contour surfaces G(a), and to characterize the shapes of the formal molecular bodies enclosed by these contours. If the size of the cubes is characterized by the uniform edge length s, then by gradually decreasing s and increasing the number n of cubes in P_n, one can approximate the formal molecular body B(a) to any desired accuracy. A similar approach for shape characterization of planar domains enclosed by Jordan curves is described in another chapter of this volume [70].

Given a molecular contour surface G(a), any finite polycube P_n can fit within G(a) if the size s of the cubes is chosen small enough. In this study we do not consider orientation constraints and we assume that the contour surface G(a) and polycube P_n may be rotated with respect to one another, that is, the relative orientation of G(a) and mesh $M(P_n)$ is not fixed. Note, however, that the polycube method of shape analysis can be extended to the case with orientation constraints.

For a given molecular contour surface G(a) and size s there exists a countable family F(G(a),s) of polycubes which fit within G(a). The polycubes of this family are the *inscribed polycubes* of size s. For a given contour surface G(a) and size s there exists a maximum number n(G(a),s) of cubes for inscribed polycubes. A small change of cube size s does not necessarily change the value n(G(a),s), hence n(G(a),s) is invariant for some size range s(G(a),n). Polycubes P_n with the maximum number n=n(G(a),s) of cubes for the given range s(G(a),n) are the n-cube *interior filling polycubes* of the contour surface G(a). The interior filling polycubes have been suggested for the quantification of similarity of shapes of three-dimensional objects, such as formal molecular bodies [71].

A polycube P_n is an interior filling polycube of the contour surface G(a) if and only if no polycube P_{n+1} of the same cube size s can be inscribed in G(a). We shall use the $P_n(G(a))$ notation for the interior filling n-cubes of contour surface G(a).

A polycube P_n may be chiral or achiral. In our present model the identity of a polycube is independent of its orientation. Any two polycubes P_n and P'_n which can be superimposed on one another by translation and rotation in 3D space are regarded identical,

$$P_n = P'_n . \tag{48}$$

That is, we do not pay attention to any actual embedding property of the polycube within a cubic lattice. Consequently, the usual condition of chirality applies: a polycube P_n is achiral if and only if P_n can be superimposed on its mirror image $P^{\lozenge}_n$ by translation and rotation:

$$P_n = P^{\lozenge}_n \, . \qquad (49)$$

Otherwise the polycube P_n is chiral.

We shall use the chirality properties of interior filling polycubes $P_n(G(a))$ inscribed in molecular contour surfaces G(a) in order to assess the chirality properties of the G(a) contours and the formal molecular bodies B(a) enclosed by them. The size s of the cubes will be associated with a formal level of resolution at which the chirality properties of G(a) and B(a) are analysed. The concepts we propose follow some of the basic patterns of visual perception of the shape features of ordinary objects.

Consider the visual assessment of chirality of ordinary objects, such as a potato, at various levels of resolution. At very low resolution, e.g., from a great distance, the object appears as a single dot, hence it appears achiral. At a slightly higher resolution, it may appear as a small, achiral sphere. Somewhat higher resolution is required to observe details of its shape, and to detect the presence or the lack of chirality. An object may exhibit chirality at relatively low resolution, while some others manifest their chirality only at a much higher resolution. It appears useful to introduce a measure to express how detailed a shape description is needed to detect chirality of a given object. This measure will be called the degree of chirality. Interior filling polycubes provide a tool to generate descriptions of G(a) and B(a) at various levels of resolution: for large size parameter s the description is crude, and gradually higher levels of resolutions, hence, finer descriptions can be obtained by decreasing the size parameter s.

We shall use the chirality properties of interior filling polycubes to introduce the concept of the *degree of chirality* of molecular contour surfaces and formal molecular bodies.

The chirality properties of the interior filling polycubes $P_n(G(a))$ for each value n provide information on the chirality of the molecular body B(a) at the given level of resolution. In order to define the level of resolution, scaled relative to the molecular size, we shall not use the absolute size parameter s directly. A given size s can provide few details of a small object of delicate features, but can be sufficiently descriptive for a large object of cruder features. The characterizations are more comparable if both small and large objects are described by the same number of cubes. Consequently, each level of resolution is defined by n, which depends on the relative size of the object as compared to the cube size s .

One might expect that the interior filling polycubes $P_n(G(a))$ of a chiral molecular contour surface $G(a)$ are also chiral. However, this is not necessarily the case for small n values. For example, for large enough s, the *achiral* P_1 is an interior filling polycube of many *chiral* contour surfaces. That is, on the level of resolution represented by $n=1$ and the actual s value, the given molecular body $B(a)$, when approximated with an interior filling polycube, does not appear chiral. A higher level of resolution with a greater number n of cubes for an interior filling polycube $P_n(G(a))$ is required to detect the chirality of $G(a)$ and $B(a)$. In general, the more prominent the chirality of $B(a)$, the more likely that the interior filling polycubes P_n are already chiral for smaller n values.

The smallest number n_1 for which there exists a chiral interior filling polycube $P_n(G(a))$ is not a suitable measure for the degree of chirality, since even achiral contour surfaces $G(a)$ may have chiral interior filling polycubes. In order to introduce a more appropriate measure, we shall consider a chirality index for each $G(a)$, defined as a critical cube number $n_o(G(a))$, above which all interior filling polycubes $P_n(G(a))$ are chiral.

We say that $G(a)$ is chiral at and above cube number n_c if each interior filling polycube $P_n(G(a))$ is chiral if $n \geq n_c$. The *chirality index* $n_o(G(a))$ is the smallest n_c value above which all interior filling polycubes $P_n(G(a))$ are chiral,

$$n_o(G(a)) = \begin{cases} \min\{n_c : P_n(G(a)) \text{ is chiral if } n \geq n_c\}, & \text{if the minimum exists} \\ \infty & \text{otherwise.} \end{cases} \tag{50}$$

We use the chirality index $n_o(G(a))$ to define the degree of chirality:

Definition. The degree of chirality $\chi(G(a))$ of a molecular contour surface $G(a)$ is

$$\chi(G(a)) = 1 / (n_o(G(a))-3). \tag{51}$$

The inclusion of number three on the right hand side of the defining equation is motivated by the fact that the smallest chiral polycube has four cubes, arranged as the cubes C, C', C", and C'" in condition (ii), and we assign the highest degree of chirality (chirality manifested at the lowest level of resolution) to such a P_4 polycube. Hence, if a molecular contour surface G(a) has this polycube P_4 as an interior filling polycube, and all other interior filling polycubes P_n of G(a) with $n \geq 4$ are also chiral, then G(a) belongs to the family of contours of the highest possible degree of chirality (objects that appear chiral at the lowest possible level of resolution where chirality may appear at all). Hence, it is natural to assign a maximum degree

$$\chi(G(a)) = 1 / (n_o(G(a))-3) = 1 / (4-3) = 1 \quad (52)$$

to this contour. The choice of the denominator provides a scale where the extreme value $\chi(G(a)) = 1$ corresponds to the case of most pronounced chirality, whereas $\chi(G(a)) = 0$ corresponds to an achiral molecular contour surface G(a) .

Appendix.

A brief review of the background and statements of the vertical point symmetry theorem and the catchment region point symmetry theorem are presented below, based on the description given in ref.[17].

For a given stoichiometry of N nuclei, consider the corresponding nuclear configuration space M that contains all possible nuclear arrangements. Choose any surface B that divides the nuclear configuration space M into two parts, M_1 and M_2 , where set M_1 of nuclear configurations contains surface B as its boundary (that is, M_1 is a closed set).

Some symmetry elements are present for all nuclear configurations K' along the boundary B, and we denote a family of such symmetry elements, $R'_1, R'_2, \ldots R'_p$, by $\boldsymbol{R'}$,

$$\boldsymbol{R'} = \{ R'_1, R'_2, \ldots R'_p \} . \quad (A1)$$

Choose a nuclear configuration K from set M_1, and denote by $\boldsymbol{R}$ a family of symmetry elements, $R_1, R_2, \ldots R_q$ that are present at point K :

$$\boldsymbol{R} = \{ R_1, R_2, \ldots R_q \} . \tag{A2}$$

If one takes all common symmetry elements from B, then the family of symmetry operators corresponding to the family $\boldsymbol{R'}$ of symmetry elements is a group. Similarly, if one extends family $\boldsymbol{R}$ to all symmetry elements of nuclear configuration K, then the corresponding symmetry operators form a group, the point symmetry group of nuclear configuration K. The results, however, are valid even if one includes only some of the eligible symmetry elements in families $\boldsymbol{R'}$ and $\boldsymbol{R}$.

The vertical point symmetry theorem of nuclear configuration spaces states the following:

Theorem.

If

(i) no configuration along B possesses the family $\boldsymbol{R}$ of symmetry elements,

or, if

(ii) configuration K does not have all the symmetry elements of family $\boldsymbol{R'}$,

then the family M1 of configurations must contain at least one critical point for the potential energy surface of each electronic state (of each possible overall electronic charge). The proof can be found in ref.[17].

Note that if either one of conditions (i) and (ii) is fulfilled, then K must be an interior point of M_1, that is, K cannot fall on the boundary B.

The following is the catchment region point symmetry theorem:

Theorem.

Within each catchment region $C(\lambda,i)$, the nuclear configuration corresponding to the critical point $K(\lambda,i)$ has the highest point symmetry.

A proof of this theorem is given in ref.[17].

Note that within a catchment region $C(\lambda,i)$, there always exists a point symmetry group (that of the critical point $K(\lambda,i)$) which contains as subgroups all other point symmetry groups occurring in $C(\lambda,i)$. Hence, within a catchment region, it is meaningful to refer to the "highest" point symmetry. The theorem does not imply that the critical point $K(\lambda,i)$ is the only point within C(λ,i) that has the highest point symmetry. Other points, and possibly all points of the catchment region C(λ,i) may have the same, highest point symmetry. Note that each point symmetry group is considered to be one of its own subgroups.

These and other related theorems of ref.[17] provide a link between point symmetry and molecular stability. The point symmetries of various nuclear configurations are purely geometrical properties, not directly dependent on energy, whereas the arrangements of catchment regions and their critical points, being the properties of the potential energy surface, are dependent on energy relations. One should also note that the point symmetry of a fixed nuclear configuration is independent of the net charge and the electronic state of the molecular system. Consequently, the above results are general for the potential energy surfaces and catchment regions of all electronic states of the neutral and ionic species of the given stoichiometric family. The point group symmetry of nuclear configurations provides a condition that interrelates the catchment regions and other energy properties of various electronic states of neutral and ionic species.

Acknowledgements. The author would like to acknowledge stimulating discussions with Prof. Frank Harary on the relations between polycubes and lattice animals that has led to the concept of the degree of three-dimensional chirality, and interesting conversations with Prof. Stuart Whittington and Prof. Chris Soteros on embedding problems on cubic lattices. This work was supported by both strategic and operating research grants from the Natural Sciences and Engineering Research Council of Canada. The support of The Upjohn Company is gratefully acknowledged.

References.

1 J.H. van 't Hoff, Bull. Soc. Chim. Fr. **23**, 295 (1875).
2. P. Curie, J. Phys. (Paris), **3**, 393 (1894).
3 R.S. Cahn, C.K. Ingold and V. Prelog, Angew. Chem., Internat. Ed. Engl., **5**, 385 (1966).
4 E. Ruch and A. Schönhofer, Theor. Chim. Acta **10**, 91 (1968).
5 E. Ruch, Theor. Chim. Acta **11**, 183 (1968).
6 R.W. Robinson, F. Harary and A.T. Balaban, The numbers of chiral and achiral alkanes and monosubstituted alkanes. Tetrahedron **32**, 355 (1976).
7 D.J. Klein and A.H. Cowley, J. Am. Chem. Soc., **100**, 2593 (1978).
8 R.L. Flurry Jr., J. Am. Chem. Soc.,**103**, 2901 (1981).
9 E. Ruch and D.J. Klein, Theor. Chim. Acta **63**, 447 (1983).
10 F.A.L. Anet, S.S. Miura, J. Siegel and K. Mislow, J. Am. Chem. Soc., **105**, 1419 (1983).
11 K. Mislow and J. Siegel, J. Am. Chem. Soc., **106**, 3319 (1984).
12 D.M. Walba, Stereochemical Topology, in Chemical Applications of Topology and Graph Theory, Ed. R.B. King, Elsevier, Amsterdam, 1983.
13 D.M. Walba, Tetrahedron, **41**, 3161 (1985).
14 P.G. Mezey, J. Amer. Chem. Soc., **108**, 3976 (1986).
15 P.G. Mezey, Topology of Molecular Shape and Chirality, in New Theoretical Concepts for Understanding Organic Reactions, J. Bertrán and I.G. Csizmadia (Eds.), Kluwer Academic Publ., Dordrecht,1989, pp 77-99.
16 P.G. Mezey, Potential Energy Hypersurfaces. Elsevier, Amsterdam, 1987.
17 P.G. Mezey, J. Amer. Chem. Soc., **112**, 3791 (1990).
18 P.G. Mezey, Theor. Chim. Acta, **63**, 9 (1983).
19 P.G. Mezey, Int. J. Quantum Chem., **26**, 983 (1984).
20 E.B. Wilson Jr., J.C. Decius, and P.C. Cross, Molecular Vibrations, McGraw-Hill, New York, 1955.
21 R.A. Marcus, J. Chem. Phys., **45**, 4493 (1966).
22 (a) J.N. Murrell and K. Laidler, Trans. Faraday Soc.,**64**, 371 (1968), (b) J.N. Murrell and G.L. Pratt, Trans. Faraday Soc.,**66**, 1680 (1970).
23 K. Fukui, J. Phys. Chem., **74**, 4161 (1970).

24 D.G. Truhlar and A. Kuppermann, J. Amer. Chem. Soc., **93**, 1840 (1971).
25 P. Pechukas, J. Chem. Phys., **64**, 1516 (1976).
26 P.G. Mezey, Progr. Theor. Org. Chem., **2**, 127 (1977).
27 A. Tachibana and K. Fukui, Theor. Chim. Acta, **49**, 321 (1978).
28 A. Tachibana and K. Fukui, Theor. Chim. Acta, **51**, 189 (1979).
29 K. Müller, Angew. Chem. Int. Ed., **19**, 1 (1980).
30 W.H. Miller, N.C. Handy, and J.E. Adams, J. Chem. Phys., **72**, 99 (1980).
31 P.G. Mezey, Theor. Chim. Acta, **58**, 309 (1981).
32 P.G. Mezey, Reaction Topology and Quantum Chemical Molecular Design on Potential Surfaces, in New Theoretical Concepts for Understanding Organic Reactions, J. Bertrán and I.G. Csizmadia (Eds.), Kluwer Academic Publ., Dordrecht,1989, pp 55-76.
33 P.G. Mezey, Differential and Algebraic Topology of Chemical Potential Surfaces, Chapter 19, in "Mathematics and Computational Concepts in Chemistry" Ed. N. Trinajstic, Ellis Horwood Publ. Co., Chichester, U.K., 1986, pp 208-221.
34 C.A. Cayley, Philos. Mag., **18**, 264 (1859).
35 J.C. Maxwell, Philos. Mag., **40**, 233 (1870).
36 M.R. Hoare, Adv. Chem. Phys.,**40**, 49 (1979).
37 P.G. Mezey, Theor. Chim. Acta, **67**, 115 (1985).
38 E.H. Spanier, Algebraic Topology, McGraw-Hill, New York, 1966.
39 I.M. Singer and J.A. Thorpe, Lecture Notes on Elementary Topology and Geometry, Springer-Verlag, New York, 1976.
40 P.G. Mezey, From Geometrical Molecules to Topological Molecules: A Quantum Mechanical View, in Molecules in Physics, Chemistry and Biology, Vol. II, J. Maruani, (Ed.), Reidel, Dordrecht, 1988, pp 61-81.
41 P.G. Mezey, Theor. Chim. Acta, **73**, 221 (1987).
42 J. Maruani and P.G. Mezey, Compt. Rend. Acad. Sci. (Paris), **305**, SerII, 1051, (1987), **306**, SerII, 1141, (1988).
43 P.G. Mezey and J. Maruani, Mol. Phys., **69**, 97 (1990).
44 P.G. Mezey, J. Math. Chem., in press (1990).
45 P.G. Mezey, Int. J. Quant. Chem., in press (1990).
46 P.G. Mezey, Fivefold Symmetry in the Context of Potential Surfaces, Molecular Conformations and Chemical Reactions, in "Quasicrystals, Networks, and Molecules with Fivefold Symmetry", Ed. I. Hargittai, VCH Publishers, New York, 1990, pp 223-238.

47 L.A. Zadeh, Theory of Fuzzy Sets, in Encyclopedia of Computer Science and Technology, Marcel Dekker, New York, 1977.

48 A. Kaufmann, Introduction à la Théorie des Sous-Ensembles Flous, Masson, Paris, 1973.

49 M.M. Gupta, R.K. Ragade, and R.R. Yager, Eds., Advances in Fuzzy Set Theory and Applications, North-Holland Publ. Co., Leyden, 1979.

50 D. Dubois and H. Prade, Fuzzy Sets and Systems: Theory and Applications, Academic Press, New York, 1980.

51 E. Sanchez and M.M. Gupta, Eds., Fuzzy Information, Knowledge Representation and Decision Analysis, Pergamon Press, London, 1983.

52 E. Prugovecki, Found. Phys., **4**, 9, 1974.

53 E. Prugovecki, Found. Phys., **5**, 557, 1975.

54 E. Prugovecki, J. Phys. A, **9**, 1851, 1976.

55 S.T. Ali and H.D. Doebner, J. Math. Phys., **17**, 1105, 1976.

56 S.T. Ali and E. Prugovecki, J.Math. Phys., **18**, 219, 1977.

57 P.G. Mezey, Three-dimensional Topological Aspects of Molecular Similarity, in "Concepts and Applications of Molecular Similarity", Eds., G.M. Maggiora and M.A. Johnson, Wiley Interscience, New York, 1990.

58 P.G. Mezey, Molecular Surfaces, in "Reviews in Computational Chemistry", Eds. K.B. Lipkowitz and D.B. Boyd, VCH Publ., New York, 1990, pp 265-294.

59 P.G. Mezey, J. Math. Chem., **2**, 299 (1988).

60 P.G. Mezey, Nonvisual Molecular Shape Analysis: Shape Changes in Electronic Excitations and Chemical Reactions, in "Computational Advances in Organic Chemistry", Eds. C. Ögretir and I.G. Csizmadia, Kluwer Academic Publ., Dordrecht, The Netherlands, 1990.

61 G.A. Arteca, P.G. Mezey, Configurational Dependence of Molecular Shape, in Reports in Molecular Theory, Eds. G. Náray-Szabó and H. Weinstein, CRC Press, to be published.

62 P.G. Mezey, Int. J. Quant. Chem. Quant. Biol. Symp., **12**, 113 (1986).

63 P.G. Mezey, J. Comput. Chem., **8**, 462 (1987).

64 P.G. Mezey, J. Math. Chem., **2**, 325 (1988).

65 F. Harary and P.G. Mezey, J. Math. Chem., **2**, 377 (1988).

66 G.A. Arteca and P.G. Mezey, Int. J. Quant. Chem., **34**, 517 (1988).

67 G.A. Arteca and P.G. Mezey, J. Phys. Chem., **93**, 4746 (1989).

68 G.A. Arteca and P.G. Mezey, J. Math. Chem., **3**, 43 (1989).

69 G.A. Arteca, G.A. Heal, and P.G. Mezey, Theor. Chim. Acta, **76**, 377 (1990).

70 F. Harary and P.G. Mezey, Chiral and Achiral Square-Cell Configurations; The Degree of Chirality, in this volume.
71 P.G. Mezey, J. Math. Chem., **6**, xxx (1991).

INDEX

Understanding Chemical Reactivity

---

1. Z. Slanina: *Contemporary Theory of Chemical Isomerism.* 1986 ISBN 90-277-1707-9
2. G. Náray-Szabó, P. R. Surján, J. G. Angyán: *Applied Quantum Chemistry.* 1987 ISBN 90-277-1901-2
3. V. I. Minkin, L. P. Olekhnovich and Yu. A. Zhdanov: *Molecular Design of Tautomeric Compounds.* 1988 ISBN 90-277-2478-4
4. E. S. Kryachko and E. V. Ludeña: *Energy Density Functional Theory of Many-Electron Systems.* 1990 ISBN 0-7923-0641-4
5. P. G. Mezey (ed.): *New Developments in Molecular Chirality.* 1991 ISBN 0-7923-1021-7

Kluwer Academic Publishers – Dordrecht / Boston / London